Human Biologists in the Archives

Many physical anthropologists study populations using data that come primarily from the historical record. For this volume's authors, the classic anthropological 'field' is not the glamour of an exotic locale, but the sometimes tedium of the dusty back rooms of libraries, archives and museum collections. This book tells of the way in which archival data inform anthropological questions about human biology and health. The authors present a diverse array of human biological evidence from a variety of sources including the archeological record, medical collections, church records, contemporary health and growth data and genetic information from the descendants of historical populations. The chapters demonstrate how the analysis of historical documents expands the horizons of research in human biology, extends the longitudinal analysis of microevolutionary and social processes into the present and enhances our understanding of the human condition.

D. Ann Herring is Associate Professor in the Department of Anthropology at McMaster University in Hamilton, Canada. Her current interests center on the anthropology of infectious disease, historical demography and epidemiology, environmental health and Aboriginal health in Canada. She has co-authored *Aboriginal Health in Canada: Historical, Cultural and Epidemiological Perspectives* (1995) with James Waldram and T. Kue Young, and co-edited *Strength in Diversity: A Reader in Physical Anthropology* (1994) with Leslie K.-Y. Chan, and *Grave Reflections: Portraying the Past Through Cemetery Studies* (1995) with Shelley R. Saunders.

Alan C. Swedlund is Professor of Anthropology and Research Associate in the Social and Demographic Research Institute at the University of Massachusetts, Amherst. He is also a Research Associate at the New Mexico State Museum of Indian Arts and Culture/Laboratory of Anthropology. His research interests are in anthropological demography, historical epidemiology, and the history of physical anthropology.

Cambridge Studies in Biological and Evolutionary Anthropology

Series Editors

HUMAN ECOLOGY
C. G. Nicholas Mascie-Taylor, University of Cambridge
Michael A. Little, State University of New York, Binghamton
GENETICS
Kenneth M. Weiss, Pennsylvania State University
HUMAN EVOLUTION
Robert A. Foley, University of Cambridge
Nina G. Jablonski, California Academy of Science
PRIMATOLOGY
Karen B. Strier, University of Wisconsin, Madison

Consulting Editor
Emeritus Professor Derek F. Roberts

Selected titles also in the series

Human Biologists in the Archives

Demography, Health, Nutrition and Genetics in Historical Populations

EDITED BY

D. ANN HERRING
McMaster University

and

ALAN C. SWEDLUND
University of Massachusetts, Amherst

CAMBRIDGE UNIVERSITY PRESS
Cambridge, New York, Melbourne, Madrid, Cape Town, Singapore, São Paulo

Cambridge University Press
The Edinburgh Building, Cambridge CB2 2RU, UK

Published in the United States of America by Cambridge University Press, New York

www.cambridge.org
Information on this title: www.cambridge.org/9780521801041

First published 2003
This digitally printed first paperback version 2005

A catalogue record for this publication is available from the British Library

Library of Congress Cataloguing in Publication data

Human biologists in the archives : demography, health, nutrition and genetics
in historical populations / edited by D. Ann Herring & Alan C. Swedlund.
   p.   cm. – (Cambridge studies in biological and evolutionary
anthropology ; 34)
Includes bibliographical references (p.).
ISBN 0-521-80104-4
1. Medical anthropology – Archival resources – Congresses.   2. Physical
anthropology – Archival resources – Congresses.   I. Herring, Ann, 1951–
II. Swedlund, Alan C.   III. Series.
GN296 .H86   2003
306.4′61 – dc21   2003024452

ISBN-13 978-0-521-80104-1 hardback
ISBN-10 0-521-80104-4 hardback

ISBN-13 978-0-521-02011-4 paperback
ISBN-10 0-521-02011-5 paperback

Contents

Contributors

Sylvia Abonyi
Saskatchewan Population Health and Evaluation Research Unit (SPHERU),
Department of Community Health and Epidemiology, University of
Saskatchewan, Saskatoon, Saskatchewan S7N 5E5, Canada

Stacie D. A. Burke
Department of Anthropology, McMaster University, Hamilton, Ontario L8S 4L9,
Canada

Alison K. Donta
Massachusetts Institute for Social and Economic Research, University of
Massachusetts at Amherst, Amherst, MA 01003-4805, USA

Anne L. Grauer
Department of Sociology and Anthropology, Loyola University Chicago, 6525 N
Sheridan Road, Chicago, IL 60626, USA

D. Ann Herring
Department of Anthropology, McMaster University, Hamilton, Ontario L8S 4L9,
Canada

Rosanne L. Higgins
Department of Health Sciences, Cleveland State University, 1983 East 24th Street,
Cleveland, OH 44115-2440, USA

Robert D. Hoppa
Department of Anthropology, University of Manitoba, Winnipeg, Manitoba R3T
5V5, Canada

John R. Johnson
Department of Anthropology, Santa Barbara Museum of Natural History, Santa
Barbara, CA 93105, USA

Lynette Leidy Sievert
Department of Anthropology, Machmer Hall, University of Massachusetts at
Amherst, Amherst, MA 01003-4805, USA

Michael A. Little
Department of Anthropology, Binghamton University, State University of
New York, Binghamton, NY 13902-6000, USA

Lorena Madrigal
Department of Anthropology, College of Arts and Sciences, University of South Florida, 4202 E. Fowler Avenue SOC 107, Tampa, FL 33620, USA

James H. Mielke
Department of Anthropology, University of Kansas, 1415 Jayhawk Blvd, Lawrence, KS 66045-7556, USA

Shawn M. Phillips
Department of Geography, Geology, and Anthropology, Holmstedt Hall 001-A, Indiana State University, Terre Haute, IN 47809, USA

John H. Relethford
Department of Anthropology, State University of New York College at Oneonta, Oneonta, NY 13820, USA

Lisa Sattenspiel
Department of Anthropology, 107 Swallow Hall, University of Missouri, Columbia, MO 65211, USA

Lawrence A. Sawchuk
Department of Anthropology, University of Toronto at Scarborough, 1265 Military Trail, Scarborough, Ontario M1C 1A4, Canada

Sydel Silverman
Graduate School, City University of New York, 365 Fifth Avenue, New York, NY 10016, USA

Malcolm T. Smith
Department of Anthropology, 43 Old Elvet, University of Durham, Durham DH1 3HN, UK

Alan C. Swedlund
Department of Anthropology, Machmer Hall, University of Massachusetts at Amherst, Amherst, MA 01003-4805, USA

Phillip L. Walker
Department of Anthropology, University of California, Santa Barbara, CA 93106, USA

Foreword

History embodies knowledge, tradition, and identity, which are at the core of the human condition. Anthropology has access to a vital record of history represented by the past study of cultures that no longer endure, of languages now extinct, of earlier conditions of health and human biology, and, in general, of the material and written evidence of patterns of existence from the near and distant past. That record is as essential to our understanding of humankind as is the ongoing data collection of the discipline today.

The historical records that incorporate anthropological knowledge consist of many things beyond the finished products that appear in publications: the raw data of research projects; the process of analysis and interpretation that led to published conclusions, contained in notes and worksheets and written drafts; the personal papers of the anthropologists themselves, which give context to the research and document the biographical and social realities of the researchers' lives; and the vast array of materials created by others and for other purposes that anthropologists discover they can mine for use in their own work. What all these things have in common is that they are 'records' only by virtue of the fact that someone has saved them and deposited them in archives.

Individual researchers in all the subfields of anthropology have long made use of such records, but until recently the discipline as a whole has had a certain ambivalence toward them. In cultural anthropology, fieldwork with living populations was always accorded greater value than archival research. Moreover, the unpublished papers produced by anthropologists in the course of their professional lives, including field notes, were considered by many to be of little significance and often discarded, on the mistaken assumption that everything important would have been published. An added complication was fieldworkers' sensitivity about their notes, whether because of unease over the prospect of someone else examining their methods and data or because of ethical concerns, warranted or not.

Other subfields have also undervalued unpublished records, for a variety of reasons. Linguistic anthropology has struggled against time to describe and analyze the large number of the world's rapidly disappearing languages, and there have been minimal resources available for preservation of the paper slips and sound recordings that represent the original data. In archeology, the pressure

S. Silverman & M.A. Little

has long been toward excavating ever more sites and the reward system has favored new data, while unanalyzed collections and the paper records that support them have accumulated and often lain neglected. Where collections have been deposited in museums, priority is given to the care of artifacts (especially the more dramatic ones), whereas the documents that describe their collection and accession – without which the artifacts make no sense – are considered expendable. In biological anthropology there is a wealth of data just begging for new analysis and for comparative studies; on a positive note, some early twentieth century anthropometric databases have been computerized, and researchers have been increasingly inclined to seek information from archival and historical sources.

About ten years ago, a small group of anthropologists began some conversations to see how the historical records of the discipline could be more fully valued and better conserved. With the support of the Wenner-Gren Foundation for Anthropological Research, these conversations grew into a series of conferences and workshops, attended by an ever expanding network of anthropologists from all the subfields, along with archivists and librarians responsible for anthropological records. Recognizing the need for an institutional structure that would keep their efforts alive over time, these concerned scholars and practitioners created the Council for the Preservation of Anthropological Records (CoPAR).

The purpose of CoPAR is to identify, encourage the preservation of, and foster the use of the records of anthropological research. Its activities span four areas: (1) awareness of and education about preservation needs; (2) records location and access, including the development of a comprehensive database identifying significant materials in repositories; (3) consulting and technical assistance to individuals and organizations on preservation issues; and (4) advocacy for existing anthropological-archival facilities. CoPAR is also dedicated to encouraging appropriate use of the archived records by researchers, students, members of the communities in which the records originated, and other interested groups. (The background and initial efforts of CoPAR are described in Silverman and Parezo (1995); see also the CoPAR web page[1] at: http://copar.asu.edu/.)

Much progress has already been made in raising the consciousness of those who produce or hold records about the importance of archival preservation; in devising strategies and providing guidelines to assist with preservation; and in

[1] The publisher has used its best endeavors to ensure that the URLs for external websites referred to in this book are correct and active at the time of going to press. However, the publisher has no responsibility for the websites and can make no guarantee that a site will remain live or that the content is or will remain appropriate.

developing tools for more effective utilization of records. For example, concerted efforts to survey biological anthropologists about the disposition of their papers not only yielded the beginnings of a compilation of locations but also succeeded in saving valuable biographical material from imminent destruction. An extensive computer file of archival locations of the papers of anthropologists in all fields is now available through the National Anthropological Archives web page (http://www.nmnh.si.edu/naa/).

Once records are deposited in archives, they serve two major functions. First, as primary data, they can be returned to again and again to clarify, critically evaluate, or reinterpret the research for which they were gathered; to elucidate the development of approaches to a topic; and, especially, to answer new questions. When they refer to peoples or cultures no longer extant, they may be our only link to those parts of the record of human activity and human diversity. Data records are therefore more than merely of historical interest; they will remain current as long as the anthropological enterprise continues. Second, these records contain the intellectual and social history of anthropology in its own right and as part of the history of science. It is only through the documents of anthropologists' lives and those of anthropological institutions and organizations that we can come to understand the processes through which our discipline, its work and its ideas, unfolded and the directions in which it is moving.

Human Biologists in the Archives is directed at the first of these functions: the use of records as primary data to solve scientific problems, including those that may not have been conceptualized at the time of the original data collection. It explores the rich sources of archival records from the past, compiled by anthropologists and non-anthropologists alike, that can be applied to new problems or to old questions rephrased in new ways. It finds pay dirt in sources ranging from church registers of vital events to government documents on emigration, from measures of the health status of schoolchildren to accounts of the devastating effects of epidemics and famine. The collection suggests that research based on archival materials might signal a newly emerging area of biological anthropology, *human biohistory*: the historical reconstruction of the human biology of past populations.

Skeletal biologists have been studying human history and prehistory and attempting to reconstruct the health and biology of human populations from skeletal remains for nearly a century. For example, paleodemography investigates past age/sex population distributions, mortality, and fertility; paleopathology considers diseases that are apparent from skeletal features; paleonutrition deals with the combined evidence from archeology and skeletal biology to reconstruct dietary status and health; and paleoepidemiology is likely to incorporate both skeletal and archeological evidence and historical and archival sources.

Burial grounds, cemeteries, battlefields, and other sites where skeletal populations may be found provide evidence for human biohistorical reconstruction and problem solving. Skeletal biologists have always drawn on historical documents, when available, to support their interpretations of hard-tissue evidence. However, the traditional means of reconstructing human health and biology from skeletal remains is now being augmented by a variety of new and renewed approaches that allow for a broader historical understanding of human population biology. It is worth considering briefly some of these studies, the sources of data they have used and the innovative approaches they have applied in their analyses.

Demographic data are abundant in the archival record, but compiling them for research purposes is often a tedious and time-consuming activity. In Mexico and Peru, parish records of births, deaths, and marriages may date back to the late sixteenth or early seventeenth centuries; in Europe records may go back much further. Other demographic data may be found in government censuses, actuarial data, military records, and government health records. Most of us are familiar with the migration study conducted during the early twentieth century by Franz Boas (1911), in which a demographic pattern (migration) was used to demonstrate a biological phenomenon (plasticity). Fewer may be aware of the fact that these data were published and are available for reanalysis (Boas 1928). Archival demographic data are most valuable when they are applied to interesting problems, including evolutionary problems. Differential fertility and mortality are the primary means of selection; out-migration is a response to deteriorating environmental, social, economic, and/or health conditions; genetic drift is a function of population size; and gene flow is tied to social and ethnic boundaries, mating practices, physical migration, and a host of other variables that have evolutionary implications. Basic data on such processes are often found in previously unexplored archives.

Historical information on child growth and maturation can be discovered in unusual places. It has for many years been recognized that the size of children is greater and maturation earlier in present times than in the past. This is known as the *secular trend* in growth (Tanner 1962: 143–55); it has been attributed to a number of factors, but principally it can be ascribed to improved health and nutritional status. Historical data from the past two centuries have been used to document these trends. Now, we are beginning to draw on archival records to provide the fine resolution in growth variation, variation most often linked to socioeconomic differentials. Increasing evidence is bringing to light cyclical patterns of growth and maturation linked to economic cycles, warfare, ethnic persecution, modernization, and social reform.

One study using archival records may serve as an example of just what can be done with imagination and good research design. Although there is abundant evidence that girls reach menarche earlier today than in the past, puberty indicators in boys are more difficult to document. Voice changing or 'breaking' is one indicator of maturation in boys, and where better to find this documented than in baroque boys' choirs of the eighteenth century? Daw (1970) used voice breaking or *cambiata* (transfer from treble to lower voice status) in boys who were members of J. S. Bach's Leipzig Thomasschule choir from 1727 to 1749 as an inferred measure of the transition through puberty. He found that the estimated age of maturation of these boys was in the late teens, compared with the early teens among contemporary boy choristers.

Health and epidemiological data can be found in numerous sources, including colonial archives (such as the Kenya Archives), public health records, military chronicles, and parish records of mortality. Reconstruction of catastrophic epidemics is not only a challenging intellectual venture in population history; these historical examples may also provide important lessons about social, environmental, and health patterns of epidemic disease transmission. A case in point is the awakened interest in the worldwide influenza epidemic that struck at the end of World War I. Hundreds of thousands of Americans died during the latter months of 1918, and few families remained untouched by this major pandemic. Again, a finer-grained resolution of those infected with the 1918 flu strain by socioeconomic level, ethnicity, and residence should be possible through careful work with archival sources. There are many problems of this nature that can only be solved with anthropological and human biological expertise.

The most challenging archival explorations are likely to be those that require integrated approaches in human biology. Archival demographic data can be used in studies of health, disease, growth, evolution, and social change, and to reconstruct longitudinal trends. Each problem will call for a variety of sources of historical evidence. Even individual biographical evidence can be drawn on to provide biobehavioral information about the past. Hints about marriage practices in the past, such as degrees of consanguinity permitted, can be found in biographical and historical documents; for instance, the prevalence of first-cousin marriages may be detected, as in the case of Charles Darwin and Emma Wedgwood. Another example of archival records being used to approach problems of contemporary interest is the study of Floud and colleagues (1990) of heights of boys in United Kingdom military schools between 1750 and 1980. They found remarkable variations in height, reflecting economic and health conditions at specific times. Finally, an interesting integrated problem might be approached with Peace Corps records, good evidence that has never been explored. These records include heights and weights of tens of thousands of volunteers over more than three decades of international (and often rigorous) experiences. A question that could be asked is: why do young men

xvi   *S. Silverman & M.A. Little*

lose body weight and young women gain body weight during their Peace Corps years? Among the variables that might be germane are level of maturation, body composition, health status, stress response, Peace Corps duties, and physical activity.

Human Biologists in the Archives is an important first step toward demonstrating the value of archival records for the exploration of contemporary problems in biological anthropology. Anthropologists have skills for understanding social behavior, population processes, evolution, development, and genetics through the integration of perspectives on human biology and on human society and culture. Abundant sources of archival data are available that can be exploited for such integrated analysis. This kind of research will surely grow as new records are discovered and as more anthropologists recognize their importance.

Sydel Silverman and Michael A. Little

References

Boas, F. (1911). Changes in the Bodily Form of Descendants of Immigrants. Senate Document 208, 61st Congress. Washington: U.S. Government Printing Office. (Published by Columbia University Press, New York, 1912.)

Boas, F. (1928). *Material for the Study of Inheritance in Man.* Columbia University Press, New York.

Daw, S.F. (1970). Age of boys' puberty in Leipzig, 1727–49, as indicated by voice breaking in J. S. Bach's Choir members. *Human Biology* **42**: 87–9.

Floud, R., Wachter, K. and Gregory, A. (1990). *Health, Height and History: Nutritional Status in the United Kingdom, 1750–1980.* Cambridge Studies in Population, Economy and Society in Past Time 9. Cambridge University Press, Cambridge (England).

Silverman, S. and Parezo, N. J. eds. (1995). *Preserving the Anthropological Record,* second edition. Wenner-Gren Foundation for Anthropological Research, Inc., New York.

Tanner, J. M. (1962). *Growth at Adolescence: With a General Consideration of the Effects of Hereditary and Environmental Factors Upon Growth and Maturation from Birth to Maturity,* 2nd edition. Blackwell Scientific Publications, Oxford.

Acknowledgements

The editors wish to express their appreciation to a number of individuals without whom this volume would not have been possible. Dr Michael Little encouraged us to work up for publication the proceedings of a symposium, *Human Biology in the Archives*, jointly sponsored by the Human Biology Association and American Association of Physical Anthropologists. Dr Kathryn Denning managed the project during the first year, giving insightful editorial comments and gently keeping us on track. Karen Mason provided invaluable assistance with copy editing and indexing. We are also grateful to Tracey Sanderson, Carol Miller, and Lynn Davy at CUP for expert production and editorial assistance.

Cover: Family Register from the collection of the Pocumtuck Valley Memorial Association, Memorial Hall Museum, Deerfield, Massachusetts.

1 Human biologists in the archives: demography, health, nutrition and genetics in historical populations

ALAN C. SWEDLUND AND D. ANN HERRING

Introduction

A few years ago we began a conversation about how the use of archival data in biological anthropology was changing. Whereas in the past archival research was conducted by a very small number of us (see Foreword, this volume), that number seemed to be growing steadily. In addition, questions, themes, and approaches appeared to be emerging that gave archival research its own character and rationale. Rather than simply being an adjunct, less desirable alternative, or even an afterthought to field or laboratory work, archival research had become the approach of choice for some researchers and some questions. Today, a significant number of physical anthropologists engage in research on populations whose data come primarily from the historical record. In addition, distinctive approaches have emerged from these investigations that can be differentiated from the work of colleagues in, for example, historical demography or the history of medicine (see chapter 14). We also observed that there has been little opportunity to assemble a body of this work or to reflect on its common themes or its contributions to theory and method in physical anthropology. The purpose of this volume is to provide a forum in which researchers who are actively engaged in historical projects present their current research and at the same time more explicitly consider its place in physical anthropological theory and method.

The articles in this volume are new contributions to the physical anthropology literature. They are based on updated and expanded papers presented in a symposium entitled *Human Biology in the Archives*, held in Columbus, Ohio, in April, 1999. The symposium was jointly sponsored by the Human Biology Association and the American Association of Physical Anthropologists. For these authors, the classic anthropological 'field' is not the glamour of research in an exotic locale; rather, it often involves long hours consulting collections curated in the dusty back rooms of libraries, laboratories, and museums. Data collection can be tedious anywhere, but these repositories rarely offer the compensatory experiences that can be found in many field locations. What archives do provide are rich sources of information and multiple lines of evidence that can be used to reconsider traditional questions or address new concerns in biological anthropology. By virtue of their historical context, they also invite another layer of questions about time, place, and purpose surrounding their original assemblage and the circumstances related to their eventual preservation.

The contributors to this volume present a diverse array of archived human biological evidence from a variety of sources, including the archeological record, skeletal collections, and school, government and church documents. This collection showcases a variety of approaches, and includes investigators who are beginning their careers in archival research as well as senior scholars who have been engaged in this kind of investigation for many years. The chapters demonstrate the ways in which the analysis of archived historical documents and biological remains expands the horizons of research in human biology, fills gaps in the chronologies of anthropological populations, extends the longitudinal analysis of microevolutionary and social processes into the present, and enhances our understanding of the human condition.

What is an archive?

In conventional terms most of us probably think of an archive in a somewhat restricted, dictionary sense, as a collection of written records. But the archives of human biology are much more than that. As we look over the use of archival approaches that are being employed in biological anthropology today we see that they include the published and unpublished data of numerous studies in anthropometry, growth and development, genetics, nutrition, demography, paleoanthropology, and so forth. However, they also increasingly employ digitized files and manipulations of these derived forms of the original data. Importantly, they also include repositories of fossils, skeletal remains, photographs, blood and tissue samples, and even cell lines used in DNA research. For those of us who investigate human biology in an archival context, therefore, the definition of archive is broad and extends well beyond written documents. While it was not possible to explore the full range of data sources for this volume, we want to acknowledge and encourage the continuing experimentation with diverse types of archives. What is central to us is that these be seen for their historical significance and that researchers are encouraged to preserve their original data for the use of future investigators, as called for in the Foreword to this volume.

Many of the archives traditionally used in research by physical anthropologists are 'accidental' data sets. That is, the data were gathered for purposes quite different from the ways in which we deploy and interrogate them today. A most obvious example would be the census records and vital statistics marshaled by states, churches, and municipalities and which make up the great body of archives used in historical demography and population structure. Another might be the anthropometric data amassed on men who were historically being surveyed for eligibility for military service. In these examples the empowered agencies desired this information primarily to maintain surveillance over, tax, marry, conscript and bury their citizenry. However, now a very large body of literature exists using these same kinds of data for studies in mortality, growth, nutrition, population structure/genetics, and epidemiology (see, for example, Damon 1968; Tanner 1981; chapter 14, this volume).

A second type of archive is one that has been created as a result of questions that can be considered explicitly anthropological/biological, such as anthropometric data gathered during anthropological surveys. These archives provide the opportunity to revisit and reevaluate previous research in light of new technologies, approaches and methodologies, as well as to analyze data anew to address contemporary questions (Jantz *et al.* 1992; chapter 3, this volume). With the penetration of globalization, the increasing cosmopolitan-ness of indigenous groups, the independence of formerly colonized peoples, and the increasingly difficult conditions under which fieldwork is conducted in the twenty-first century, these kinds of archives will not only be useful in the years ahead, but we would argue that they will be essential to the continued existence of biological anthropology and to the future of the biocultural enterprise as a whole (see Goodman and Leatherman 1998; Smith and Thomas 1998). Whether the area of inquiry be human genetics, bioarcheology, or paleoanthropology, in our estimation, new data collected specifically for studies of human biology will become increasingly difficult to acquire. Legitimate concern by Aboriginal groups in Canada and the United States, for instance, about the removal, study and reburial of the skeletal remains of indigenous people has effectively limited analysis of newly discovered sites (Waldram *et al.* 1995; Thomas 2001); resistance by indigenous people to the collection, analysis and curation of samples of their DNA is another example of the same phenomenon. The positive sides of this are that there are so many useful archives now available (Foreword, this volume), and that many productive collaborations between First and Third World research teams and between scientists and indigenous groups are being fostered.

Biological anthropologists seldom find time to reflect, or at least write, on the power and privilege that has allowed us to conduct research in many Third World and colonial situations. The historical archive, because it is often a product of colonial administrations, does not have to be a military census or measurements taken on institutionalized individuals to remind us that the original agenda that led to the creation of the archive might be one very different from our current, scientific purpose. In being so reminded the archival researcher may find herself addressing questions somewhat differently than she would otherwise. Several contributions in this volume illustrate how different historical archives – be they from a state-run asylum in New York (chapter 6) or a Roman Catholic mission in California (chapter 4) – can be used to study the biological consequences of political, religious and medical regimes.

One of the other benefits of researching archival data has been that it has often focused attention on 'ourselves': namely, Western and Westernized populations from whom most biological anthropologists working today descend. 'We' become 'the other.' Since these populations, primarily in Europe and North America, have such a long record of documenting and recording numerical data, it is these populations that now make up the bulk of archival sources. Indeed, many of the major studies on historical population structure originated with populations in the United Kingdom and continental Europe (see chapter 14 for examples).

One such project, the Otmoor–Oxfordshire study conducted in England by Drs Geoffrey Harrison and Anthony Boyce (see, for example, Boyce *et al.* 1967) was a primary catalyst to one of us (Swedlund) choosing to do historical research. Another project, on the British colony of Gibraltar by Dr Larry Sawchuk (see, for example, Sawchuk 1980; Sawchuk and Flanagan 1979), sparked the interest in this kind of research for the other author (Herring). The Otmoor study also represents a very good example of how archival data on the history of a population can also be linked to research on contemporary human biological variation. The fact that, in the popular imagination, physical anthropologists are usually only recognized as scholars of 'old bones' and 'exotic' populations is illustrated by an anecdote regarding the Otmoor study. While one of us (Swedlund) was attempting to order a copy of Harrison's recent compendium titled *The Human Biology of the English Village* (Harrison, 1995), the clerk remarked, 'That is a curious title!' When asked how she meant 'curious' she indicated that the title was 'odd' or 'strange'. Indeed, we suspect someone from a university press or bookstore would find nothing odd in a title like '*The Human Biology of the Peruvian Village*' or '*The Human Biology of the Australian Aborigine*,' but somehow the topic seemed curious when applied to a contemporary British village and its recent historical records. Dr Harrison is to be commended for this title and its topic, as it demonstrates the way in which investigations of contemporary Western societies tends to democratize our research questions, whether they are archival or not.

Explorations in archival research

Much of the early work of human biologists in the archives is perhaps best characterized as genetic demography (see, for example, Crow and Mange 1965;

Küchemann *et al.* 1967; Lasker 1977; Roberts 1971). Archival data were used primarily to develop genealogical histories of populations with a view to understanding the operation of microevolutionary processes of human adaptability, gene flow, genetic drift and natural selection. As more research was undertaken, it became increasingly clear that archival projects lead to questions outside of this traditional biological domain and demand an exploration of social processes, historical moments, and the lives of individuals represented in the records (see chapter 14). Human biologists studying archival data are inevitably confronted with interpretive dilemmas around issues of epistemology, representation, the large and small tragedies of daily life, and the history of physical anthropological theory and method.

While archival data continue to offer an important and effective means for examining the genetic history of populations, the results are increasingly interpreted in terms of historical contingency, political economy, and the complexity of the sociocultural context that shapes biological and genetic processes (chapters 2, 3 and 14). Historical archival research provides added depth and new data with which to address questions about the way in which, for example, episodes of infectious disease affect the growth and development of children (chapter 7), about the sociopolitical circumstances that influence the course of epidemics (chapters 8, 9 and 10), how epidemics spread from place to place (chapter 11), and how gender relations affect vulnerability to disability, disease and death (chapters 4, 6 and 12). It also affords the opportunity to examine the way in which particular health issues, such as malnutrition, have been understood as the field of human biology itself has developed (chapter 13). Skeletal and cemetery records are being carefully linked to show how each informs or compensates for weaknesses in the other and how the tissue and documentary records themselves have been formed by social circumstances (chapters 4, 6 and 12).

The chapters in this volume do not fit comfortably within any identifiable subfield of physical anthropology. We would argue that this is in the nature of archival research. The areas of potential investigation are only limited by the availability of data in some archival form, and by the imagination of the investigator. In organizing the chapters for this volume we therefore have avoided assigning the contributions into standard categories such as osteology, growth and development, or genetics. Rather, we see the contributions as loosely structured around four main themes that cross-cut the classic boundaries within physical anthropology: population history (chapters 2, 3 and 4), the biological consequences of institutional living (chapters 5, 6 and 7), the impact of demographic and epidemiological crises (chapters 8, 9 and 10), and methodological and epistemological implications of archival research to human biological inquiry (chapters 11–14). Even these groupings are not mutually exclusive, but they serve to draw the reader's attention to the primary research question of each contribution. In addition, the reader will note that the articles cover a range of geographical locations (Costa Rica, Ireland, Canada, Gibraltar, USA, Finland, England) and historical periods from the sixteenth to the twentieth centuries.

Contributions

Physical anthropologists are carving out distinctive niches in archival research if the chapters in this volume can be considered representative of the area as a whole. This is beautifully illustrated in Smith's masterful review of central tendencies in the development and practice of archival anthropology (chapter 14). He makes the case that, whereas anthropologists, historians and geographers often explore similar archival terrain and employ similar methods (and perhaps feel that they share more intellectual ground with each other than they do with colleagues in their disciplines), the work of human biologists in the archives is indeed distinctive. For Smith, elucidation of evolutionary mechanisms and investigation of the components of the genetic structure of populations fall unequivocally within the domain of biological anthropology and sets it apart from other areas of inquiry. The boundaries with other disciplines become blurred, however, when that research touches on questions relating to fertility, mortality and disease, as it inevitably must. Here human biologists occupy common ground with social and medical historians, demographers, historical geographers, and even economists and political scientists, and each informs the perspective of the others in both small-scale local analyses, and in developing global-scale models of human behavior. Smith believes that an interdisciplinary approach, coupled with the integration of fine-grain local studies, is necessary for any systematic analysis of the effects of broader social processes and historical contingency on human biology and microevolution.

The crucial place that context and contingency occupy in understanding human biology through time is not always made explicit in research in physical anthropology. Often the study 'subjects' are treated in isolation of the dominant populations with whom they interact – what Smith and Thomas (1998: 460) call 'the biological cocoon.' Yet that dynamic figures prominently in many of the chapters in this volume. This can be seen clearly in Herring and colleagues' chapter (13) on the deleterious impacts of the fur trade and government policies on health and nutrition among the Moose Factory Cree of northern Canada. In Walker and Johnson's study (chapter 4) we observe the destructive effects of Spanish colonial policy, and religious and political agendas, on the health and reproductive patterns of the Chumash of coastal California. Higgins and Phillips (chapters 5 and 6) examine the health of institutionalized people who are in a very different social and medical relationship to the dominant society than their non-institutionalized counterparts. Higgins' study of nineteenth century infant death in the Erie County Almshouse (New York State) lends support to the idea that the underlying motive for establishing such institutions was to deter all those but the most desperate from seeking relief. Phillips demonstrates how nineteenth century labor therapy at the Oneida County Asylum for the Mentally Ill (New York State) – which was ostensibly designed to control the inmates, reduce the costs of running the institution, and 'recreate a normal life' – resulted in extreme skeletal robusticity among the inmates as they were literally 'worked to the bone'. Leidy Sievert (chapter 7) looks at a very different kind of institutionalized population, one that is statistically at the other end of the spectrum, that is, privileged and economically well-cared-for boarding school students in a twentieth century New England middle school. This population permits her to investigate questions related to nutrition and infectious disease in a controlled, natural experiment where the positive impacts of adequate nutrition and high socioeconomic status are much more easily observed.

Relethford's contribution (chapter 3) is noteworthy not only for the intriguing inferences he makes about the population history of Ireland, but also for his creative use of anthropometric data for adult Irish men and women, collected by researchers at Harvard University during the 1930s and by researchers affiliated with the Anthropometric Laboratory of Trinity College, Dublin, in the early 1890s. His comments on the historical context of these data make explicit the political nature of this research in its earlier manifestation. Whereas these data originally formed the basis for antiquated questions about racial difference, here Relethford uses them instead to look at a contemporary question about how anthropometric variation can inform our understanding of population history in a manner similar to variation in DNA. In his re-analysis of the data, he argues that spatial variation in anthropometric patterns reflects social processes by which communities in Ireland were formed, especially the influence of Viking and English invasions, settlement and admixture.

Madrigal (chapter 2) focuses her cool critical eye on the parish records from Escazú, Costa Rica. Her chapter presents a classic analysis of population structure revealed through fertility, mortality and marriage records. The results confirm observations made in several historical populations, namely, that the evidence for the effects of gene drift are weaker than those of gene flow through high mobility of marriage partners. She finds that variation in fertility is probably less important than differential mortality in accounting for biodifferentiation.

Mielke (chapter 10), Sawchuk and Burke (chapter 9), and Swedlund and Donta (chapter 8) explore mortality under the stressful conditions of war, crowding and epidemics. Historically, these three factors are often intertwined, as Mielke demonstrates in his study of regional patterns of death on the

Åland Islands, an archipelago located between Sweden and Finland, during the War of Finland (1808–09). His analysis of the extraordinarily rich archival resources for the archipelago helps to disentangle the impacts of outright hostility and aggression from other social and biological effects that accompany military movements and activities. Mielke's study illustrates very clearly the ways in which elements in the 'epidemiological landscape' (Dobson 1992), in conjunction with sex and age, shape spatial variation during mortality crises.

Sawchuk and Burke examine the impact of cholera on the wonderfully well documented civilian population of Gibraltar during the nineteenth century. Their analysis of the differential impact of cholera in the various districts of the town illustrates how inequalities in wealth, status and residential location are strong determinants of vulnerability to morbidity and mortality (cf. Farmer 1996). Where Madrigal (chapter 2) would suggest that differential mortality is a function of natural selection in Escazú, Sawchuk and Burke (chapter 9) emphasize the socioeconomic and cultural risk factors contributing to the risks of contracting and dying from cholera in Gibraltar. In Swedlund and Donta's chapter (8) the epidemic of interest is the scarlet fever pandemic (*Streptococcus pyogenes*) of the second half of the nineteenth century. Their analysis of scarlet fever deaths during 1858–59 and 1867–68 is directed toward four communities studied by the Connecticut Valley Historical Demography Project. They observe that a number of deaths occurred in households considered to be of middle-to-high socioeconomic status, exemplifying that 'democratic diseases' like scarlet fever flow easily across class and other social boundaries, in contrast to 'undemocratic diseases' like cholera (Porter and Ogden 1998). Not only do they evaluate the demographic, economic and cultural factors that contributed to the likelihood of death from scarlet fever, but they also raise questions about the virulence of the pathogen itself. They conclude that enhanced pathogenic virulence may have been a major determinant of this local manifestation of the larger epidemic, a timely reminder that in seeking explanations for epidemics, we ignore at our peril the significance of evolutionary shifts in the pathogens themselves (Ewald 1991a,b).

Grauer and Sattenspiel's contributions (chapters 12 and 11) caution that all is not rosy under the archival sky. Each of these chapters deals with aspects of missing data, and provides approaches and suggestions of how to deal with data deficiencies. In Grauer's chapter, the case is made by comparing English Medieval history, as represented in documents, to another, different history exposed through the analysis of skeletal remains from cemeteries from the period. She addresses the problem of the under-representation of women in written documents, relative to the abundance of their remains in cemetery samples. The problem is not uniquely English, Medieval, or even historical, but is amplified in historical data because the socioeconomic status of males has traditionally meant a greater emphasis on the recording of their lives, events and activities. Grauer makes the case that coupling skeletal data with historical records can begin to redress this imbalance (see also Grauer 1995; Saunders and Herring 1995).

Sattenspiel (chapter 11) takes on the broader issue of how archival taphonomy and sampling processes can leave us with data that are inadequate for the questions we wish to ask. By using mathematical distributions known to represent demographic and epidemiological processes fairly well, and by applying mathematical and computer modeling approaches, she makes the case that it is often possible to provide estimates for missing parameters that, in turn, make new inferences possible. She shows how effective models can be developed and illustrates the process by way of her research on the spread of the 1918–19 influenza epidemic in Cree fur trapping communities in the Canadian subarctic. Sattenspiel argues that mathematical modeling enhances traditional approaches to archival research because it provides direction, focus, and a powerful tool for identifying factors that are most influential, less important or even insignificant in demographic and epidemiological patterns.

Finally, we have aimed to give the reader a cross section of archival studies that shows the varieties of themes and approaches that are currently being investigated. There are many other fine examples emerging in the literature and, taken together, we believe that these justify the argument that human biology in the archives is a growing field with a promising future. We would like to add that we find in many of these studies something that is characteristic of the best traditions in biological anthropology, no matter where undertaken or what time period is under consideration. That is, virtually all of these studies deal with communities or populations in a way that enriches our understanding of them beyond the simple reporting of numbers and events. The best archival research strives to bring life to the aging documents and museum collections on which we depend. In doing so, not only is it possible to carry out more interesting science, but also to represent and give voice to the individuals whose lives are represented in the archival record.

References

Boyce, A.J., Küchemann, C.F. and Harrison, G.A. (1967). Neighbourhood knowledge and the distribution of marriage distances. *Annals of Human Genetics* **30**, 335–8.

Crow J.F. and Mange, A.P. (1965). Measurement of inbreeding from the frequency of marriages of persons of the same surname. *Eugenics Quarterly* **12**, 199–203.

Damon, A. (1968). Secular trend in height and weight within old American families at Harvard, 1870–1965. *American Journal of Physical Anthropology* **29**, 45–50.

Dobson, M.J. (1992). Contours of death: Disease, mortality and the environment in early modern England. In *Historical Epidemiology and the Health Transition*. ed. J. Landers. *Health Transition Review* **2** (suppl.), 77–95.

Ewald, P.W. (1991a). Waterborne transmission and the evolution of virulence among gastrointestinal bacteria. *Epidemiology and Infection* **106**, 83–119.

Ewald, P.W. (1991b). Transmission modes and the evolution of virulence, with special reference to cholera, influenza and AIDS. *Human Nature* **2**, 1–30.

Farmer, P. (1996). Social inequalities and emerging infectious diseases. *Emerging Infectious Diseases* **2**(4), 259–69.

Goodman, A.H. and Leatherman, T.L. (1998). Traversing the chasm between biology and culture: an introduction. In *Building a New Biocultural Synthesis: Political-economic Perspectives on Human Biology*. ed. A.H. Goodman and T.L. Leatherman, pp. 3–41. Ann Arbor: The University of Michigan Press.

Grauer, A. (ed.) (1995). *Bodies of Evidence: Reconstructing History Through Skeletal Analyses*. New York: Wiley-Liss.

Harrison G.A. (1995). *The Human Biology of the English Village*. Oxford: Oxford University Press.

Harrison, G.A. and Boyce, A.J. (eds) (1972). *The Structure of Human Populations*. Oxford: Clarendon Press.

Jantz, R.L., Hunt, D.R., Falsetti, A.B. *et al.* (1992). Variation among North Amerindians: Analysis of Boas's anthropometric data. *Human Biology* **64**, 435–61.

Küchemann, C.F., Boyce, A.J. and Harrison, G.A. (1967). A demographic and genetic study of a group of Oxfordshire villages. *Human Biology* **39**, 251–76.

Lasker G.W. (1977). A coefficient of relationship by isonymy: a method for estimating the genetic relationship between populations. *Human Biology* **49**, 489–93.

Porter, J.D.H. and Ogden, J.A. (1998). Social inequalities in the re-emergence of infectious disease. In *Human Biology and Social Inequality*, ed. S.S. Strickland and P.S. Shetty, pp. 96–113. Cambridge: Cambridge University Press.

Roberts, D.F. (1971). The demography of Tristan da Cunha. *Population Studies* **25**, 465–79.

Saunders, S. R. and Herring, A. (eds) (1995). *Grave Reflections: Portraying the Past Through Cemetery Studies*. Toronto: Canadian Scholars' Press.

Sawchuk, L.A. (1980). Reproductive success among the Sephardic Jews of Gibraltar: evolutionary implications. *Human Biology* **52**, 731–52.

Sawchuk, L.A. and Flanagan, L.E. (1979). Infant mortality among the Jews of Gibraltar, 1869 to 1977. *Canadian Review of Physical Anthropology* **2**, 63–72.

Smith, G.A. and Thomas, R.B. (1998).What could be: biocultural anthropology for the next generation. In *Building a New Biocultural Synthesis: Political-economic Perspectives on Human Biology*, ed. A.H. Goodman and T.L. Leatherman, pp. 451–74. Ann Arbor: The University of Michigan Press.

Tanner, J.M. (1981). *A History of the Study of Human Growth*. Cambridge: Cambridge University Press.

Thomas, D.H. (2001). *Skull Wars: Kennewick Man, Archaeology, and the Battle for Native American Identity*. New York: Basic Books.

Waldram, J., Herring, D.A. and Young, T.K. (1995). *Aboriginal Health in Canada: Historical, Cultural and Epidemiological Perspectives*. Toronto: University of Toronto Press.

2 The use of archives in the study of microevolution: changing demography and epidemiology in Escazú, Costa Rica

LORENA MADRIGAL

Introduction

Archival data constitute a rich source of information for microevolutionary studies, a primary focus of human biology and biological anthropology research. Particularly important is that, by virtue of their ubiquity, archives allow researchers to study changes in demography and epidemiology in multiple cultures and time settings. Thus, it is possible to compare microevolutionary processes across cultures at different time periods, and across time within a group. For example, Scott and Duncan (2001) look at the behavior of different epidemics throughout Europe, and at different time periods. Scott and Duncan also (1998) provide a cross-cultural review of the impact of disease on human demography.

Several archival approaches to microevolutionary studies are possible, all of them of interest, and all of them depending on the data available. However, in the best of situations, it could be possible to study mortality and fertility through death and birth/baptism archives, and population structure and gene flow through marriage records. This chapter presents a case study of changes in demography and epidemiology in the historical population of Escazú, Costa Rica, through archival research, which incorporates the mortality, birth, and marriage patterns of the population.

The specific question addressed about mortality is whether it had a seasonal distribution, as has been described in similar human populations. Seasonality of deaths has been researched in recently collected data sets (Bako et al. 1988; Bowie and Prothero 1981; Dzierzykray-Rogalski and Prominska 1971; Hajek et al. 1984; Kalkstein and Davis, 1989; Malina and Himes 1977; Mao et al. 1990; Schwartz and Marcus 1990; Shumway et al., 1988; Stroup et al. 1988), as well as in longer series (Barrett 1990), some of which were started in previous centuries (Breschi and Bacci 1986; Galloway 1988; Landers 1986; Lin

11

and Crawford 1983; Marcuzzi and Tasso 1992; Mielke *et al.* 1984). Studies of long mortality time series have also determined the presence of long-term trends, and have frequently reported a decrease in seasonal variation of deaths (Barrett 1990; Lin and Crawford 1983). It is likely that a decline in mortality fluctuations results from a decrease in the importance of season-specific diseases, and from an increase in the importance of degenerative ones (Barrett 1990; Lin and Crawford 1983). Diseases with strong seasonal fluctuations include gastrointestinal maladies, more frequent in the wet and warm months, and upper respiratory diseases during the winter months (Barrett 1990; Kalkstein and Davis 1989; Landers 1986; Malina and Himes 1977).

This chapter also looks at a related question about the fertility data for Escazú: do births follow a seasonal distribution? Seasonality of human births appears to be an almost universal phenomenon in agricultural groups. It has been reported in populations from different geographical and cultural areas as well as from different historical periods (Lam and Miron 1991; Leslie and Fry 1989). Lam and Miron (1991) summarized the most commonly reported explanations of birth seasonality, namely: 1, weather (which may affect the probability of conception, or the frequency of coitus); 2, agricultural cycles (since labor demand may affect the desired timing of births or the absence of spouses from home); 3, other economic variables (such as labor migration); 4, holidays (since festivities may affect the frequency of sexual activity); and 5, marriage seasonality (in communities where marriages are seasonal). It is of interest to study whether Escazú experienced fertility cycles because such cycles appear to be more noticeable in peasant societies (Huss-Ashmore 1988), and because they have not been studied in Central American populations from the nineteenth century.

To address the issue of seasonal variation of births and deaths, climatic data (rainfall and temperature) were also collected, looking for an impact of climate on vital events.

The marriage records were analyzed with the aim of computing the level of inbreeding, as well as analyzing the frequencies of surnames new to the records, by gender. Moreover, the marriage records allow an examination of population structure by showing if there were any systematic barriers to the flow of genes within the group.

The population and its records

The Parish of San Miguel de Escazú (Escazú for short) is 15 km from San José, Costa Rica's capital. Both archeological and historical records support the notion that Escazú was not an Amerindian settlement, but that it was a settlement founded during colonial times, sometime in the late 1600s (Sibaja

1970). A few civil and ecclesiastical documents from the early 1700s mention the emerging settlement, and indicate that the area was inhabited part of the year by families who raised corn, sugar cane and plantains. These families were likely the result of Spanish and Indian admixture, as is most of the Costa Rican population (Sibaja 1970).

During the 1700s, the Spanish Empire authorities passed and enforced laws to settle into communities the very dispersed and rural Costa Rican population. Thus, edicts were issued to the effect that the rural population had to coalesce into the (then) emerging cities. Interestingly, the inhabitants of the Escazú area resisted these orders, and issued a letter to the governor of the province in 1755 in which they presented their case for staying in the area. The Costa Rican governor asked the colonial government in Guatemala to rule on the appeal. The Guatemalan seat allowed some community members to stay in Escazú, as long as they could build houses in San José, where they would spend holy days. For the rest of the year, they were allowed to permanently settle Escazú. Otherwise, individuals who could not afford two houses were ordered to relocate to San José. Thus, Sibaja (1970) suggests that after this edict, a core of families remained in Escazú, and that the population of San José received a number of families who did not have the wealth to remain in the rural areas.

After this mandated depopulation, the population of Escazú increased enough during the second part of the 1700s (presumably out of this core of families) for the settlers to request from the Church the establishment of a parish with a permanent priest. In response to this request, the Catholic Church established a parish in 1799. The population continued to increase, with the 1864 national census reporting 2533 inhabitants (Ministerio de Economia y Hacienda 1964). Migration into the community was also probably an important force in the population growth.

Costa Rica became a free republic in 1821, and the forced urbanization laws enacted during the first part of the 1700s were discontinued. González-Garcia (1983) notes that there was a veritable demographic explosion in Costa Rica at the end of the 1700s, especially in the Central Valley, where the four largest cities (including San José) and Escazú are located. As a result of this population increase, there were population movements out of these four cities to rural regions, particularly those within the central valley (Chaves-Camacho, 1969). This population movement was supported by a governmental effort to distribute freely, or for a small price, unused land in the Central Valley, in an effort to promote coffee farming by small family homesteads (Sáenz-Maroto 1980). Several sources indicate that Escazú, like other rural regions in the Central Valley surrounding the four major cities, received migrants from the urban centers after 1800. Escazú remained a distinct population from San José until recently. Currently, it is virtually a suburb of the capital.

Since its foundation in 1799, the Parish has kept excellent vital event certificates. During the 1800s, Catholic priests were appointed civil servants, with the obligation of registering all vital events, even if non-Catholics were involved. For example, Madrigal (1992, 1993, 1994) has noted that the deaths of persons who refused to take, or were denied the last sacraments, were routinely recorded (with an explanation of the situation). It is fairly certain that all vital events were recorded (P. Howell-Castro, priest of Escaźu, personal communication 1995); in addition, some priests included further information such as detailed descriptions of the deaths (Madrigal 1992). Besides the Church records, there are two civil documents (one already mentioned) from the second part of the 1700s, which list the names of community members. These documents allow a reconstruction of the earliest family names in the region.

Although the church records are of excellent quality, they are unfortunately missing for the decade of 1840. Thus, all of the analysis represented here will exclude by necessity that decade. The specific years covered by the certificates analyzed here are: mortality 1851–1921, baptismal 1851–1901, and marriage 1800–1839 and 1850–1899.

In order to study the seasonality of vital events, climatic data consisting of monthly temperatures (for 1888–1901) and rainfall (1888–1921) averages for

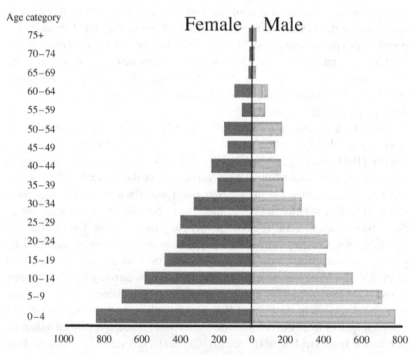

Fig. 2.1. Population pyramid based on the 1864 census.

San José were collected. Because San José and Escazú are only 15 km apart and have almost the same elevation (1100 m in San José, slightly higher in Escazú), they have virtually the same climatic pattern (Costa Rican National Meteorological Institute, personal communication 1986). The climate of San José and Escazú can be described as mild, with small temperature fluctuations (an average of 18.85 °C in January and an average of 20.12 °C in July for 1888–1901), but with marked rainfall fluctuations (averaging 217.16 mm in July compared with 40.58 mm in December during 1888–1921).

Mortality patterns

This section considers whether mortality was seasonally distributed, and whether mortality changed through time in Escazú. Based on the three censuses taken during the 1800s, Arriaga (1976) computed the life expectancy in Costa Rica for both sexes as 27.7 (1864), 28.9 (1883) and 30.5 (1892). Population pyramids based on the 1864 and 1892 censuses were developed for Escazú, and are shown in Figs. 2.1 and 2.2. It is very clear that the earlier pyramid is one of a population with very high fertility and large infant and childhood mortality. In contrast, the 1892 pyramid is less steep, indicating that the population had

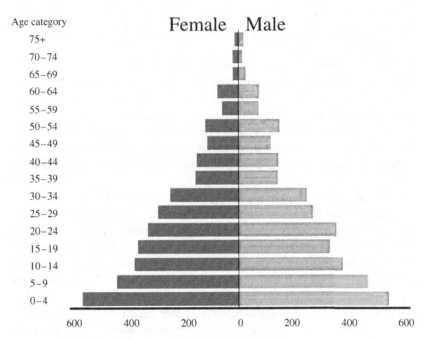

Fig. 2.2. Population pyramid based on the 1892 census.

Table 2.1. *Descriptive statistics of mortality (data adjusted for calendar differences) 1851–1921*

Month	Number of deaths	Mean	Standard deviation
January	505.16	7.11	4.99
February	544.28	7.66	4.92
March	579.67	8.16	4.69
April	588.00	8.28	4.62
May	713.22	10.04	6.58
June	829.00	11.67	6.94
July	801.29	11.28	6.81
August	600.96	8.46	5.58
September	524.00	7.38	4.45
October	530.32	7.46	4.33
November	464.00	6.53	3.68
December	460.64	6.48	3.72
χ^2	280.04		
d.f.	11.00		
p	<0.0001		
Freedman	6.47		
n	3678.716		
p	<0.01		

experienced some degree of change as predicted by the demographic transition model (Molnar 1998).

Table 2.1 shows descriptive statistics of the mortality series by month during 1851–1921 (data adjusted for calendar differences; see Madrigal 1994). The mean and sum of deaths is highest during the early rainy season (May–July), and lowest during the dry months of November, December and January. Table 2.1 also shows the results of the χ^2 and Freedman tests (Freedman 1979). Both tests yield highly significant statistics, rejecting the null hypothesis of equal number of deaths per month (see Appendix for a brief discussion of Freedman's test).

Although the wet months are also the warmer ones, and the dry ones also the coldest, rainfall and not temperature is most likely the main precipitating factor of the mortality seasonality, since temperature fluctuates very little in this region. An analysis of the monthly distribution of respiratory-related and gastrointestinal-related deaths by month was employed to determine if either had a seasonal distribution, and is presented in Table 2.2. Cause of death information only began to be maintained in the records in 1867. The number of gastrointestinal-related deaths was highly significantly different by month whereas the number of respiratory-related deaths by month was not significantly

Table 2.2. *Number of gastrointestinal- and respiratory-related deaths (data adjusted for calendar differences) 1867–1921*

Month	Gastrointestinal ailments	Respiratory ailments
January	43.548	18.387
February	43.929	19.286
March	52.258	20.323
April	53	15
May	68.71	10.645
June	100	20
July	115.161	15.484
August	72.581	15.484
September	45	18
October	54.194	25.161
November	41	13
December	31.935	18.387
χ^2	115.15	9.15
d.f.	11	11
p	0.0001	n.s.
Freedman	4.22	0.76
n	721.316	209.157
p	0.01	n.s.

different from that proposed by the null hypothesis of equal distribution. The table shows that, whereas gastrointestinal diseases showed a peak during the wet months, the number of upper respiratory diseases remained fairly stable through the year. This analysis of cause of death strongly suggests that gastrointestinal maladies (associated with waterborne pathogens and hence rainfall) were the main cause of mortality seasonality in Escazú. Upper respiratory tract diseases (associated with temperature fluctuations), in contrast, did not show any discernible seasonal pattern.

In order to determine whether the long-term seasonal distribution of mortality changed in the time series under study, a time-series analysis employing the Box–Jenkins methodology was applied (see Appendix for a discussion of the Box–Jenkins technique). After repeated attempts, the entire series could not be modeled, but had to be split as suggested by Jenkins (1979). The first part (1851–1891) was successfully modeled including a seasonal parameter. In contrast, the second part of the series (1892–1921) did not incorporate a seasonal parameter. (The Appendix includes a discussion of the models, and an interpretation of their components.)

That different models were needed to fit the two sections of the time series indicates that the mortality pattern of the population changed through time;

specifically, it became non-seasonal. This is a clear example of the underlying structure of a time series changing through time, as discussed by Jenkins (1979). The data suggest that the ecological and cultural characteristics that allowed pathogens to have a cyclical behavior in Escazú began to disappear sometime during the late 1800s. This could be simply the result of basic hygienic practices, although there are no records available to document this suggestion.

Fertility patterns

Seasonality of human births appears to be an almost universal phenomenon in agricultural groups. It has been reported in populations from different geographical and cultural areas as well as from different historical periods (Lam and Miron 1991; Leslie and Fry 1989). In contrast, lack of fertility cycles has rarely been reported (Arcury *et al.* 1990). Indeed, Arcury *et al.* (1990) report that only one published study was identified in which no seasonality of births was demonstrated. It was of interest to determine whether births were seasonally distributed just as mortality was found to be, and whether fertility could be tied to climatic variables, just as mortality could.

In Escazú during the time period under consideration, the priests did not record births, only baptisms. According to the current Parish head (P. Howell-Castro, personal communication 1995), the baptismal records are an accurate reflection of the actual number of births in the population: he estimated that all infants were baptized within 2 or 3 days of their birth. Indeed, Madrigal corroborated that even infants who appeared to be in imminent danger of dying soon after birth (as stated in the baptismal record) were baptized, and their deaths recorded soon after (Madrigal, 1993).

The calendar-adjusted number of baptisms by month for the entire period under analysis (1851–1901) is shown in Table 2.3. The sum, mean number and standard deviation of baptisms across all months are remarkably similar, with a high of 19.39 in April and a low of 16.13 in December. Although a plot of the baptisms across months did not imply any clear seasonality (not shown), the slight increase in April did cause the hypothesis of equal distribution of baptisms across months to be rejected at the $p = 0.05$ level. A Freedman test was not attempted because the distribution of baptism was not clearly seasonal, as was that of mortality (see Appendix).

The Box–Jenkins approach was used to fit the baptism series, which was successfully fitted with a non-seasonal model (see Appendix). It is of note that, whereas the mortality series was fitted with two models because its behavior changed throughout the years, the baptismal series was fitted with a single

Table 2.3. *Descriptive statistics of baptisms (data adjusted for calendar differences) 1851–1901*

Month	Number of baptisms	Mean	Standard Deviation
January	861.29	16.89	5.61
February	864.64	16.95	5.54
March	930.97	18.25	5.53
April	989.00	19.39	5.44
May	924.19	18.12	4.80
June	882.00	17.29	4.92
July	896.13	17.57	5.35
August	915.48	17.95	5.36
September	936.00	18.35	5.74
October	886.45	17.38	5.21
November	898.00	17.61	5.68
December	822.58	16.13	5.10
χ^2	22.51		
d.f.	11.00		
p	<0.05		

model. Thus, the fertility behavior did not seem to change as did the mortality pattern.

Marriage, surnames and population structure

Marriage archives which specify the consorts' last names provide a rich source of information for the human biologist. In fact, Lasker (1988b) notes that the distribution of surnames reflects the effect of mate selection on genetic structure. It is commonly agreed that last names behave as neutral alleles and thus provide a means to study gene flow in historical populations (Koertvelyessy *et al.* 1990; LaFranchi *et al.* 1988; Lasker 1988a; Relethford 1986; Wijsman *et al.* 1984; Zei *et al.* 1983, 1986). It has been noted that surnames may be used to study migration diachronically in a population, and may provide an opportunity to determine the effect of gene flow on a population's structure and evolution (Lasker and Kaplan, 1985). Since surnames behave like genetic markers (Guglielmino *et al.* 1991), it is assumed that new surnames appear as a result of in-migration. Surnames have also proven to be of great utility in the historical study of population structure. Specifically, a Wahlund effect would be evident in marriage records if there were more repeated-surname (RP) marriages than those expected by chance (Lasker 1988a; Lasker and Kaplan 1985). A Wahlund effect is the presence of a subdivision within a larger population, or a set of subdivided populations resulting from an isolation mechanism such as preferential mating among families (Ridley 1996).

Surnames in marriage archives have also provided invaluable data to study inbreeding levels. To study inbreeding in historic populations, human biologists have relied on the method of isonymy, based on the frequency of same-surname marriages in a population (Calderón *et al.* 1993; Crawford 1980; Ellis and Starmer 1978; Koertvelyessy *et al.* 1990; Kosten and Mitchell 1990; Mielke and Swedlund 1993; Pollitzer *et al.* 1988). Most, if not all, of the isonymic investigations cite as their primary source the seminal work of Crow and Mange (1965).

This section describes the work on the Escazú archives which investigates the structure of the population and inbreeding. To investigate population structure, the Lasker and Kaplan (1985) RP (repeated pairs) statistic is used to study preferential mating among families in Escazú. The random component or RP_r as derived by Chakraborty (1985, 1986) and the standard errors of RP and RP_r are also computed according to Chakraborty (1985, 1986). Finally, a test of the null hypothesis that $RP = RP_r$ as proposed by Relethford (1992) is applied here (Madrigal and Ware 1999).

To study inbreeding, Crow and Mange's (1965) formulae for computing the total inbreeding (Ft), the non-random inbreeding (Fn), and the random inbreeding (Fr) are applied to the frequencies of grooms' and brides' last names per decade. A second type of isonymy analysis can also be undertaken because in Iberoamerican populations, each individual inherits two surnames, one from the father and one from the mother. Therefore, the methodology of Pinto-Cisternas *et al.* (1985) is used to compute these three inbreeding coefficients for the records in which the two surnames are present for both bride and groom. Because of the large number of records which do not include the four surnames (presumably because of illegitimacy), less than half of the total number of records is thus analyzed. To differentiate both types of isonymy methodology, the Crow and Mange methodology is referred to as C&M, and the Pinto-Cisternas *et al.* methodology is referred to as IB (Iberoamerican), as done by Pinto-Cisternas *et al.* (1985). According to common practice (see, for example, Pettener 1990; Pollitzer *et al.* 1988) minor spelling differences of the same last name were considered to represent only one name. Fortunately, ecclesiastical dispensations are also available in Escazú, and provide yet another way to measure the average inbreeding coefficient in the population. However, only during the 1860, 1870, 1880 and 1890 decades are there enough dispensations to compute the coefficient (Bodmer and Cavalli-Sforza 1976; Madrigal and Ware 1997).

During the entire time period, 255 surnames were recorded, with most new ones appearing before the 1820 decade. A simple χ^2 test indicated that the frequency of surnames introduced by males (146 surnames) is significantly greater than that introduced by females (111 surnames) ($\chi^2 = 4.56$, d.f. $= 1$, $p < 0.05$).

Twenty-seven of the 255 surnames found in the entire period under analysis were present in the archives in all decades and are thus referred to as core surnames. Most of these 27 surnames were listed in the 1755 and 1786 civil documents, indicating that the core surnames were already present before the parish was officially established in 1799. These core families contributed a great deal to the breeding population. Out of the total 4142 people who got married in the community in the entire period of analysis, 2327 consorts carried one of these last names, indicating that 56% of all marriages involved people who belonged to one of these 27 core families.

The repeated-pair analysis of all marriages every two decades indicated that most of the surname repetition is random (Table 2.4). Thus, the population did not seem to be engaging in preferential mating for most of the time period analysis. The hypothesis that the observed level of RP is the same as the random RP was tested as proposed by Relethford (1992). The z scores testing this hypothesis are also shown in Table 2.4 and are not significant for the entire period, or for three out of the five 'generations' studied. However, the hypothesis is rejected for the periods 1850–1869 and 1870–1889. This is important because the level of inbreeding (as researched by isonymy with two or four surnames, and as researched with ecclesiastical dispensation) increased as well in the decade of 1850 (Madrigal and Ware 1997). Figure 2.3 shows the inbreeding coefficients estimated though both isonymy methods, and Table 2.5 displays them. Although their behavior is not exactly the same, they indicate that during the latter part of the century, consanguinity increased. This increase in consanguineous marriages is also indicated by the dispensation coefficient, which consistently rises for the four decades in which it could be computed.

That consanguinity did increase during the latter part of the century is also supported by an analysis of the random and non-random components of the inbreeding coefficients computed through both the Crow and Mange and the Pinto-Cisternas methods. After 1850, and particularly after 1860, the non-random components increased in value, suggesting that the increase in total isonymy was not a random fluctuation but was actually due to non-random forces (see Table 2.5).

In conclusion, there is strong evidence from different methodological approaches that there was an increase in consanguineous marriages in Escazú, as well as in repeated-pairs marriages (the latter evidence of preferential mating among families) during the 1850s.

Table 2.4. Repeated-pair analysis of all marriages

	All years	1800–1819	1820–1839	1850–1869	1870–1889	1890–1899
No. of couples	2071	243	520	666	595	47
No. of unique pairs of names	1887	234	484	592	530	47
No. of non-unique pairs of names	184	9	36	74	65	0
No. of repetitions	470	18	84	204	164	0
Proportion of repetitions (RP)	3.43×10^{-4}	3.06×10^{-4}	3.11×10^{-4}	4.61×10^{-4}	4.64×10^{-4}	0
Random RP (RP_r)	2.93×10^{-4}	3.38×10^{-4}	2.72×10^{-4}	3.08×10^{-4}	3.46×10^{-4}	3.20×10^{-4}
$RP - RP_r$	4.97×10^{-5}	-3.21×10^{-5}	3.95×10^{-5}	1.53×10^{-4}	1.18×10^{-4}	-3.20×10^{-4}
$((RP - RP_r)/RP_r) \times 100\%$	16.94	−9.50	14.54	49.67	34.17	−100.00
$(RP_r/RP) \times 100\%$	85.51	110.50	87.31	66.81	74.53	—
SE RP	0.0000326	0.0004107	0.0001505	0.0001283	0.0001397	0.0043890
SE RP_r	0.0000148	0.0000610	0.0000302	0.0000299	0.0000331	0.0002451
H0: RP = RP_r	3.3552446	−0.5269935	1.3102988	5.1084952	3.5749078	−1.3058283

Table 2.5. *The random, non-random and total inbreeding coefficients computed by Crow and Mange's methodology (Fr, Fn, Ft) and by Pinto-Cisternas' methodology (FrIB, FnIb, FtIB)*

Average α calculated from the dispensations.

Decade	Fr	Fn	Ft	FrIb	FnIB	FtIB	Average α
1800	0.004375	−0.004450	−0.000058	0.0047	−0.0034	0.00130	—
1810	0.005100	−0.003410	0.001696	0.0047	−0.0032	0.00150	—
1820	0.003700	−0.001420	0.002122	0.0044	−0.0013	0.00310	—
1830	0.004000	−0.003180	0.000836	0.0044	−0.0032	0.00120	—
1840	—	—	—	—	—	—	—
1850	0.004175	0.002280	0.006457	0.0028	−0.0029	−0.00009	—
1860	0.004450	0.000290	0.004751	0.0052	0.0045	0.00970	0.0036
1870	0.004350	0.001765	0.006104	0.0046	0.0031	0.00770	0.004
1880	0.005350	0.004750	0.010076	0.0062	0.0070	0.00930	0.0051
1890	0.005200	0.000116	0.005321	0.0062	0.0062	0.01240	0.0056

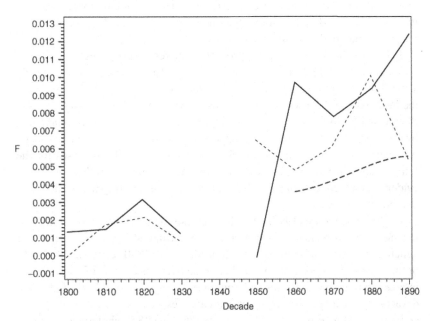

Fig. 2.3. Total inbreeding coefficients computed by the methods of Crow and Mange (dashed line) and Pinto-Cisternas *et al.* (solid line). Dotted line indicates α as computed by ecclesiastical dispensations.

Discussion

The analysis of surnames indicated that Escazú had a steady core of 27 families which provided more than 50% of the breeding individuals for the century. The majority of these core families were present before the parish was founded, since they are mentioned in the two existing 1700s documents discussed earlier. Given the long history of the 27 core families, and their important contribution to the breeding population, a Wahlund's effect separating these from the other families could have been expected, but none was found. Moreover, as Lasker (1988b) mentions, the more traditional a population, the higher the level of RP (repeated-pair surname marriages) that can be expected. The results for Escazú probably indicate that the population size of this community was too small to allow for such exclusion of potential mates. Moreover, since the new immigrants acquired free or very cheap land as part of the government's policy, they were not without a means of living. Thus, the socioeconomic difference between immigrants and local residents was probably not sufficiently large to prevent them from marrying. Therefore, although the 27 core families provided 56% of all consorts for the entire period, they did not preferentially mate among themselves. Indeed, it was found that the proportion of random repetitions is 85.51% of the total repeated pairs.

It is very interesting that the proportion of repeated-pair marriages (RP) was significantly different from the random RP for the entire sample during 1850–1889. This is significant in light of the fact that inbreeding increased in the latter part of the 1800s. This conclusion was reached with both Crow and Mange and Pinto-Cisternas' methodologies for the computation of inbreeding by isonymy. This increase in consanguineous marriages is also indicated by the dispensation–inbreeding coefficient, which consistently increases for the four decades in which it could be computed. In conclusion, there is strong evidence that there was an increase in consanguineous marriages and a decrease in random mating.

Madrigal and Ware (1997, 1999) have proposed that the increase in inbreeding and repeated-pairs marriages is a result of the devastating cholera epidemic which Costa Rica suffered in 1856. Indeed, Pérez (1980) estimates that 10% of the Costa Rican population perished in that year as a result of cholera. It is interesting to note that Bittles and Smith (1994) report that in Ireland after the devastating potato famine, isonymy increased, especially as measured by total and non-random coefficients. It is possible that the 1856 cholera epidemic in Costa Rica had a similar effect on marriage patterns as did the potato famine in Ireland: an increase in isonymous marriages perhaps as a result of a reduction in the number of family names, or as a mechanism for maintaining precious resources within the family. It would be very interesting to determine if, cross-culturally, other populations react in a similar manner to such catastrophes. The author is currently examining the records of second marriages to determine, for example, if there is any evidence for preferential sister-in-law to brother marriages (L. Madrigal, work in progress).

The lack of birth seasonality in Escazú is unexpected because such fluctuations have been interpreted as the dominant pattern in human fertility. Indeed, seasonality of births has been reported in populations from a diverse geographic and cultural background, and has been considered to be an almost universal phenomenon (Lam and Miron 1991). It is possible that the lack of birth seasonality in Escazú is a result of the very stable and mild climate of the area and of a fairly constant food supply (Madrigal 1993). In other words, the climate in the region is one that allows for a constant food supply. As Rosetta (1993) mentions, seasonal food shortages may regulate reproductive function in women, and hence, births. It would be of interest to study populations under similar conditions to determine whether Escazú is still unique, or whether comparable communities also show no birth seasonality. However, these results suggest that seasonal patterns of birth distribution are not ubiquitous in human populations, as the literature may suggest (Lam and Miron 1991; Leslie and Fry 1989).

In contrast, strong mortality cycles were evident in the first part of the 1800s, though the seasonality of the cycles disappeared in the more recent years. The cause of death analysis indicated that gastrointestinal-related deaths were the likely cause of the seasonal distribution of mortality as well as the cause of the decline in mortality seasonality in the more recent years of the series. The change in mortality behavior suggests that, in Escazú, infectious diseases caused by water-borne pathogens started to decrease in importance as a major selective force. As Tomkins (1993) notes, a decrease in the prevalence of seasonal infections does not necessarily result from hygienic measures, but could result from better housing or water supply. Tomkins (1993) also notes that such a decrease in seasonal infections is obvious in the industrialized countries over the 1800s. Costa Rica, which now has health standards comparable with industrialized countries (see *http://www.who.int/inf-pr-2000/en/pr2000-life.html*), might have experienced similar changes which decreased the importance of seasonal infections. It is reasonable to speculate that the action of natural selection in this population changed from an emphasis on differential mortality to a greater importance of differential fertility.

That an epidemiological change was detected over a few generations in this time series demonstrates the usefulness of archival research for human biologists interested in demographic and epidemiological changes. In the case of the Escazú project, archival work has allowed Madrigal (1992, 1993, 1994) an entry into the community, whose members and clergy have shown great interest in the project. Thus, archival research could be the start of further biological research of the descendants of the individuals researched in the community.

Acknowledgements

The author thanks the personnel of the Escazú Parish. Specifically, the support, enthusiasm and understanding of Father Howell and of S. Belén are gratefully acknowledged. Mrs Barbara Ware's contribution to the long-term Escazú project is also acknowledged.

Appendix. A note on Freedman's test and the Box–Jenkins method

The appropriate methodology for testing the presence of vital-events seasonality has been discussed in the physical anthropological literature (McCullough 1985; Madrigal 1994; O'Brien and Holbert 1987; Reijneveld, (1990). In this chapter, the presence of mortality seasonality in Escazú was tested with the Freedman and the χ^2 tests, since McCullough (1985), O'Brien and Holbert (1987) and Reijneveld (1990) convincingly show them to be the tests best suited to such a purpose. The Freedman statistic is compared with the estimated percentiles of its distribution provided by Freedman (1979).

The Box–Jenkins analysis of time series offers several advantages for the study of mortality seasonality. It incorporates the correlation among the data set's points, its models are constructed empirically from the data without any preconceived ideas, and it allows the description of highly complex data sets with a few parameters (Jenkins 1979).

Two drawbacks of the Box–Jenkins methodology for the study of mortality seasonality can be identified. The series has to be relatively long for the model parameters to be estimated with accuracy. At the same time, if a series is too long, changing circumstances may cause the structure of the series to change slowly over time. Therefore, a single model may not be estimated for the entire series. In this situation, Jenkins (1979) suggests that the series be split into two series of roughly equal size, and a model estimated for each. This is the case for the Escazú mortality series, since the whole series could not be fitted with one model, but had to be split into two parts, each fitted by different models. The first part (1851–1891) was successfully modeled as follows.

$$Y_t = \{[(1 - 0.17B)(1 - 0.88B)(1 - 0.88B^{12})]/(1 - B)(1 - B^{12})\}a_t,$$

where Y_t is the time series, a_t represents the white noise, $1 - B$ is the backward shift operator, $1 - B^{12}$ is the seasonal backward shift operator, 0.17 is the autoregressive (AR) parameter and 0.88 is the seasonal moving average (MA) parameter.

By incorporating non-seasonal AR and MA parameters, the model indicates that the number of deaths in a month was related to the number of deaths in the previous month, as well as to the previous month's random error or shocks (such as the specific strains of a disease which caused the deaths). The MA seasonal parameter indicates that the number of deaths in a specific month y_t is related to random error or shocks of the y_{t-s} datum, where s is the number of periods in the season (twelve in this case). In other words, the factors which influenced the number of deaths at time $t - s$ such as season-specific strains or season-specific climatic variables also influenced the number of deaths at time t.

The second part of the series (1892–1921) was modeled as follows.

$$Y_t = [(1 - 0.43B - 0.11B^2)]^a t,$$

where Y_t is the time series, a_t represents the white noise, 0.43 is the first autoregressive parameter, and 0.11 is the second autoregressive parameter.

The model indicates that a month's number of deaths is related to the number of deaths registered during the previous two months. This model describes a non-seasonal rather than a seasonal mortality distribution: if mortality is strongly seasonal, the number of deaths in a month is more likely to be influenced by the number of deaths in the previous, not in the next-to-the-previous and the previous months. Moreover, the model does not incorporate a seasonal parameter. Therefore, the number of deaths at time t is not correlated with the number of deaths at time $t - s$.

The differenced baptismal series was successfully fitted with the following model.

$$Y_t = \frac{1 - 0.91B}{(1 - B)} a_t,$$

where Y_t is the time series, a_t represents the white noise, $1 - B$ is the backward shift operator, and 0.91 is the moving average parameter.

Thus, the baptism series is successfully fitted with a (0,1,1) model, clearly indicating that the series does not have a seasonal component.

References

Arcury, T.A., Williams, B.J. and Kryscio, R.J. (1990). Birth seasonality in a rural U.S. county, 1911–1979. *American Journal of Human Biology* **2**, 675–89.

Arriaga, E.E. (1976). *New Life Tables for Latin American Populations in the 19th and 20th Centuries: Mortality Decline and its Demographic Effects in Latin America*. Population Monograph Series 3. Westport, CT: Greenwood Press.

Bako, G., Ferenczi, L., Hill, G.B. and Lindsay, J. (1988). Seasonality of mortality from various diseases in Canada 1979–83. *Canadian Journal of Public Health* 7, 388–9.

Barrett, R.E. (1990). Seasonality in vital processes in a traditional Chinese population. *Modern China* 16, 190–225.

Bittles, A.H. and Smith, M.T. (1994). Religious differentials in postfamine marriage patterns, Northern Ireland, 1840–1915. 1. Demographic and isonymy analysis. *Human Biology* 66, 59–76.

Bowie, C. and Prothero, D. (1981). Finding causes of seasonal diseases using time series analysis. *International Journal of Epidemiology* 10, 87–92.

Bodmer, W.F. and Cavalli-Sforza, L.L. (1976). *Genetics, Evolution, and Man*. San Francisco: W.H. Freeman and Company.

Breschi, M. and Bacci, M.L. (1986). Effet du climat sur la mortalite infantile. *Population* 41, 1072–4.

Calderón, R., Pena, J.A., Morales, B. and Guevara, J.I. (1993). Inbreeding patterns in the Basque country (Alava province), 1831–1980. *Human Biology* 65, 743–770.

Chakraborty, R. (1985). A note on the calculation of random RP and its sampling variance. *Human Biology* 57, 713–17.

Chakraborty, R. (1986). Erratum. *Human Biology* 58, 991.

Chaves-Camacho, J. (1969). Evolucion Demografica de la Poblacion de Costa Rica. *Revista de la UCR* 27, 37–42.

Crawford, M.H. (1980). The breakdown of reproductive isolation in an Alpine genetic isolate: Acceglio, Italy. In: *Population Structure and Genetic Disorders*, ed. A.W. Eriksson and S.J. Stiftelse, pp. 57–70. 7th Sigrid Juselius Foundation symposium, Mariehamn, Åland Islands, Finland. New York: Academic Press.

Crow, J.F. and Mange, A.P. (1965). Measurement of inbreeding from the frequency of marriages between persons of the same surname. *Eugenics Quarterly* 12, 199–203.

Dzierzykray-Rogalski, T. and Prominska, E. (1971). Seasonal rhythms of mortality in Sanuris. *Africana Bulletin* 14, 59–68.

Ellis, W.S. and Starmer, W.T. (1978). Inbreeding as measured by isonymy, pedigrees, and population size in Torbel, Switzerland. *American Journal of Human Genetics* 30, 366–76.

Freedman, L.S. (1979). The use of a Kolmogorov-Smirnov type statistic in testing hypotheses about seasonal variation. *Journal of Epidemiology and Community Health* 33, 223–8.

Galloway, P.R. (1988). Basic patterns in annual variations in fertility, nuptiality, mortality, and prices in pre-industrial Europe. *Population Studies* 42, 275–303.

González-Garcia, Y. (1983). *Continuidad y Cambio en la Historia Agraria de Costa Rica (1821–1880)* . Dissertation, Université Catholique de Louvain.

Guglielmino, C.R., Zei, G. and Cavalli-Sforza, L.L. (1991). Genetic and cultural transmission in Sicily as revealed by names and surnames. *Human Biology* 63(5), 607–28.

Hajek, E.R., Gutiérrez, J.R. and Espinosa, G.A. 1984. Seasonality of mortality in human populations of Chile as related to a climatic gradient. *International Journal of Biometeorology* 28, 29–38.

Huss-Ashmore, R. (1988). Seasonal patterns of birth and conception in rural highland Lesotho. *Human Biology* **60**, 493–506.

Jenkins, G.M. (1979). *Practical Experiences with Modeling and Forecasting Time Series.* Jersey, UK: GJP Publications.

Kalkstein, L.S. and Davis, R.E. (1989). Weather and human mortality: an evaluation of demographic and interregional responses in the United States. *Annals of the Association of American Geographers* **79**, 44–64.

Koertvelyessy, T., Crawford, M.H., Pap, M. and Szilagyi, K. (1990). Surname repetition and isonymy in northeastern Hungarian marriages. *Human Biology* **62**(4), 515–24.

Kosten, M. and Mitchell, R.J. (1990). Examining population structure through the use of surname matrices: Methodology for visualizing nonrandom mating. *Human Biology* **62**, 319–35.

LaFranchi, M., Mitchell, R.J. and Kosten, M. (1988). Mating structure, isonymy and social class in late nineteenth century Tasmania. *Annals of Human Biology* **15**(5), 325–336.

Lam, D.A. and Miron, J.A. (1991). Seasonality of births in human populations. *Social Biology* **38**, 51–78.

Landers, J. (1986). Mortality, weather and prices in London 1675–1825: a study of short-term fluctuations. *Journal of Historical Geography* **12**, 347–64.

Lasker, G.W. (1988a). Repeated surnames in those marrying into British one-surname "lineages:" an approach to the evaluation of population structure through analysis of the surnames in marriages. *Human Biology* **60**(1), 1–10.

Lasker, G.W. (1988b). Application of surname frequency distributions to studies of mating preferences. In *Human Mating Patterns*, ed. C.G.N. Mascie-Taylor and A.J. Boyce, pp. 101–14. Cambridge: Cambridge University Press.

Lasker, G.W. and Kaplan, B.A. (1985). Surnames and genetic structure: repetition of the same pairs of names of married couples, a measure of subdivision of the population. *Human Biology* **57**(3), 431–40.

Leslie, P.W. and Fry, P.H. (1989). Extreme seasonality of births among Nomadic Turkana pastoralists. *American Journal of Physical Anthropology* **79**, 103–15.

Lin, P.M. and Crawford, M.H. (1983). A comparison of mortality patterns in human populations residing under diverse ecological conditions: a time series analysis. *Human Biology* **55**, 35–62.

Madrigal, L. (1992). Differential sex mortality in a 19th Century rural population: Escazú, Costa Rica. *Human Biology* **64**, 199–213.

Madrigal, L. (1993). Lack of birth seasonality in an agricultural 19th Century population: Escazú, Costa Rica. *Human Biology* **65**, 255–71.

Madrigal, L. (1994). Mortality seasonality in Escazú, Costa Rica: 1851–1921. *Human Biology* **66**, 433–52.

Madrigal, L. and Ware, B. (1997). Inbreeding in Escazú, Costa Rica (1800–1840, 1850–1899): isonymy and ecclesiastical dispensations. *Human Biology* **69**, 703–14.

Madrigal, L. and Ware, B. (1999). Mating pattern and population structure in Escazú, Costa Rica: a study using marriage records. *Human Biology* **71**, 963–75.

Malina, R.M. and Himes, J.H. (1977). Differential age effects in seasonal variation of mortality in a rural Zapotec-speaking municipio, 1945–1970. *Human Biology* **49**, 415–28.

Mao, Y., Semenciw, R., Morrison, H. and Wigle, D.T. (1990). Seasonality in epidemics of asthma mortality and hospital admission rates, Ontario, 1979–86. *Canadian Journal of Public Health* **81**, 226–8.

Marcuzzi, G. and Tasso, M. (1992). Seasonality of death in the period 1889–(1988) in the Val di Scalve. *Human Biology* **64**, 215–22.

McCullough, J.M. (1985). Application of the Kolmogorov-Smirnov test to seasonal phenomena may be inappropriate. *American Journal of Physical Anthropology* **68**, 393–4.

Mielke, J.H. and Swedlund, A.C. (1993). Historical demography and population structure. In *Research Strategies in Human Biology: Field and Survey*, ed. G.W. Lasker and C.G.N. Mascie-Taylor, pp 140–85. Cambridge: Cambridge University Press.

Mielke, J.H., Jorde, L.B., Trapp, P.G., Anderton, D.L., Pitkanen, K. and Eriksson, A.W. (1984). Historical epidemiology of smallpox in Aland, Finland: 1751–1890. *Demography* **21**, 271–95.

Ministerio de Economía, Industria y Comercio. (1974). *Censo de Población: 1892*. San José: Dirección General de Estadística y Censos.

Ministerio de Economía, Industria y Comercio. (1975). *Censo de Población: 1883*. San José: Dirección General de Estadística y Censos.

Ministerio de Economia y Hacienda. (1964). *Censo de Población. 1864*. San José: Dirección General de Estadística y Censos.

Molnar, S. (1998). *Human Variation*, 4th ed. New York: Prentice-Hall.

O'Brien, K.F. and Holbert, D. 1987. Note on the choice of statistic for testing hypotheses regarding seasonality. *American Journal of Physical Anthropology* **72**, 523–4.

Pérez, H. (1980). *Las variables demográficas en las economías de exportación: el ejemplo del Valle Central de Costa Rica (1800–1950)*. Avances de Investigación (irregular series). San José: University of Costa Rica.

Pettener, D. (1990). Temporal trends in marital structure and isonymy in S. Paolo Albanese, Italy. *Human Biology* **62**, 837–51.

Pinto-Cisternas, J., Pineda, L. and Barrai, I. (1985). Estimation of inbreeding by isonymy in Iberoamerican populations: and extension of the method of Crow and Mange. *American Journal of Human Genetics* **37**, 373–85.

Pollitzer, W.S., Smith, M.T. and Williams, W. R. (1988). A study of isonymic relationships in Fylingdales Parish from marriage records from 1654 through 1916. *Human Biology* **60**, 363–82.

Reijneveld, S.A. (1990). The choice of a statistic for testing hypotheses regarding seasonality. *American Journal of Physical Anthropology* **83**, 181–4.

Relethford, J. (1986). Microdifferentiation in historical Massachusetts: a comparison of migration matrix and isonymy analyses. *American Journal of Physical Anthropology* **71**, 365–75.

Relethford, J. (1992). Analysis of marital structure in Massachusetts using repeating pairs of surnames. *Human Biology* **64**, 25–33.

Ridley, M. (1996). *Evolution*. Oxford: Blackwell Science.

Rosetta, L. (1993). Seasonality and fertility. In *Seasonality and Human Ecology*, ed. S.J. Ulijaszek and S.S. Strickland, pp. 65–75. Symposia of the Society for the Study of Human Biology. Cambridge: Cambridge University Press.

Sáenz-Maroto, A. (1980). *Algunos Aspectos Basicos Agrologicos de Costa Rica.* San José: Universidad de Costa Rica Press.

Schwartz, J. and Marcus, A. (1990). Mortality and air pollution in London: a time series analysis. *American Journal of Epidemiology* **131**, 185–94.

Scott, S. and Duncan, C.J. (1998). *Human Demography and Disease.* Cambridge: Cambridge University Press.

Scott, S. and Duncan, C.J. (2001). *Biology of Plagues.* Cambridge: Cambridge University Press.

Shumway, R.H., Azaari, A.S. and Pawitan, Y. (1988). Modeling mortality fluctuations in Los Angeles as functions of pollution and weather effects. *Environmental Research* **45**, 224–41.

Sibaja, L.F. (1970). Los origenes de Escazú. *Revista de la UCR* **28**, 97–106.

Stroup, D.F., Thacker, S.B. and Herndon, J.L. (1988). Application of multiple time series analysis to the estimation of pneumonia and influenza mortality by age 1962–1983. *Statistics in Medicine* **7**, 1045–59.

Tomkins, A. (1993). Environment, season and infection. In *Seasonality and Human Ecology*, ed. S.J. Ulijaszek and S.S. Strickland, pp. 123–34. Cambridge: Cambridge University Press.

Wijsman, E., Zei, G., Moroni, A. and Cavalli-Sforza, L.L. (1984). Surnames in Sardinia: computation of migration matrices from surname distributions in different periods. *Annals of Human Genetics* **48**, 65–78.

Zei, G., Guglielmino Matessi, R., Siri, E., Moroni, A. and Cavalli-Sforza, L. (1983). Fit of frequency distributions for neutral alleles and genetic population structure. *Annals of Human Genetics.* **47**, 329–52.

Zei, G., Piazza, A., Moroni, A. and Cavalli-Sforza, L.L. (1986). Surnames in Sardinia: the spatial distribution of surnames for testing neutrality of genes. *Annals of Human Genetics* **50**, 169–80.

3 Anthropometric data and population history

JOHN H. RELETHFORD

Introduction

This paper deals with the use of anthropometric data in reconstructing the history of human populations. Anthropometric measures of the body and head have been collected on many populations over the past few centuries. Although data collection and analysis were frequently done within the outdated context of racial classification, new models and methods allow us to revisit anthropometric variation from the perspective of current evolutionary theory. In these days of rapidly developing means of directly assessing DNA variation, anthropometric data are often likely to be seen as old-fashioned. However, despite their antiquity, anthropometric data still have a lot to offer us in terms of insight into population structure and history. The purpose of this paper is two-fold: (1) to describe in general terms the use of anthropometric data in the study of population history, and (2) to present summaries of two case studies using anthropometric data from Ireland. In both cases, the results show the clear influence of historical events relating to differential immigration on the pattern of biological variation within Ireland.

What do anthropometric data have to do with the focus of this volume – archival data? Many anthropometric data are historical, in the sense of having been recorded at an earlier time. Some measurements, notably height and weight, were routinely recorded in the past, particularly in military records. More comprehensive anthropometric data collections, including other measures of the body and head, stem from the nineteenth and twentieth centuries, and sometimes earlier. As such, they form part of the archival record representing the 'biological present' from past decades. Retrieval and analysis of such data provide us with a window into the recent past. One or more centuries may not seem like much at first, but consider that data on genetic markers and DNA sequences are much more recent. Even the ABO blood group, the first discovered genetic marker, has only been around a century, and protein polymorphisms and DNA sequences are much more recent discoveries. The ability to go back in time even a century or so is important given the short time we have been looking at the microevolution of human populations with genetic markers.

The importance of historical data even a few generations old should not be minimized. Changes in the modern world, particularly in population growth and migration, can rapidly change preexisting genetic structure. The ability to peer back in time even a few decades offers us an opportunity to look at a world that no longer exists. In addition, the methods developed for anthropometric data can also be extended to osteometric and craniometric variation from skeletal data, providing us with the potential to look at human variation and evolution over thousands of years, or longer.

There is another connection between anthropometric data and other archival materials. Anthropometric data, as well as genetic markers, are used to make inferences about the history of populations. Historical relationships and events leave their signature on patterns of genetic variation, but it is often difficult to determine the underlying population history. If, for example, we look at a matrix of genetic distances and find that one population is distinctly different, we are left with several possible explanations for that difference, such as genetic drift due to small population size, cultural or geographic isolation, or admixture with distant populations, among others. As Felsenstein (1982) has noted, different population histories can give rise to the same observed patterns of genetic distances. This problem is particularly acute in the modern human origins debate, where the underlying history is not known, and is the question we seek to answer (Relethford 1998, 2001). Studies of genetics and population history in the more recent past can rely on supporting historical data to resolve issues of interpretation; indeed, the most common use of biological data in this context is to help test hypotheses of population history derived from other analyses of the historical record.

Collecting and analyzing historical anthropometric data

Historical anthropometric data can be collected in two ways: either from published data or from the retrieval of original data forms. In both cases, we must depend entirely on whatever was recorded at the time, and deal with potential problems that might arise given that such data may have been collected for a different purpose. Such data generally include information on sex, age, and geographic location. Some data sets might provide further information, such as birthplace of individuals and parents or surnames, thus providing data that can be used for migration matrix and isonymy studies with which to compare anthropometric analyses (Rogers and Harpending 1986;

Relethford 1988b). An excellent example of the use of archival anthropometric data, other than the case studies presented here, is the recent computerization and analysis of the anthropometric data collected by Franz Boas on North American Indian populations in the late nineteenth century (Jantz *et al.* 1992).

Since anthropometrics reflect phenotypic variation, we must guard where possible against non-genetic influences on variation, including sex differences, ageing effects, and environmental plasticity. Sex and age effects are easily controlled for by appropriate sample selection and statistical adjustment. The potential influence of the environment is usually harder to deal with, although restricting the analysis to a limited geographic region might control for some environmental differences. Interobserver error can also be a potential problem, particularly for multivariate analysis, where data from more than one observer should be avoided, or at least viewed with considerable caution (Jamison and Zegura 1974).

Studies of population history based on anthropometric data most frequently use some sort of distance measure for summarizing the average difference across a set of variables. Distances are computed between all pairs of populations. The final distance matrix is usually used to derive a picture, or 'map,' showing a graphic representation of the relationships between populations, usually obtained using cluster analysis, principal coordinates analysis, or multidimensional scaling.

Two different measures of biological distance have been used in most studies of anthropometrics and population history. One is Euclidean distance, which is simply the squared difference between population means averaged over all variables. In order to avoid the problem of measures with large values (e.g. stature) having more effect than measures with small values (e.g. nose breadth), all means are first standardized. Euclidean distances do not account for intercorrelation of variables, and for that reason the multivariate distance measure, Mahalanobis' D^2, is preferred. Of course, this measure requires access to the raw data; when not available, Euclidean distances can be computed simply from group means.

Euclidean distances and Mahalanobis' D^2 are closely related to many genetic distance measures typically used with genetic marker data (Morton 1975; Williams-Blangero and Blangero 1989; Relethford *et al.* 1997). Such measures express the genetic distance between two populations as a function of the squared difference in allele frequencies. Since phenotypic means are directly proportional to allele frequencies under an equal and additive effects model of quantitative inheritance, Euclidean distances and Mahalanobis' D^2 have theoretical justification, and are not simply *ad hoc* measures of population affinity.

Precise estimation of genetic distances and other population-genetic parameters is possible using methods of analyzing quantitative traits based on the extension of **R** matrix theory to quantitative genetics. An **R** matrix is the standardized variance–covariance matrix of allele frequencies among a set of populations. For a given allele and g populations, the g-by-g **R** matrix has elements

$$r_{ij} = \frac{(p_i - \bar{p})(p_j - \bar{p})}{\bar{p}\,(1 - \bar{p})}, \tag{3.1}$$

where p_i and p_j are the allele frequencies of populations i and j, respectively, and $\bar{p}$ is the mean allele frequency over all g populations, weighted by population size. The final **R** matrix is derived by averaging over all alleles. By definition, pairs of populations with positive r_{ij} values are more closely related than average (where $r_{ij} = 0$), and pairs of populations with negative r_{ij} values are less closely related than average. The average diagonal element $(i = j)$ of the **R** matrix, weighted by population size, is an estimate of Wright's F_{ST}, giving an index of differentiation among populations. In addition, the **R** matrix can be transformed into a matrix of genetic distances using the equation

$$d_{ij}^2 = r_{ii} + r_{jj} - 2r_{ij} \tag{3.2}$$

(Harpending and Jenkins 1973).

R matrices are easily derived from quantitative traits such as anthropometrics (Williams-Blangero and Blangero 1989; Relethford and Blangero 1990). After Z-score standardization of the data matrix, the g-by-g codivergence matrix **C** is computed, with elements

$$c_{ij} = (x_i - \mu)'\mathbf{P}^{-1}(x_j - \mu), \tag{3.3}$$

where x_i and x_j are vectors of the means for groups i and j, μ is the vector of total means averaged over all groups, **P** is the pooled within-group phenotypic covariance matrix, $\mathbf{P}^{-1}$ is its inverse, and the prime $(')$ indicates matrix transposition. The **R** matrix is then computed as

$$\mathbf{R} = \frac{\mathbf{C}(1 - F_{ST})}{2t}, \tag{3.4}$$

where t is the number of traits and

$$F_{ST} = \sum_{i=1}^{g} w_i r_{ii} = \frac{\displaystyle\sum_{i=1}^{g} w_i c_{ii}}{2t + \displaystyle\sum_{i=1}^{g} w_i c_{ii}}. \tag{3.5}$$

Genetic distances can then be derived using equation 3.2. It is important to note that the mean vector (μ) and the variance–covariance matrix (**P**) are both computed as weighted averages across groups, and the weighting is by population size (not sample size!). **R** matrix theory describes deviations from panmixis, a situation where all members of all populations are part of the same breeding population, and the **R** matrix must therefore be derived by weighting all figures by population size. When population sizes are equal, the genetic distances will be proportional to Mahalanobis' distances; when population sizes are different, the distances derived from the **R** matrix are more appropriate. As defined in equations 3.3–3.5, F_{ST} and the **C**, **R**, and genetic distance matrices are actually minimum values obtained under the assumption that all traits are completely heritable. If estimates of heritability are available for each trait, the genotypic covariance matrix can be substituted for the phenotypic covariance matrix in equation 3.3. If not, the minimum genetic distance matrix can still be used for comparative analysis, since it will be proportional to the true genetic distance matrix if heritabilities are equal across populations, a reasonable assumption in many cases of local populations within the same environmental regime or with a fairly narrow geographic area. Standard errors for these matrices can also be easily computed (see Relethford *et al.* 1997) and minimum genetic distances can be derived after controlling for variation in population size, and hence, variation in genetic drift (Relethford 1996). A computer program (RMET) that performs all of these calculations is available upon request (requires the Windows™ 95/98 operating system).

Historical anthropometry and studies of Irish population history

The remainder of this chapter reviews two case studies of minimum genetic distances derived from anthropometric data, both from Ireland (as used here, the term 'Ireland' refers to the entire island, consisting today of the Republic of Ireland and the UK province of Northern Ireland). The history of Ireland makes it interesting for studies of human biology and population history. The initial prehistoric settlement of Ireland occurred in several waves, followed by a history of numerous invasions and settlements, including Viking invasions as well as immigrants from England, Scotland, and Wales. The demographic history of Ireland is also very interesting in terms of population size undergoing a classic 'boom–bust' pattern of change. Following the introduction of the potato, the population size of Ireland increased dramatically during the eighteenth and early nineteenth centuries, because the potato crop allowed continued subdivision of existing land into small economically viable family plots, thus allowing an increase in marriage rates and a reduction in the average age of marriage. Consequently, the birth rate and population size increased, reaching a peak of about 8 million people by the middle of the nineteenth century. The repeated failure of the potato crop over a five-year period (1846–1851) led to a rapid decrease in population size due to increased mortality and emigration (Connell 1950; Kennedy 1973).

My own interest in the human biology and population history of Ireland began in graduate school. As is often the case, my entry into this area was a function of serendipity. My graduate advisor, Francis Lees, had developed an interest in Ireland in part because of the interest of his graduate advisor, Michael Crawford, who had conducted genetic and demographic research among a group known as the Irish Tinkers (Crawford 1975; Crawford and Gmelch 1975). In 1977, I was in graduate school wandering (literally) through the hallways thinking about potential dissertation topics, when Frank asked me if I would like to participate that coming summer in fieldwork in Ireland. That fieldwork focused on a preliminary study of child growth in Ireland based on anthropometrics (Relethford *et al.* 1978) and several analyses of skin color (Lees *et al.* 1979; Relethford *et al.* 1985), as well as a survey of migration rates of parents who returned to Ireland after initially emigrating (Relethford and Lees 1983b). Initial thoughts about returning to Ireland for additional fieldwork leading to a dissertation on human growth were abandoned for a variety of reasons.

My dissertation topic shifted due to several events. At the time, Mike had obtained the original coding forms from a very large anthropometric survey originally collected as part of an anthropological survey conducted by Harvard University during the 1930s. C. Wesley Dupertuis collected anthropometric data on almost 9000 adult Irish men living throughout the island, and Helen Dawson collected similar data on almost 2000 adult Irish women living in the west coast counties (Hooton 1940; Hooton and Dupertuis 1951; Hooton *et al.* 1955). As a graduate student, I was involved with Frank and Mike in some initial analyses of a subset of these data. At the same time, my interests in biological anthropology were being shaped to a large extent by the influence of two books. The first, *Patterns of Human Variation*, by Jon Friedlaender (1975) dealt with the comparison of biological, geographic, and linguistic distances for understanding the population structure and history of Bougainville Island. The second book, *Multivariate Morphometrics* by Blackith and Reyment (1971), dealt with the application of multivariate statistical analyses in understanding patterns of variation based on quantitative traits. The influence of these two works, combined with the availability of the 1930s anthropometric data, led to a dissertation (Relethford 1980b) focusing on the relationship of anthropometric variation to geographic, demographic, and cultural factors among 12 towns in western Ireland. Various parts of this study were later published in several journals and a book chapter (Relethford 1980a; Relethford *et al.*

1980, 1981; Lees and Relethford 1982; Relethford and Lees 1983a; Relethford 1984).

During my dissertation research I became aware of an earlier anthropometric study of the west coast of Ireland, conducted in the early 1890s by the Anthropometric Laboratory of Trinity College, Dublin. The original reports were published in several issues of the *Proceedings of the Royal Irish Academy* from 1893 through 1900 (Browne 1893, 1895, 1896, 1898a, 1900; Haddon and Browne 1893), and included the original anthropometric measurements as well as background information on ethnography, history, and demography. My first use of these data focused on the published surname lists, using frequencies of isonymy (same surname) to address geographic and demographic aspects of population structure (Relethford 1982, 1985).

Up to this point, my primary focus on Ireland was more on local population structure and less on regional population history, other than a short re-analysis of published data (Relethford 1983). My focus shifted in the mid-1980s, when I became more interested in using anthropometric data to address questions of Irish history using the two data sets described above. My first set of analyses used the published anthropometric measurements from Browne's nineteenth century survey of the west coast (Relethford 1988a, 1991; Relethford and Blangero 1990). I then turned to the broader questions concerning the population history of the entire island by returning to the data set originally collected by Dupertuis and Dawson in the 1930s. After receiving a research grant from the National Science Foundation, and with the assistance of Mike Crawford and graduate students at the University of Kansas (Ravi Duggarali and Kari North), several analyses were completed that focused on the genetic impact of historical events (Relethford and Crawford 1995, 1998; Relethford *et al.* 1997; North *et al.* 1999).

The following two case studies provide a summary of these two anthropometric data sets and the relationship of patterns of anthropometric variation to differential migration and gene flow in Irish history. In each case, I describe the purpose of the original studies and contrast it with the inferences made from my analyses.

Case study: English admixture in western Ireland

The first case study is an example of analyzing data from the published literature using data described above that were published in the *Proceedings of the Royal Irish Academy*. These reports are descriptive accounts of the 'ethnography' of various communities in western Ireland, including descriptions of the physical terrain, history, lists of anthropometric measures, summaries of vital statistics, surname lists, and observations of folklore, customs, occupations, and other aspects of culture. As noted by Browne, the purpose of this research was to increase anthropological knowledge of Ireland:

> ... the physical anthropology of Ireland might almost be said to have been an untrodden field. Little or no systematic work has been undertaken in that direction, and yet there was no part of the United Kingdom which promised a richer harvest for the investigator. Anyone who has traveled through the country districts of Ireland must be familiar with the very different types which are presented by the inhabitants. It therefore occurred to us that we might employ the anthropometric methods for the purpose of giving assistance to the anthropologist in his endeavors to unravel the tangled skein of the so-called Irish race.
>
> (Browne 1898b: 269)

Despite this stated goal, the anthropometric analyses were purely descriptive, with only minimal comparisons with other European data. The descriptive nature of these reports is not surprising given their publication dates, well before the development of statistics and population genetics.

The data used here represent seven populations located on or near the western coast of Ireland in County Galway and County Mayo (Fig. 3.1). Three populations are located within 13 km of one another in County Galway: the small

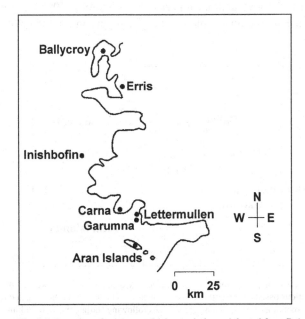

Fig. 3.1. Location of west coast Irish populations. Adapted from Relethford (1988a).

village of Carna and two small islands, Garumna and Lettermullen. The Aran Islands are a group of three islands located about 20 km away in Galway Bay. Another island, Inishbofin, is located further north, and the populations of Ballycroy and Erris are even further north in County Mayo. All seven populations lie within 119 km of one another.

In my analyses, data on ten measurements for 259 adult men were used for analysis of population affinities: head length, head breadth, bizygomatic breadth, bigonial breadth, nose height, nose breadth, head height, stature, hand length, and forearm length. These data were age adjusted using stepwise polynomial regression (Relethford 1988a).

A common pattern in many studies of genetic distances between human populations is a moderate to strong correlation with geographic distance, a pattern known as isolation by distance (Jorde 1980). Because geographic distance acts to limit migration and gene flow, we expect to see a positive correlation between genetic and geographic distance. If isolation by distance were the only factor affecting genetic similarity in the west coast Irish populations, we would expect two distinct clusters corresponding to the two counties sampled, with Inishbofin lying somewhere between them. We might also expect some additional isolation of the Aran Islands and Inishbofin owing to the possible limiting effect of water on migration.

Genetic distances between the seven populations were derived by using the methods described above and are reported in Relethford (1991). Figure 3.2 shows the principal coordinates plot of the genetic distance matrix with each

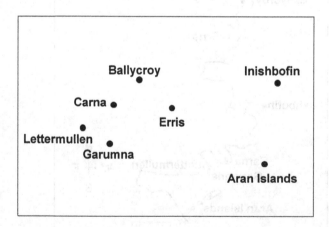

Fig. 3.2. Principal coordinates plot of the genetic distances between west coast Irish populations based on ten anthropometric measures. Each axis has been scaled by the square root of its corresponding eigenvalues following Harpending and Jenkins (1973). The **R** matrix used to obtain the genetic distances is reported in Relethford (1991).

axis scaled by the square root of its corresponding eigenvalues as suggested by Harpending and Jenkins (1973). The first axis, accounting for 59% of the total variation in the distance matrix, clearly separates the Aran Islands and Inishbofin from the remaining populations. Looking at both axes, and ignoring the Aran Islands and Inishbofin for the moment, we can see some effect of geography: the County Galway populations tend to cluster, as do the County Mayo populations. The rank-order correlation between straight-line geographic distance and genetic distance is high and significant ($r = 0.706$, $p = 0.008$ using the Mantel permutation test). When considering all seven populations, however, the correspondence with geography disappears ($r = -0.102$, $p = 0.647$), since both the Aran Islands and Inishbofin are much more distant genetically than expected based on the isolation by distance model.

By itself, Fig. 3.2 does not show *why* the Aran Islands and Inishbofin are genetically distinct. At first glance, we might suspect that both of these populations are distinct because of their greater isolation over water. However, we would still expect greater proximity of the Aran Islands with the other populations in County Galway, and this is not apparent from the results. In addition, we also need to explain why they both plot together in Fig. 3.2, a result that suggests they have something in common. One clue came from the application of the Relethford–Blangero (1990) method, which compares observed phenotypic variation with that expected under a model of equal external gene flow to all populations. In that analysis, both the Aran Islands and Inishbofin show greater within-group phenotypic variation than expected, consistent with a model where these two groups have experienced admixture with an external source (Relethford and Blangero, 1990).

Other historical data were needed to determine the source of this admixture. Both the Aran Islands and Inishbofin have a history of English admixture dating back several centuries, resulting from the garrisoning of British soldiers. This influx began in 1588 with 20 soldiers and continued with larger numbers over several centuries, including detachments of Cromwell's armies in the seventeenth century. It has been suggested that these islands had strategic value as a possible defense of the city and harbor of Galway, and against the possible maritime invasion of Ireland from the west. Hackett and Folan note that there is a popular belief in the west of Ireland that 'some of the ancestors of the Aran islanders (and also the people of Inishbofin) were men of "Cromwell's garrisons."' (1958: 251). The influx of soldiers suggests that the gene pool of these islands may have experienced English admixture. Hackett and Folan summarize this hypothesis:

> There were military garrisons placed on Aran in the 16th and 17th centuries. The more permanent included English soldiers. They may have married island women and left descendants. The native population at the time of the

military garrison was probably small, so any such liaisons then could have made a notable genetic contribution to island stock which subsequently multiplied.

(1958: 255)

Hackett and Folan's analysis of ABO and Rhesus blood group allele frequencies supports this hypothesis for the Aran Islands, where the allele frequencies are more similar to England than is the case for other Irish populations.

Comparative analysis of anthropometric means from English populations confirmed this hypothesis, since these two populations are the most similar phenotypically to England (Relethford 1988a). Analysis of genetic distances derived from surnames did not show this effect, which is expected since surnames change so rapidly that prior history is often erased. Further historical data lends support to this view; Hackett and Folan (1958) looked at surname lists on the Aran Islands and found that the rate of surname extinction was very rapid – of 135 surnames on the island in 1821, 67% had become extinct by 1892 (Relethford 1988a). In this case, the anthropometric data reveal long-term genetic change while surnames likely reflect recent stochastic shifts.

Case study: the population history of Ireland

The above study deals with the impact of an historic event on genetic relationships among populations along part of the west coast of Ireland. This second case study focuses on population genetics and history of the entire island of Ireland (Fig. 3.3), using the data collected by Dupertuis and Dawson in the 1930s as described above. The purpose of this study was to link anthropometric and anthroposcopic variation to racial history. Although much of the original analyses focused on variation relating to geography and religious affiliation (Catholic versus Protestant), much effort was also made for racial classification, looking at the relative frequency of 'morphological types' across Ireland, such as 'Nordic,' 'Keltic,' 'East Baltic,' 'Dinaric,' 'Nordic Mediterranean,' 'Mediterranean,' and 'Nordic Alpine' (Hooton et al. 1955).

The focus on typology in the original study is a bit confusing, representing a semantic shift from 'races' to 'morphological types.' Hooton et al. describe how Hooton set up sorting criteria with the assistance of Carleton Coon in an effort to '... get away from the time-worn method of talking about "racial types" based on arbitrary abstractions from isolated means and modes within a population' (p. 141). The typological orientation, whether they were talking about 'races' or 'morphological types,' is clear. They noted that

no attempt was made to set up new 'races.' It was designed merely to apply to individuals the conventional 'racial' features more or less agreed upon by

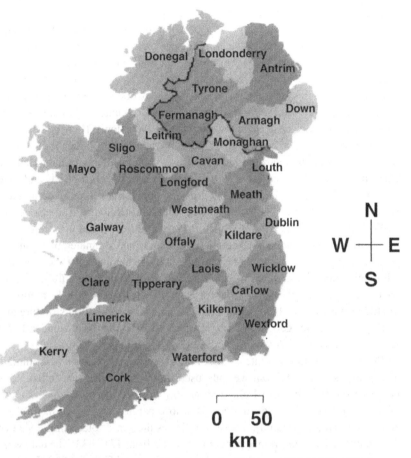

Fig. 3.3. Geopolitical map of Ireland showing the 26 counties in the Republic of Ireland and the 6 counties in Northern Ireland. The two nations are separated by the solid line. Adapted from a public domain map obtained from the Internet at *www.irelandstory.com.*

anthropologists and established in the literature. However, it was necessary to rig a scheme whereby every person fell into some 'race' or type, whether or not he conformed to classic race definitions.

(p. 141)

and further stated

The senior author does not pretend that these crudely delimited 'types' are 'races.' Perhaps they are 'subraces,' perhaps only 'breeds.' Let us not worry for the moment about the taxonomic designations, but find out rather what the sorting of these types can teach us about the Irish population.

(p. 142)

Although the emphasis was clearly on the delineation of morphological 'types,' the results obtained by Hooton *et al.* (1955) in most cases parallel those that I report here with regard to historical influences on anthropometric variation.

The analyses reported here focus on the male data only, since they were available for the entire island, whereas Dawson collected data on females from only the western counties. These data were analyzed according to the birth county of each individual. There are 26 counties in the Republic of Ireland, but one county was excluded because of low sample size (Wicklow). There are also data from the six counties in Northern Ireland, for a total of 31 counties for analysis. Seven body measures and ten head measures were available. Analyses were performed using all 17 variables, as well as separate analyses for body and head measures. The inferred historical patterns and population structure parameters were easiest to interpret using only the head measures (Relethford and Crawford 1995; Relethford *et al.* 1997), which are reported here (head circumference, head length, head breadth, head height, minimum frontal diameter, bizygomatic breadth, bigonial breadth, upper facial height, nose height, and nose breadth). Although head measures are often associated with environmental differences over large continental regions (see, for example, Beals *et al.* 1984), a number of studies have found them quite useful as indices of population structure and history over smaller geographic ranges (see, for example, Friedlaender 1975; Lees and Crawford 1976; Williams-Blangero and Blangero 1989).

Because of the multivariate nature of the analyses, only cases with complete data were used. In addition, we only used those cases where an individual's parents were both born in Ireland so as to minimize the potential influence of recent immigration (e.g. individuals whose parents were born in England and then moved to Ireland). These criteria resulted in a total sample size of 7214 adult males. Sample size per county ranges from 47 to 833. The data were age-adjusted using regression methods (Relethford and Crawford 1995).

Minimum genetic distances were obtained from the **R** matrix. The principal coordinates analysis based on the ten head measures is shown in Fig. 3.4, which shows two primary historical influences. The first principal coordinate, accounting for 38% of the total variation, clearly separates four counties from the remaining 27 counties. These four counties (Leitrim, Longford, Roscommon, and Westmeath) are all adjacent to each other in the Irish midlands near the center of the island. From a purely geographic view, we would expect centrally located populations to be the least distinct genetically. This is not the case here, as further shown by a low and non-significant correlation between geographic and genetic distance ($r = 0.050$, $p = 0.289$). The second principal coordinate, accounting for 24% of the total variation, separates the remaining populations along a strong longitudinal gradient, with the west coast counties toward the top of the plot and the east coast counties toward the bottom. The location of

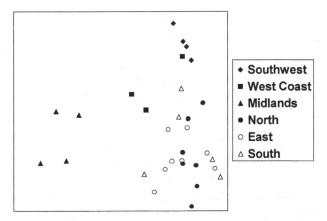

Fig. 3.4. Principal coordinates plot of the genetic distances between 31 Irish counties based on ten head measures. Each axis has been scaled by the square root of its corresponding eigenvalues following Harpending and Jenkins (1973). Counties have been labeled as belonging to one of six geographic regions: Southwest (Clare, Cork, Kerry, Limerick), West Coast (Galway, Mayo, Sligo), Midlands (Leitrim, Longford, Roscommon, Westmeath), North (Antrim, Armagh, Donegal, Down, Fermanagh, Londonderry, Tyrone), East (Carlow, Cavan, Dublin, Kildare, Louth, Meath, Monaghan, Wexford), South (Kilkenny, Laois, Offaly, Tipperary, Waterford). The first axis separates the four Midland counties from all other counties. The second axis separates counties along a west–east longitudinal gradient. The more western counties (Southwest, West Coast) tend to plot near the top, whereas the more eastern counties (North, East, South) tend to plot near the bottom. Adapted from Relethford and Crawford (1995).

the populations along the second principal coordinate shows a strong and significant correlation with the longitude of the geographic center of each county ($r = 0.813$, $p < 0.001$). Spatial autocorrelation analysis of these data has also detected an east–west gradient (North *et al.* 1999).

Both of these patterns have known historical correlates. The distinctiveness of the Irish midlands is best explained by the influx of Vikings moving into the center of the island via the Shannon River (accessible from the Atlantic Ocean). Permanent Norse settlements began in 832, consisting of many thousands of men, and including a major headquarters at Lough Ree, a lake in the Irish midlands. Historical evidence also shows Danish Vikings moving into the region during the tenth and eleventh centuries (see references in Relethford and Crawford 1995). The longitudinal gradient, found in other studies of Irish genetics (Hackett *et al.* 1956; Hackett and Dawson 1958; Dawson 1964; Tills *et al.* 1977), has been explained in terms of the history of Irish settlement. Hooton *et al.* (1955) suggested that early settlers in Ireland were successively pushed westward by subsequent population settlements from the east coast. The earliest Irish settlement dates to roughly 6000 BC, and the first Celtic migrants arrived in at least four waves starting in the sixth century BC (North *et al.* 1999, 2000). It is also likely that more recent historical events contributed to the contrast between west and east coast populations, particularly the Anglo-Norman invasion of the twelfth century and the immigration of settlers from England and Wales starting in the sixteenth century. Tills *et al.* (1977) have suggested that immigrants from England and Wales heavily settled the eastern counties of Ireland, while the northern counties received many immigrants from England and Scotland.

The phenotypic distinctiveness of the Irish midlands and the longitudinal gradient were examined further by comparing Irish means with published anthropometric means. As described by Relethford and Crawford (1995), means for several anthropometric measures were taken from the published literature and compared with the Irish means using Euclidean distance analysis (since only the means were published and not the original data, multivariate distance methods could not be used). For this analysis, the Irish data were subdivided into six geographic clusters for comparison with mean values from England, Denmark, and Norway. The principal coordinates plot of the distance matrix is shown in Fig. 3.5. The Irish midlands are closest to Denmark and Norway, as expected from the Viking influx hypothesis. The northern, southern, and eastern regions lie closest to England, as expected from recent differential settlement from England into the northern and eastern parts of Ireland. The Viking admixture hypothesis has been recently been supported by genetic marker analysis (North *et al.* 2000).

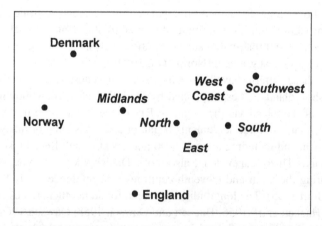

Fig. 3.5. Principal coordinates plot of the Euclidean distances between six geographic regions of Ireland (in italics) and Denmark, Norway, and England based on comparative anthropometric data. Each axis has been scaled by the square root of its corresponding eigenvalues following Harpending and Jenkins (1973). The Irish midlands are the closest to Denmark and Norway, and the northern and eastern regions of Ireland are closest to England. Adapted from Relethford and Crawford (1995).

Are anthropometrics useful measures of population history?

With the advent of new molecular markers of population history, even the commonly used blood polymorphisms popular during the 1960s and 1970s may seem old-fashioned, and anthropometrics might be regarded as positively archaic. While there are many advantages to the newer DNA markers, with many applications not available with other data, this does not mean that data from earlier times are useless – quite the contrary. The two case studies of Irish population history show clear interpretable patterns; biological variation in these cases is a useful mirror of underlying historical processes, particularly differential admixture and population settlement. Despite non-genetic influences and uncertainty about the mode of inheritance, anthropometric data have been quite useful in these particular studies. Other studies of human variation and population history also show anthropometric data to be valuable (see, for example, Friedlaender 1975; Lees and Crawford 1976; Jantz *et al.* 1992).

The availability of historical anthropometric data provides us with the opportunity to examine the relationship between biological diversity and population history from new perspectives and new methodologies. It is worthwhile comparing the results that my colleagues and I have obtained with those originally reported for both of the case studies. The original analysis of the 1890s data set was purely descriptive, listing data tables and a few comparisons with other European populations. To my eye, few inferences were made, again perhaps owing to the lack of appropriate population genetic models and statistical methods in the late nineteenth century. The 1930s study undertaken by Dupertuis and Dawson is a bit different. Here, great effort was invested in computing basic descriptive statistics for a variety of measures and collated on a number of different levels (e.g. geographic, religious affiliation, racial type). This type of analysis was all done before the advent of electronic computing technology, although punch cards were used for mechanical sorting (Hooton *et al.* 1955). Indeed, Hooton has been credited as being the first to use this then-new technology in anthropology (Shapiro 1981). The analyses conducted by Hooton *et al.* are summaries of univariate analyses focusing on geographic structure (counties and aggregates of different counties) and racial classification, as well as the relationship of various estimates of the frequency of racial types with geography. Despite the methodological limitations and the overriding typological orientation of the times, their results also showed the possible influence of Viking invasion and admixture, as well as the west–east pattern resulting from differential settlement and admixture from England and elsewhere. I regard the results of my own multivariate analyses as being clearer and more firmly grounded in population and quantitative genetics theory, but the general patterns were suggested as well by Hooton *et al.* I suspect that the agreement between the two approaches (racial versus microevolutionary) is the fact that Hooton *et al.*

based their conclusions on summaries of many variables together. Thus, their attempt to quantify morphological 'types' was, to some extent, a crude form of multivariate analysis. However, I do not view this as justification for the bio-logical race concept or morphological types; population genetic-based methods give a much clearer picture of the underlying population history with respect to patterns of gene flow and their historical and cultural correlates. To me, this is further demonstration that 'racial' analyses are, at best, crude attempts to summarize overall patterns of human variation. Their use was perhaps more justified in time before the widespread availability of multivariate statistics and computer methods, but this time has passed. There is no longer any need or jus-tification to aggregate data by 'races' or 'types' – we have better alternatives. Still, this does not mean that we should automatically dismiss the hypotheses raised by earlier studies; the overall pattern of historical relationships might be confirmed with more modern theory and methods. This is the case for the 1930s Irish study: my work confirmed some of the hypotheses first proposed by Hooton *et al.* (1955).

The time depth provided by anthropometric data provides another advantage in studies of population history. Anthropometric data are often available from the nineteenth and early twentieth centuries, and sometimes earlier, thus allow-ing us a window that is not available for many genetic and DNA markers. Given the rapid demographic changes in our species over the past century, even data from several decades ago provide us with an opportunity to view a world that has since changed. Further, the methods developed for anthropometric studies can also be applied to osteometric data, allowing inferences about prehistoric populations (see, for example, Steadman 1998, 2001).

Despite their antiquity relative to new genetic technologies, anthropomet-ric data are very useful in studies of population history. They can be used to reexamine old hypotheses and to test new ones. They provide us with some additional time depth and are frequently as useful as genetic markers in un-raveling population structure and history. I suggest that we be on the lookout for anthropometric data sets that are still unanalyzed, either from old published reports or residing in boxes in the back of museum storerooms.

Acknowledgements

I thank my colleagues, Michael Crawford and Francis Lees, for their support and assistance with my Irish studies. I thank Ravi Duggarali and Kari North for their assistance in entry and editing of data described in the second case study. Portions of this research were supported by National Science Foundation Grant No. DBS-9120185.

References

Beals, K.L., Smith, C.L. and Dodd, S.M. (1984). Brain size, cranial morphology, climate and time machines. *Current Anthropology* **25**, 301–10.

Blackith, R.E. and Reyment, R.A. (1971). *Multivariate Morphometrics*. New York: Academic Press.

Browne, C.R. (1893). The ethnography of Inishbofin and Inishark, County Galway. *Proceedings of the Royal Irish Academy* **3**, 317–70.

Browne, C.R. (1895). The ethnography of the Mullet, Inishkea Islands, and Portacloy, County Mayo. *Proceedings of the Royal Irish Academy* **3**, 587–649.

Browne, C.R. (1896). The ethnography of Ballycroy, County Mayo. *Proceedings of the Royal Irish Academy* **4**, 74–111.

Browne, C.R. (1898a). The ethnography of Garumna and Lettermullen, in the County Galway. *Proceedings of the Royal Irish Academy* **5**, 223–68.

Browne, C.R. (1898b). Report of the work done in the Anthropometric Laboratory of Trinity College, Dublin, from 1891 to 1898. *Proceedings of the Royal Irish Academy* **5**, 269–93.

Browne, C.R. (1900). The ethnography of Carna and Mweenish, in the Parish of Moyruss, Connemara. *Proceedings of the Royal Irish Academy* **6**, 503–34.

Connell, K.H. (1950). *The Population of Ireland, 1750–1845*. London: Oxford University Press.

Crawford, M.H. (1975). Genetic affinities and origin of the Irish Tinkers. In *Biosocial Interrelations in Population Adaptation*, ed. E.S. Watts, F.E. Johnston and G.W. Lasker, pp. 93–104. Chicago: Aldine.

Crawford, M.H. and Gmelch, G. (1975). Demography, ethnohistory, and genetics of the Irish Tinkers. *Social Biology* **21**, 321–31.

Dawson, G.W.P. (1964). The frequencies of the ABO and Rh(D) blood groups in Ireland from a sample of 1 in 18. *Annals of Human Genetics* **28**, 49–59.

Felsenstein, J. (1982). How can we infer geography and history from gene frequencies? *Theoretical Biology* **96**, 9–20.

Friedlaender, J.S. (1975). *Patterns of Human Variation: The Demography, Genetics, and Phenetics of Bougainville Islanders*. Cambridge, MA: Harvard University Press.

Hackett, E. and Folan, M.E. (1958). The ABO and Rh blood groups of the Aran Islands. *Irish Journal of Medical Science* **390**, 247–61.

Hackett, W.E.R. and Dawson, G.W.P. (1958). The distribution of the ABO and simple Rhesus (D) blood groups in the Republic of Ireland from a sample of 1 in 37 of the adult population. *Irish Journal of Medical Science* **387**, 99–109.

Hackett, W.E.R., Dawson, G.W.P. and Dawson, C.J. (1956). The pattern of the ABO blood group frequencies in Ireland. *Heredity* **10**, 69–84.

Haddon, A.C. and Browne, C.R. (1893). The ethnography of the Aran Islands, County Galway. *Proceedings of the Royal Irish Academy* **2**, 768–830.

Harpending, H.C. and Jenkins, T. (1973). Genetic distance among Southern African populations. In *Methods and Theories of Anthropological Genetics*, ed. M.H. Crawford and P.L. Workman, pp. 177–99. Albuquerque: University of New Mexico Press.

Hooton, E.A. (1940). Stature, head form, and pigmentation of adult male Irish. *American Journal of Physical Anthropology* **26**, 229–49.

Hooton, E.A. and Dupertuis, C.W. (1951). *Age Changes and Selective Survival in Irish Males. Studies in Physical Anthropology No. 2.* American Association of Physical Anthropologists and Wenner-Gren Foundation for Anthropological Research. Ann Arbor, MI: Edwards Brothers.

Hooton, E.A., Dupertuis, C.W. and Dawson, H. (1955). *The Physical Anthropology of Ireland. Papers of the Peabody Museum Vol. 30, Nos. 1–2.* Cambridge, MA: Peabody Museum.

Jamison, P.L. and Zegura, S.L. (1974). A univariate and multivariate examination of measurement error in Anthropometry. *American Journal of Physical Anthropology* **40**, 197–204.

Jantz, R.L., Hunt, D.R., Falsetti, A.B. and Key, P.J. (1992). Variation among North Amerindians: Analysis of Boas's anthropometric data. *Human Biology* **64**, 435–61.

Jorde, L.B. (1980). The genetic structure of subdivided human populations: A review. In *Current Developments in Anthropological Genetics*, Volume 1: *Theory and Methods*, ed. J.H. Mielke and M.H. Crawford, pp. 135–208. New York: Plenum Press.

Kennedy, R.E. Jr. (1973). *The Irish: Emigration, Marriage, and Fertility.* Berkeley: University of California Press.

Lees, F.C., Byard, P.J. and Relethford, J.H. (1979). New conversion formulae for light-skinned populations using Photovolt and E.E.L. reflectometers. *American Journal of Physical Anthropology* **51**, 403–8.

Lees, F.C. and Crawford, M.H. (1976). Anthropometric variation in Tlaxcaltecan populations. In *The Tlaxcaltecans: Prehistory, Demography, Morphology, and Genetics*, ed. M.H. Crawford, pp. 61–80. *University of Kansas Publications in Anthropology No. 7.* Lawrence, KS: University of Kansas.

Lees, F.C. and Relethford, J.H. (1982). Population structure and anthropometric variation in Ireland during the 1930s. In *Current Developments in Anthropological Genetics*, Volume 2: *Ecology and Population Structure*, ed. M.H. Crawford and J.H. Mielke, pp. 385–428. New York: Plenum Press.

Morton, N.E. (1975). Kinship, information and biological distance. *Theoretical Population Biology* **7**, 246–55.

North, K.E., Crawford, M.H. and Relethford, J.H. (1999). Spatial variation of anthropometric traits in Ireland. *Human Biology* **71**, 823–45.

North, K.E., Martin, L.J. and Crawford, M.H. (2000). The origins of the Irish travelers and the genetic structure of Ireland. *Annals of Human Biology* **27**, 453–65.

Relethford, J.H. (1980a). Bioassay of kinship from continuous traits. *Human Biology* **52**, 689–700.

Relethford, J.H. (1980b). *Population structure and anthropometric variation in rural Western Ireland.* Ph.D. dissertation, Department of Anthropology, State University of New York at Albany.

Relethford, J.H. (1982). Isonymy and population structure of Irish isolates during the 1890s. *Journal of Biosocial Science* **14**, 241–7.

Relethford, J.H. (1983). Genetic structure and population history of Ireland: A comparison of blood group and anthropometric analyses. *Annals of Human Biology* **10**, 321–34.

Relethford, J.H. (1984). Morphological size and shape variation in human populations. *Journal of Human Evolution* **13**, 191–4.

Relethford, J.H. (1985). Examination of the relationship between inbreeding and population size. *Journal of Biosocial Science* **17**, 97–106.

Relethford, J.H. (1988a). Effects of English admixture and geographic distance on anthropometric variation and genetic structure in 19th-century Ireland. *American Journal of Physical Anthropology* **76**, 111–24.

Relethford, J.H. (1988b). Estimation of kinship and genetic distance from surnames. *Human Biology* **60**, 475–92.

Relethford, J.H. (1991). Genetic drift and anthropometric variation in Ireland. *Human Biology* **63**, 155–65.

Relethford, J.H. (1996). Genetic drift can obscure population history: Problem and solution. *Human Biology* **68**, 29–44.

Relethford, J.H. (1998). Genetics of modern human origins and diversity. *Annual Review of Anthropology* **27**, 1–23.

Relethford, J.H. (2001). *Genetics and the Search for Modern Human Origins.* New York: John Wiley & Sons.

Relethford, J.H. and Blangero, J. (1990). Detection of differential gene flow from patterns of quantitative variation. *Human Biology* **62**, 5–25.

Relethford, J.H. and Crawford, M.H. (1995). Anthropometric variation and the population history of Ireland. *American Journal of Physical Anthropology* **96**, 25–38.

Relethford, J.H. and Crawford, M.H. (1998). Influence of religion and birthplace on the genetic structure of Northern Ireland. *Annals of Human Biology* **25**, 117–25.

Relethford, J.H., Crawford, M.H. and Blangero, J. (1997). Genetic drift and gene flow in post-Famine Ireland. *Human Biology* **69**, 443–65.

Relethford, J.H. and Lees, F.C. (1983a). Correlation analysis of distance measures based on geography, anthropometry, and isonymy. *Human Biology* **55**, 653–65.

Relethford, J.H. and Lees, F.C. (1983b). Genetic implications of return migration. *Social Biology* **30**, 158–61.

Relethford, J.H., Lees, F.C. and Byard, P.J. (1978). The use of principal components in the analysis of cross-sectional growth data. *Human Biology* **50**, 461–75.

Relethford, J.H., Lees, F.C. and Byard, P.J. (1985). Sex and age variation in the skin color of Irish children. *Current Anthropology* **26**, 396–7.

Relethford, J.H., Lees, F.C. and Crawford, M.H. (1980). Population structure and anthropometric variation in rural western Ireland: Migration and biological differentiation. *Annals of Human Biology* **7**, 411–28.

Relethford, J.H., Lees, F.C. and Crawford, M.H. (1981). Population structure and anthropometric variation in rural western Ireland: Isolation by distance and analysis of the residuals. *American Journal of Physical Anthropology* **55**, 233–45.

Rogers, A.R. and Harpending, H.C. (1986). Migration and genetic drift in human populations. *Evolution* **40**, 1312–27.

Shapiro, H.L. (1981). Earnest A. Hooton, 1887–1954 *in Memoriam cum amore.* *American Journal of Physical Anthropology* **56**, 431–4.

Steadman, D.W. (1998). The population shuffle in the central Illinois valley: A diachronic model of Mississippian biocultural interactions. *World Archaeology* **30**, 306–26.

52   *J.H. Relethford*

Steadman, D.W. (2001). Mississippians in motion? A population genetic analysis of interregional gene flow in west-central Illinois. *American Journal of Physical Anthropology* **114**, 61–73.

Tills, D., Teesdale, P. and Mourant, A.E. (1977). Blood groups of the Irish. *Annals of Human Biology* **4**, 23–34.

Williams-Blangero, S. and Blangero, J. (1989). Anthropometric variation and the genetic structure of the Jirels of Nepal. *Human Biology* **61**, 1–12.

4 For everything there is a season: Chumash Indian births, marriages, and deaths at the Alta California missions

PHILLIP L. WALKER AND JOHN R. JOHNSON

Introduction

The demographic history of the Chumash Indians who lived in the Santa Barbara Channel area of southern California is better documented than that of almost any other North American Indian population. An exceptionally complete archeological record makes it possible to chart population growth and changing settlement patterns during the long prehistory of this area. More importantly from the perspective of this book, an extraordinarily complete set of ecclesiastical records survives that allows the demographic consequences that European contact had for the Chumash to be studied in great detail.

A key element of Spanish colonial policy was removal of the Chumash from their native villages to mission communities where priests could give them religious instruction and vocational training. In addition to providing the local Indians with the perceived benefits of Christianity, this colonial strategy strengthened Spain's grasp on the northern borderlands of its New World holdings. These religious and political goals were pursued using Indian laborers to build a series of missions and garrisoned forts at strategic locations along the California coast (Fig. 4.1). Spain viewed these bastions of its colonial presence as essential for preventing other nations, especially Russia, from seizing control of Alta California. To measure the success of their proselytizing efforts, the priests at each mission were required to keep careful records of births, deaths, baptisms, marriages, and other important events in the lives of the neophytes (baptized Indian people) under their control. This was a task most of them performed both conscientiously and with considerable zeal. These records of the so-called 'harvest of souls' in the Santa Barbara Channel provided many insights into demographic processes responsible for the collapse of the Chumash population.

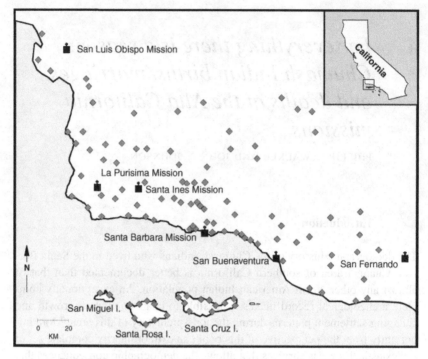

Fig. 4.1. Map of the Chumash Territory. Diamonds mark the locations of historic period Chumash villages.

We know that people have lived in the Chumash territory for more than 10 000 years (Johnson *et al.* 2000; Erlandson *et al.* 1996). Archeological evidence indicates that there was a long-term trend toward population increase, which accelerated markedly beginning around 2000 years ago (Breschini *et al.* 1996; Lambert and Walker 1991) (Fig. 4.2). This late prehistoric period population growth was associated with increased sedentism. It culminated in a non-agricultural economic system based upon fishing and intensive wild-plant resource collection that emphasized local craft specialization and intervillage economic exchange (King 1976; Walker and Hudson 1993).

The arrival of Spanish colonists abruptly terminated this long-term trend toward population growth. The earliest historically documented account of contact between the Chumash and Europeans occurred in 1542 when the explorer Juan Rodríguez Cabrillo sailed into the Santa Barbara Channel (Bolton 1916). Although Cabrillo's voyage was followed by the naval expeditions of Unamuno in 1587, Cermeño in 1595, and Vizcaíno in 1602, the Chumash had very little

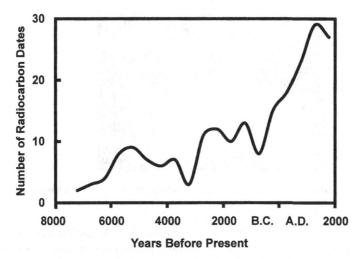

Fig. 4.2. Number of radiocarbon-dated prehistoric sites or site components in the Santa Barbara Channel area per 500-year interval. Sites with multiple dates were counted once per interval.

contact with Europeans until 1769 when an overland expedition headed by Captain Gaspar de Portolá, the governor of Baja California, and Father Junípero Serra, the newly appointed President of the then non-existent Alta California missions, passed through the Chumash territory (Smith and Teggart 1909). Based on the descriptions of villages observed by Portolá and other early explorers, it seems likely that 15 000–18 000 people speaking Chumashan languages were living in the Santa Barbara Channel area in 1772 when the first Chumash mission was established at San Luís Obispo (Brown 1967; Cook 1976b: 37–8; Walker and Johnson 1992, 1994).

Mission records

The Franciscan priests in charge of converting the Chumash to Catholicism kept careful records of important events in the lives of the neophytes under their control (Fig. 4.3). Each mission maintained registers of births, deaths, baptisms, confirmations, and marriages. A *padrón*, or census register, was also maintained that served as an up-to-date list of the people associated with each mission and their marital status (Johnson 1988b). Finally, the priests were required to make annual reports to church authorities. These reports contain demographic data on the neophyte population of each mission, as well as statistics on crop yields

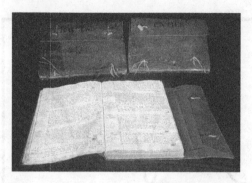

Fig. 4.3. Baptismal, marriage, and burial registers from Mission Santa Bárbara.

and livestock holdings. Fortunately, the primary ecclesiastical records for most of the Alta California missions have been preserved and were microfilmed in the 1960s and 1970s (Barton 1975; Beers 1979). Bound facsimile volumes of nearly all mission registers may be consulted by researchers at the Santa Barbara Mission Archive-Library. With their increased accessibility, ethnohistorians systematically have been gathering and analyzing information from mission records pertinent to the study of California Indians, and an effort is under way to unite separate computer databases that now have been created for most missions into a master database that will facilitate further research (Johnson 1999b).

The baptismal registers are the most important of the mission documents from the perspective of demographic research. They contain the names of all people converted from the native population, and also record all births that occurred among the mission neophytes. Along with data on the sex and approximate age of Indian converts, the baptismal registers also usually give additional information such as the person's native name, village of origin, godparents or sponsors, family relationships, and notes on their social standing within Chumash society. Sometimes there are also interesting comments on physical disabilities and health status. Owing to the priest's belief that baptism was essential for the redemption of an infant's soul, baptisms were conducted as soon after birth as possible. Early baptism was especially important among the Chumash because, as we shall see, the chances of neonatal death were high. This means that, for our demographic purposes, baptism dates of mission-born children can be treated as their birth dates. Every person baptized was assigned a unique identification number in the baptismal register that the priests often cross-reference when making entries in other registers.

The marriage registers include the date of the marriage, the names of the bride and groom, whether they were widowed or single, and the names of witnesses.

The burial registers list the date of death, information on the administration of last rites, and occasional references to the causes and circumstances of death. The *padrones* include much of the same information as the baptismal and marriage registers. This makes them useful for crosschecking the other registers for clerical errors.

The *padrones* sometimes contain entries for deaths, marriages, and baptisms that are unrecorded elsewhere. Omissions in the registers increased markedly following secularization of the missions by the Mexican government in 1834–36. During secularization, government commissioners were put in charge of each mission, and the remaining neophytes were freed from church control. As a result, the mission system began to disintegrate and only a few surviving Chumash families remained in the vicinities of each mission (Johnson and McLendon 1999).

To explore the demographic processes responsible for the historic period decline of the Chumash population, databases of baptismal and burial register entries have been created for the six missions that received Chumash converts: San Luís Obispo, Santa Bárbara, La Purísima, Santa Inés, and San Fernando. When inconsistencies and omissions were found, they could often be resolved through crosschecking names and dates in the *padrones* and other sources (Johnson 1988a, 1999a). Burial entries are unavailable for about 10% of the Chumash converts with baptismal records. Many of these people probably died in native villages after one of the early mission period mass baptisms. Such deaths may well have gone unnoticed by the priests. Other deaths undoubtedly were unrecorded because the person was either a mission fugitive whose fate was unknown, or someone who died at the end of the mission period after systematic record keeping became more difficult.

Demographic collapse

Although the demographic consequences of the earliest Spanish naval expeditions to the Santa Barbara Channel area are uncertain, it is known that the crews of these ships occasionally interacted with the Chumash while provisioning themselves. Such contacts would provide opportunities for the transmission of European diseases (Erlandson and Bartoy 1995). Evidence that infectious diseases were spread to the Chumash at this time comes from Santa Rosa Island skeletons dating to around the time of Cabrillo's 1542 voyage. Two crania from this period have lesions very similar to those produced by the venereal syphilis that was rampant throughout Europe during the sixteenth century (Walker *et al.* 2003). Around the time of Cabrillo's voyage, Spanish explorers were beginning to increasingly interact with Indians living in the puebloan area and it is conceivable that epidemic diseases spread from the southwest into the Chumash territory at this time (Dobyns 1983; Preston 1996).

A total of 9972 Chumash converts are recorded in the mission records (Johnson 1999a). This means that about two thirds of the 15 000 people estimated to have been living in the Santa Barbara Channel area at the time of the earliest Spanish overland expeditions entered a mission. These population size estimates incorporate the assumption that some epidemic-induced depopulation occurred before the missions were established (Brown 1967; Cook 1976b). Nevertheless, it is clear that many Chumash people who were alive at the beginning of the mission period did not enter the missions system. There is no evidence that large numbers of Indians fled into the interior to avoid Spanish contact; this strategy would have been very difficult to pursue because the interior was already fully occupied and the resources were simply not available in this marginal, semiarid environment to support large numbers of people from coastal settlements (Johnson 1999a).

A much more likely explanation is that introduced diseases and outbreaks of epidemic-induced warfare resulted in many Chumash deaths before the missionization process was fully under way. In the 1770s typhus and measles epidemics reportedly occurred in Baja California (Jackson 1981) and these could have been transmitted along native trade routes to the Santa Barbara Channel area. Chumash oral histories also record epidemics just before the arrival of European colonists in which 'people went about feeling sick until they fell backwards, dead' (Librado et al. 1977: 11). The missionaries knew that many deaths were occurring among unconverted Indians living outside the mission system; one measles epidemic on the Channel Islands reportedly killed more than 200 people (Johnson 1982: 63). Other ethnohistoric records suggest that intervillage warfare was common around the time the missions were being founded. This would have contributed to deaths among non-baptized Indians. The Chumash themselves described some of this killing as retaliation for the spread of epidemics by Indian enemies (Brown 1967: 75–6; Engelhardt 1932: 7; Geiger and Meighan 1976: 93, 113, 139; Johnson 1988a; Walker and Hudson 1993).

Whatever the demographic consequences were of these early contacts, we do know from documentary sources that the Chumash population began a precipitous decline soon after the Spanish settled on the mainland coast (Fig. 4.4). Within about 50 years, all of the residents of the once thriving Chumash villages described by members of the Portolá expedition had either died or relocated to one of the missions. The first of these, Mission San Luís Obispo, was founded in 1772 and others soon followed: Mission San Buenaventura in 1782, Mission Santa Bárbara in 1786, Mission La Purísima in 1787, Mission San Fernando

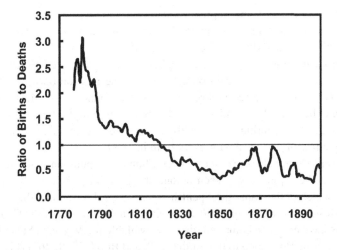

Fig. 4.4. Ratio of births to deaths for neophytes at the six Chumash missions for which birth and death records are available. The ratio was calculated based on a running 5-year mean of the number of births and the number of deaths.

in 1798, and finally Mission Santa Inés in 1804. As they did elsewhere in the borderlands of their North American holdings, the Spanish implemented the colonial strategy of *reducción* in which the Indians were relocated from their native villages to mission communities where their activities could be more easily supervised by priests (Walker 2002). The resulting concentration of many people in one place had devastating health consequences for the Chumash. It disrupted their traditional subsistence practices and greatly abetted the maintenance and spread of newly introduced infectious diseases.

The number of Chumash residing at the missions increased between 1772 and 1805. Unfortunately, this growth is explained almost entirely by the recruitment of Indians fleeing the rapidly deteriorating conditions in their native villages, not births among the neophytes already at the missions. At its zenith in 1805, the mission Chumash population consisted of about 6000 neophytes. During the next 30 years the number of Indians at the missions steadily declined, except for brief increases in 1814 and 1816 when most of the remaining Channel Island Chumash were brought to the missions. By 1832, the Chumash population had been reduced to 1182 people. Conditions at the missions deteriorated markedly after they were secularized in 1834. The Catholic Church was stripped of its holdings, and the missions were turned over to government administrators who were often more concerned with personal financial gain than improving the living conditions of their Indian charges (Engelhardt 1963: 147; Librado *et al.*

1979: 32). As a result, the population decline continued and by the time of the first California state census in 1852, fewer than 600 members of the once flourishing Chumash tribe survived (Johnson 1993).

From a demographic standpoint, the Chumash missions were failures almost from their inception. The birth-to-death ratio in the mission-born population began to decline soon after the first missions were founded (Fig. 4.4). During the decade between 1781 and 1791 it dropped from a high of 3 to 1.4 births for every death. It continued to steadily decrease for the next 60 years, finally reaching a low of two deaths for every birth in 1850.

This reproductive failure of the mission Chumash population stands in sharp contrast to the rapid growth seen among the *gente de razón*, or 'people of reason,' which is the name the Spanish colonists used to refer to themselves. Although statistics on the *gente de razón* living in the vicinity of the Chumash missions have not been compiled, they are available for Missions Santa Clara and San Carlos Borromeo to the north (Cook and Borah 1979: 263) (Fig. 4.5). At the beginning of the mission period the ratio of 2.5 births for every death among these colonists was almost identical to that of the Chumash. Within a few years, however, the two groups began to diverge because of a decrease in Chumash births and an increase in Chumash deaths. The birth to death ratio of the colonists, in contrast, remained high throughout the nineteenth

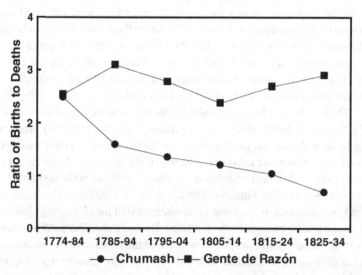

Fig. 4.5. Ratios of births to deaths for the mission Chumash and Spanish-Colonial (*gente de razón*) populations of California between 1774 and 1834. Data on the Spanish Colonists are those reported by Cook and Borah (1979: 263) for Missions Santa Clara and San Carlos Borromeo.

century and this fueled a rapid increase in California's Hispanic population (Fig. 4.5).

The demographic consequences European contact had for the Chumash is graphically illustrated by the 1822 population pyramid we have reconstructed through cross-referencing baptism and burial entries (Fig. 4.6). With the exception of having many more adult females than adult males and a shortage of children under the age of 5, the 1782 Chumash population pyramid has a triangular form similar to that of many eighteenth and nineteenth century European populations. Forty years later, the 1822 population pyramid shows the devastating effects of mission life; high infant mortality is reflected in its narrow base, which, in a growing population, would be broad because of the presence of many young children. The 1822 population is anomalous in another respect: its sex ratio is markedly skewed in favor of males with a striking deficit of young women of reproductive age and an excess of middle-aged men.

Causes of population decline

An exceptionally high infant mortality rate clearly provides part of the explanation for the mission Chumash population decline. Unfortunately for the Indians, most of their babies died during the first four years of life. This is in striking contract to the much lower infant mortality experienced by the contemporaneous Spanish colonists. Among the *gente de razón* associated with the Santa Clara and San Carlos Borromeo missions, deaths of *párvulos* (children about 7–8 years old or younger) occurred at a rate of 200 or less per 1000 births throughout the mission period (Cook and Borah 1979: 263) (Fig. 4.7). Among the mission Chumash, in contrast, the number of *párvulos* deaths per 1000 births started at four times that level at the beginning of the mission period and climbed to shocking 900 *párvulos* deaths per 1000 births by the time of secularization.

The reproductive problems of the Chumash experienced at the missions are explained in part by a high death rate among young women of childbearing age (Figure 4.8). In 1820, Father Mariano Payeras, a priest at Mission La Purísima, remarked upon this problem in a letter in which he observed that when the Chumash moved to the missions 'they become extremely feeble, lose weight, get sick and die. This plague affects the women particularly, especially those who have recently become pregnant' (Payeras 1995: 225). Our analysis of sex differences in deaths during the 1806 measles epidemic supports Payeras' observation that young female neophytes were especially vulnerable to disease (Walker and Johnson 1994: 116–18). This epidemic killed many women in their prime reproductive period between the ages of 20 and 34. Mortality among men

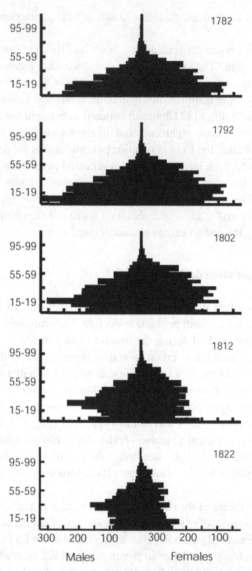

Fig. 4.6. Population pyramids of baptized Chumash. These population pyramids are based on data from baptismal and death registers of Santa Bárbara, La Purísima, Concepción, and Santa Inés. Records of a person's age at baptism and year of death were used to reconstruct the age structure of Chumash who were living at the missions as well as those who were still living in their native villages.

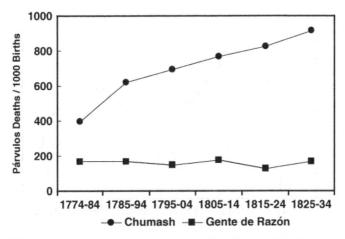

Fig. 4.7. Ratio of *párvulos* (children <8 years old) deaths per 1000 births in the mission Chumash and Spanish-Colonial (*gente de razón*) population of California between 1774 and 1834. Data on the Spanish Colonists are those reported by Cook and Borah (1979: 263) for Missions Santa Clara and San Carlos Borromeo.

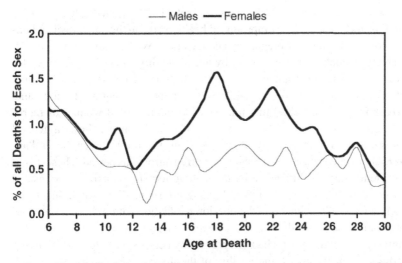

Fig. 4.8. Graph showing the deaths for each sex occurring between the ages of 6 and 30 years in the mission Chumash population as a percentage of the total number of deaths for each sex. Note the high mortality rate of young women of reproductive age relative to men of the same age and the increase in deaths of girls relative to boys around the time of puberty, which corresponds to the age at which girls started to be locked up in *monjeríos* (special convent-like buildings).

Fig. 4.9. Variation in the fertility rate of the Chumash population calculated using data from mission baptismal and death records from Mission Santa Bárbara, Santa Inés, and La Purísima. The graph plots the five-year moving average of the general fertility rate (total number of births/total population of females 15–49 years old × 1000).

of the same age was much lower. It seems likely that this high death rate among young women stems from compromised health owing to the nutritional and immunological demands of pregnancy and lactation (Walker and Johnson 1994).

The reproductive problem posed by high mortality among women of child-bearing age was compounded by the low fertility of the women who were fortunate enough to survive (Fig. 4.9). Between the founding of Mission Santa Bárbara in 1786 and 1820 the number of births per 1000 women of reproductive age (15–49 years old) among the mission Chumash shows a general downward trend from a high of about 100 births per 1000 women. Several apparent short-lived increases in fertility occurred during this period. These do not reflect temporary fertility increases among long-term mission residents; instead they correspond to intensified proselytizing efforts in 1796, 1803, and 1816, which periodically augmented the mission population with Indians who previously had remained in their native villages. Once this supply of women with relatively high fertility who had not previously been exposed to mission living conditions was exhausted, the fertility of the mission population plummeted. By the end of the mission period the population's birth rate had declined to fewer than 30 births per year for every 1000 women of reproductive age.

There is documentary evidence that venereal disease and perhaps also abortions contributed significantly to the low fertility of the mission Chumash population. The sexual mores of the Chumash differed considerably from those of the Spanish and this probably facilitated the spread of sexually transmitted diseases. The Chumash evidently had a pragmatic attitude toward sexual activity. For example, the chronicler of Vizcaíno's voyage to the Santa Barbara Channel area in 1602 remarked that a Chumash chief tried to lure the Spaniards ashore by offering each sailor ten women to 'serve and entertain them' (Wagner 1929: 240).

The sexual activities of the Chumash and other California Indians were sufficiently troubling to the Mission priests that special dormitories called *monjeríos* were built at the Missions in which girls and unmarried women were locked up at night to prevent them from having sexual relations with men (Jackson 1994: 133; Milliken 1995: 89; Webb 1983). The *monjeríos* evidently were sometimes repugnant places. For example, Diego de Borica, the Governor of California from 1794 to 1800, reported that he once attempted to enter one of these buildings at an unidentified mission, but could not endure it for even a minute because of the repellent stench that permeated it (Florescano and Gil Sánchez 1976). The health consequences of such filthy living conditions may very well have contributed to the high mortality rates documented among young Chumash women. This is strongly suggested by mission Chumash mortality profiles, which show an increase in deaths among young girls around the time that they began to be sequestered in the *monjeríos*. This increase is not seen in boys of the same age, who remained free to sleep where they pleased (Fig. 4.8).

Two of the priests at the Chumash missions recorded their concerns about the effect that syphilis was having on their charges when they responded to a questionnaire Don Ciriaco Gonzales Carvajal, Secretary of the Department of Overseas Colonies, sent them in 1812. The missionaries at Santa Barbara gave the following response to a question about the health status of the mission's neophytes: 'The sicknesses found among these Indians are those common to all mankind but the most pernicious and the one that has afflicted them most here for some years is syphilis. All are infected with it for they see no objection to marrying another infected with it. As a result births are few and deaths many so that the number of deaths exceed births by three to one' (Geiger and Meighan 1976: 74). The priests at San Luís Obispo echoed this concern and wrote that venereal disease was 'an infirmity with which generally all the Indians are infected to such an extent that any other illness during the various stages of the year kills them. Hence it is to my grief and to all who behold their unhappy fate that there are more deaths than births' (Geiger and Meighan 1976: 75).

These sexually transmitted diseases probably contributed significantly to the reproductive problems of the Chumash at the missions through producing subfertility and infertility in both men and women. Sexually transmitted infections reduce fertility by obstructing the reproductive ducts, and also increase miscarriages and neonatal death rates (Kramer and Brown 1984; Westrom 1994). Epidemiological models suggest that a 20% prevalence of untreated gonorrhea in sexually active adults can result in a 50% reduction in net population growth (Brunham *et al.* 1991; Wright 1989).

Abortions and infanticide by the Chumash may have further depressed the mission Chumash reproductive rate. In 1794 Father Lasuén observed the decline of the native population was due to their 'dominant vice of incontinence, and especially in the entire Channel region from Santa Barbara to San Luis, their inhuman and widespread practice of voluntary abortion, or on the part of the mother, of suffocating their newly born children'(Lasuén 1965: 378). José Longinos Martínez, who visited the Santa Barbara area in 1792, made a similar observation: 'In this region they have the notion that unless they have an abortion at their first pregnancy, or if the child does not die immediately, they will never conceive again. Hence they murder many babies with the efforts they make, the blows they give themselves, and the barbarous medicine they take in order to induce an abortion, so that some of the women die and others are badly injured' (Simpson 1961: 56). Elderly Chumash consultants interviewed by John Peabody Harrington at the beginning of the twentieth century verified that infanticide was practiced in the 'old days' when deformed babies were born and that abortions were sometimes performed by 'eating medicine' (Harrington 1942: 35; Walker and Hudson 1993: 107–8).

Birth seasonality

Comparisons of the reproductive patterns of the mission Chumash and Hispanic colonists of Alta California provide some insights into the environmental causes of the reproductive problems that decimated the Chumash. At the beginning of the mission period (1769–1800), Alta California's Hispanic population shows a complex pattern of birth seasonality with spring and fall peaks as well as a secondary one during mid-summer (Fig. 4.10). This pattern is atypical of an agrarian society such as this and may reflect the predominance of military personnel among these early colonists, whose activities were not directly tied to seasonal cycle of agricultural activity (Garcia-Moro *et al.* 2000). Later in the colonial period, the mid-summer birth increase disappears and births assume a strong seasonal pattern with a major spring peak and a minor early winter peak. This so-called 'European' birth pattern is typical of agricultural populations. It reflects a high frequency of July and February conception and a low conception rate during the autumn harvest season.

Throughout the mission period the pattern of Chumash birth seasonality differs significantly from that of these colonists (1774–1800: $\chi^2 = 27.91$, $p = 0.01$; 1801–1820: $\chi^2 = 26.03$, $p = 0.01$; 1821–1900, $\chi^2 = 24.67$, $p = 0.02$). Early in the mission period Chumash births show a seasonal pattern with

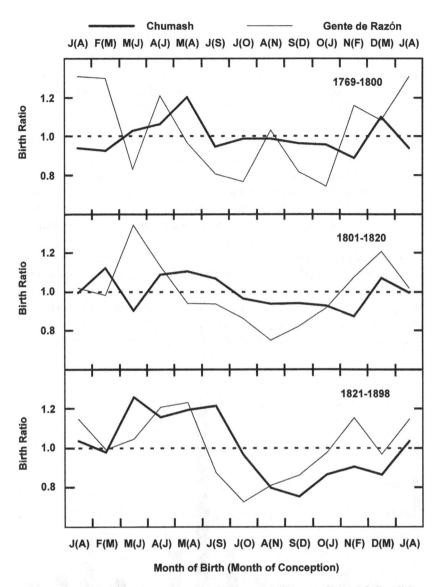

Fig. 4.10. Comparisons of patterns of birth seasonality among Chumash Indians living at the missions (*n* = 6960) and the contemporaneous Spanish colonial population (*gente de razón*) of California. The *gente de razón* data are from Garcia-Moro *et al.* (2000). The birth ratio is the number of births per day during a month divided by the average number of births per day for the entire year. A birth ratio of greater than 1.0, therefore, indicates more births during that month than the annual average.

increases in May and December, which would correspond to conception peaks during the late spring and fall. This may reflect the residual influence of seasonality in their precontact fishing and seed gathering activities, which continued to some extent at the missions. Before the arrival of Europeans, the mid-winter was a time when stores of seeds and dried fish ran low, and rain and rough seas made fishing both difficult and dangerous (Simpson 1961: 52). As the Chumash became increasingly dependent upon the mission agricultural system, this earlier pattern of birth seasonality gradually disappeared. During the first two decades of the nineteenth century, mission Chumash births show less seasonal variation, perhaps owing to the disruptive effects of the profound social and economic transformations they were undergoing (Fig. 4.10). By the end of the mission period, however, a seasonal pattern re-emerges, which is similar to that seen among contemporaneous Hispanic colonists except for a much less prominent fall birth increase among the Chumash.

A cultural preference for mid-summer marriages may have contributed to the late spring birth peak seen among both the Chumash and the colonists. Our analysis of Mission La Purísima marriages registers shows that there was a strong preference for mid-summer marriages at the missions: more than half of the weddings were held between June and August, with 32% of them performed in July (Fig. 4.11). In populations lacking effective contraceptive methods,

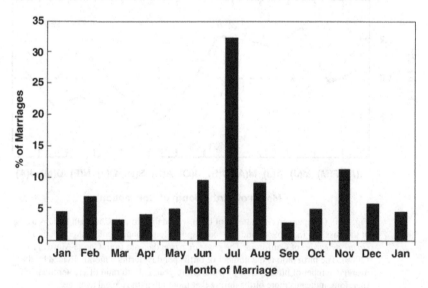

Fig. 4.11. Seasonal distribution of 219 marriages among Chumash neophytes at Mission La Purísima between 1805 and 1813. These frequencies exclude remarriages in Christian ceremonies of neophytes previously married in Chumash ceremonies.

most healthy brides will be pregnant within a few months after their marriage. This would tend to concentrate first births in March, which is ten months after the mid-summer marriage season. The spring birth increases seen among the mission Chumash and Hispanic colonists during the latter part of the mission period may, therefore, reflect a preference for July weddings.

Preferred marriage dates are not arbitrary. In agrarian societies, they reflect many practical considerations, the most important of which are accommodating the seasonal patterns of agricultural production and workload. The agricultural cycle, in turn, is directly related to climatically induced seasonal fluctuations in local environmental productivity. The mission period mid-summer marriage preference is undoubtedly in part a solution to the problem of scheduling these celebrations during a period when food was plentiful while at the same time avoiding important religious holidays (Easter and Christmas) and the period of intensive harvest-season work during late summer and early fall.

The seasonal pattern of agricultural productivity in southern California is a product of its Mediterranean climate with a cool, rainy winter season between November and March and a long warm, dry growing season between April and October. Early in the Colonial period, between 1769 and 1820, the proportion of annual births occurring each month among the Hispanic colonists shows a statistically significant positive correlation with that month's average rainfall as gauged from modern instrumental records (1769–1800: $r = 0.65$, $p = 0.02$; 1801–20: $r = 0.66$, $p = 0.02$). After 1820 this relation between births and rainfall continues to persist, but weakens to the point that it is no longer statistically significant ($r = 0.41$, $p = 0.18$). Births among the mission Chumash, in contrast, do not show a significant positive correlation with rainfall during any part of the mission period (1769–1800: $r = -0.17$, $p = 0.02$; 1801–20: $r = 0.10$, $p = 0.76$; 1821–98: $r = 0.16$, $p = 0.63$).

Death seasonality

The probability of death shows a strong age-related seasonal pattern among the mission-born Chumash. Neonates who died within the year of their birth account for 34% of the deaths recorded in the mission burial registers. Exploring death seasonality among these infants is made difficult by the fact that our death registry database contains the year of a person's death but not the day and month. Thus, if we confine our analysis to infants who die within the year of their birth, those born in January have 1–12 months of exposure to potentially lethal environmental conditions whereas children born in December have one month or less. Because of this actuarial problem, the frequency of neonatal deaths during the year of birth decreases linearly between January and

December, and this obscures any seasonal variation that might exist in neonate mortality rates. Dividing deaths during each month by the number of months of potential exposure (12 for January births, 11 for February births, and so on) does not effectively correct for this bias because the probability of death during each month of the first year of life varies dramatically, with many more deaths occurring during the first few months following birth than later in the year.

As age at death and the number of years in the pooled sample of deaths increases, the obfuscating effect of this actuarial problem rapidly diminishes. This makes it possible to examine the relation between death rate and birth season for older children whose month of death is unknown. When the proportion of deaths for each birth month of mission-born Chumash children who died between the ages of 1 and 5 years is tabulated for the early (1774–1800) and late (1801–1898) mission periods, two contrasting seasonal patterns emerge that cannot be explained by the weighting problem discussed above (Fig. 4.12). More deaths of children 1–5 years of age occur among those born during the last six months of the year early in the mission period than later in the mission period ($\chi^2 = 9.4$, $p = 0.002$). Many early mission period children who died between the ages of 1 and 5 were born during November and December. Among the later mission period children, in contrast, the birth months of children who

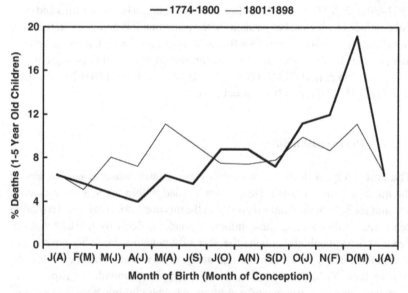

Fig. 4.12. The proportion of deaths for each birth month of mission-born Chumash children who died between the ages of 1 and 5 years during the early mission period (1774–1800) and the late mission period (1801–1898).

died are concentrated during the late spring. For both periods, there is a tendency for more childhood deaths among children born early in the year than later in the year. This is the opposite of the trend expected from the actuarial problem discussed above and suggests that its effect on the overall distribution of deaths among these older children is minimal.

Recent research suggests that prenatal and early postnatal nutrition can significantly increase the risk of death from infections contracted much later in life. This long-lasting effect of early developmental disruption provides a possible explanation for the relation between childhood mortality and the birth season of mission period Chumash children. Studies of Gambian villagers who experience marked annual cycles in food availability and disease have demonstrated a strong correlation between birth season and premature death. In this African population, young adults who die of infectious diseases are much more likely to have been born during the 'hungry season' than more favorable times of the year (Moore *et al.* 1999). Such studies suggest that nutritionally mediated intrauterine growth retardation can permanently damage a person's immunological development.

The early mission period deaths show a strong positive correlation with modern records of average rainfall nine months earlier during the probable month of conception ($r = 0.76$, $p = 0.004$). These data suggest that, during the early mission period, children conceived in the late winter had a higher probability of early childhood death than children conceived at other times. Before the mission agricultural system was fully operational, the late winter was sometimes a period of food shortage. This correlation may, therefore, reflect problems of fetal development associated with poor maternal nutrition during the winter months.

It is interesting to note that the Chumash recognized a relationship between the month of birth and adult vitality; some months were considered a healthy birth month, others not (Walker and Hudson 1993). For example, a person born during the Chumash month that corresponds to our March was considered likely to become 'sickly' in life because leaves that develop during March are often weak and sickly (Blackburn 1975: 101–2; Hudson and Underhay 1978: 126–39).

The deaths among the late mission period Chumash children do not show the correlation between month of conception and rainfall seen in the early mission period population. This perhaps reflects the fact that, for the most part, food was plentiful at the missions. After a few marginal early years when the neophytes were allowed to return to their traditional subsistence practices to compensate for mission agricultural shortfalls, food was usually superabundant, at least from a caloric standpoint, at the missions (Costello 1989; Walker *et al.* 1989). As one priest lamented: 'They live well free but as soon as we reduce them to a Christian and community life . . . they fatten, sicken, and die' (Archibald 1978: 157). Evidently the availability of large quantities of food at the missions did not result in better health.

Conclusions

The detailed documentary evidence we have for the effects that European contact had on the Chumash Indians is unusual. Many Native American groups, especially those contacted soon after Europeans discovered the New World, vanished before explorers even bothered to note their existence. The Chumash, in contrast, began to intensively interact with Europeans late in the colonial process. Alta California was the last frontier Spain colonized as part of its attempt to gain a colonial foothold in North America. Because of this, an elaborate bureaucracy was already in place that allowed the Spanish government both to provision the missionaries and soldiers assigned to remote Alta California outposts and also to monitor their activities (Perissinotto *et al.* 1998).

In Alta California, as elsewhere in the borderlands of New Spain, relations between military and clerical authorities were often less than cordial. This antipathy is good from a historian's perspective: self-serving accounts by one group are often counterbalanced by less sanguine reports of the same situation by the other.

Of greatest importance from the perspective of our demographic research is the care the Franciscan friars took to document important events in the lives of the Indians they sought to convert from paganism to Christianity. Baptism was of central importance in this regard: Catholic doctrine is uncompromising in its position that those who die without baptism, newborn infants included, are perpetually excluded from heaven. As a result, the priests were extremely assiduous in their efforts to harvest as many pagan souls as possible for the Kingdom of God. Baptism was considered of such importance that priests even received instructions for performing Cesarean sections on the bodies of pregnant women who died so that their unborn infant could be afforded the sacrament before dying in their mother's womb (Rigau-Perez 1995; Valle 1974). This baptismal zeal, in conjunction with the priests' dutiful recording of the deaths of those who passed on to heaven, affords us with extraordinarily complete picture of the process of cultural disruption and death through which the Chumash population was decimated.

A central message in this depressing story of enormous human suffering is the complexity of the social, political, and demographic processes involved. Genocidal raids by American settlers were important in contributing to the extinction of some northern California tribes during the last half of the nineteenth century (see, for example, Cook 1976a: 284; Thornton 1987: 107). Although Spanish soldiers occasionally fought with the Chumash (see, for example, Lasuén 1965: 46; Tibesar 1955: 295), we have no documentary evidence for large-scale killing. Instead, the introduction of new infectious diseases in conjunction with the inadvertent consequences of the concerted efforts of missionaries to concentrate the Chumash in mission settlements appear to be the principal causes of this rapid decline. The amount of physical and psychological abuse the Indians who came to the Alta California missions suffered at the hands of mission priests has long been the center of acrimonious scholarly debate (Castillo 1989; Costo and Costo 1987; Guest 1979). Such disputes will be difficult to resolve based on the existing documents, many of which consist of reports the priests made of their own activities. On the other hand, the baptismal and death records these friars produced are unambiguous on one point: the mission environment was exceptionally detrimental to young women and especially infants. The precise causes of the high mortality rates in these two segments of the mission Chumash population undoubtedly varied from mission-to-mission through time. Nevertheless, there is little doubt that the colonial strategy of relocating Indians from their native villages and bringing them together in dense aggregations at mission settlements created an environment that fostered infertility, infectious disease, and population decline.

Acknowledgements

We are indebted to Fr Virgilio Biasiol of the Santa Barbara Mission Archives and Msr Francis Weber of the Chancery Archives of the Archdioceses of Los Angeles for access to the mission records used in this research. Much of the database used for our research was assembled and integrated under Johnson's direction for a project funded by the National Park Service Archeology and Ethnography Program (McLendon and Johnson 1999). The contributions of many individuals were instrumental in creating this database, including Chester King, Robert López, Bob Edberg, Scott Edmondson, Sally McLendon, Elise Tripp (Santa Ynez Indian Reservation), Julie Tumamait (Ventureño Chumash), Eleanor Arellanes (Ventureño Chumash), Gilbert Unzueta (Barbareño Chumash), Virginia Ochoa (Santa Ynez Indian Reservation), Henry Cruz (Ineseño Chumash), Carmen Unzueta (Barbareño Chumash), Holly Woolson, and Dinah Crawford.

References

Archibald, R. (1978). *Economic Aspects of the California Missions*. Washington, D.C.: Academy of American Franciscan History.

Barton, N.R. (1975). Genealogical research in the records of the California Spanish missions. *Genealogical Journal* **4**(1), 11–33. Salt Lake City: Utah Genealogical Association.

Beers, H.P. (1979). *Spanish & Mexican Records of the American Southwest*. Tucson: University of Arizona Press.

Blackburn, T.C. (1975). *December's Child: A Book of Chumash Oral Narratives*. Berkeley: University of California Press.

Bolton, H.E. (1916). *Spanish Exploration in the Southwest, 1542–1706*. New York: C. Scribner's Sons.

Breschini, G.S., Haversat, T. and Erlandson, J. (1996). *California Radiocarbon Dates*. Salinas: Coyote Press.

Brown, A. (1967). The aboriginal population of the Santa Barbara Channel. *Reports of the University of California Archaeological Survey* **69**. Berkeley: University of California Archaeological Research Facility, Department of Anthropology.

Brunham, R.C., Garnett, G.P., Swinton, J. and Anderson, R.M. (1991). Gonococcal infection and human fertility in sub-Saharan Africa. *Proceedings of the Royal Society of London B* **246**, 173–7 [published erratum (1992) appears in *Proc. R. Soc. Lond. B* **247**(1320):223].

Castillo, E.D. (1989). The native response to the colonization of Alta California. In *Columbian Consequences: Archaeological and Historical Perspectives on the Spanish Borderlands West*, ed. D.H. Thomas, pp. 377–94. Washington, D.C.: Smithsonian Institution Press.

Cook, S.F. (1976a). *The Conflict between the California Indian and White Civilization*. Berkeley: University of California Press.

Cook, S.F. (1976b). *The Population of the California Indians, 1769–1970*. Berkeley: University of California Press.

Cook, S.F. and Borah, W.W. (1979). *Essays in Population History: Mexico and California*. Berkeley: University of California Press.

Costello, J. (1989). Variability among the Alta California missions: the economics of agricultural production. In *Columbian Consequences: Archaeological and Historical Perspectives on the Spanish Borderlands West*, ed. D.H. Thomas, pp. 435–49. Washington, D.C.: Smithsonian Institution Press.

Costo, R. and Costo, J.H. (1987). *The Missions of California: A Legacy of Genocide*. San Francisco: The Indian Historian Press.

Dobyns, H.F. (1983). *Their Numbers Become Thinned: Native American Population Dynamics in Eastern North America*. Knoxville: University of Tennessee Press.

Engelhardt, Z. (1932). *Mission Santa Ines, Virgen y Martir, and Its Ecclesiastical Seminary*. Santa Barbara: Mission Santa Barbara.

Engelhardt, Z. (1963). *Mission San Luis Obispo in the Valley of the Bears*. Santa Barbara: W.T. Genns.

Erlandson, J.M. and Bartoy, K. (1995). Cabrillo, the Chumash, and Old World diseases. *Journal of California and Great Basin Anthropology* **17**, 153–73.

Erlandson, J.M., Kennett, D.J., Ingram, B.L., Guthrie, D.A., Morris, D.P., Tveskov, M.A., West, G.J. and Walker, P.L. (1996). An archaeological and paleontological chronology for Daisy Cave (CA-SMT-261), San Miguel Island, California. *Radiocarbon* **38**, 355–73.

Florescano, E. and Gil Sánchez, I. (1976). *Descripciones Económicas Regionales de Nueva España. Promincias del Norte, 1790–1814.* Mexico: Instituto Nacional de Antropologia e Historia, Departamento de Investigaciones Historicas, Seminario de Historia Economica. 359 pp.

Garcia-Moro, C., Pascual, J., Toja, D.I. and Walker, P.L. (2000) Birth seasonality in the early Spanish-Mexican colonists of California (1769–1898). *Human Biology* **72**, 655–74.

Geiger, M. and Meighan, C.W. (1976). *As the Padres Saw Them: California Indian Life and Customs as Reported by the Franciscan Missionaries, 1813–1815.* Santa Barbara: Santa Barbara Mission Archive-Library.

Guest, F.F. (1979). An examination of the thesis of S.F. Cook on the forced conversion of Indians in the California missions. *Southern California Quarterly* **61**(1), 1–77.

Harrington, J.P. (1942). Culture element distributions: XIX Central California Coast. *Anthropological Records* **7**, 1–46. Berkeley: University of California Press.

Hudson, T. and Underhay, E. (1978). *Crystals in the Sky: An Intellectual Odyssey Involving Chumash Astronomy, Cosmology, and Rock Art.* Socorro, NM: Ballena Press.

Jackson, R.H. (1981). Epidemic disease and population decline in the Baja California missions, 1697–1834. *Southern California Quarterly* **63**, 308–46.

Jackson, R.H. (1994). *Indian Population Decline: The Missions of Northwestern New Spain, 1687–1840.* Albuquerque: University of New Mexico Press.

Johnson, J.R. (1982). *An ethnohistoric study of the island Chumash.* M.A. thesis, University of California, Santa Barbara.

Johnson, J.R. (1988a). *Chumash social organization: an ethnohistoric perspective.* Ph.D. dissertation, University of California, Santa Barbara.

Johnson, J.R. (1988b). Mission registers as anthropological questionnaires: understanding limitations in the data. *American Indian Culture and Research Journal* **12**(2), 9–30.

Johnson, J.R. (1993). The Chumash Indians after secularization. In *The Spanish Missionary Heritage of the United States*, ed. H. Benoist and M.C. Flores, pp. 143–64. San Antonio, Texas: National Park Service and Los Compadres de San Antonio Missions National Historical Park.

Johnson, J.R. (1999a). Chumash population history. In *Cultural Affiliation and Lineal Descent of Chumash Peoples in the Channel Islands and the Santa Monica Mountains*, ed. S. McLendon and J.R. Johnson, pp. 93–130. Washington, D.C.: National Park Service Archeology and Ethnography Program.

Johnson, J.R. (1999b). Mission registers: data base. *California Mission Studies Association* [newsletter] **16**(2), 4. Bakersfield: California Mission Studies Association.

Johnson, J.R. and McLendon, S. (1999). Chumash social history after mission secularization. In *Cultural Affiliation and Lineal Descent of Chumash Peoples in the Channel Islands and the Santa Monica Mountains*, ed. S. McLendon and J.R. Johnson, pp. 131–84. Washington, D.C.: National Park Service Archeology and Ethnography Program.

Johnson, J.R., Stafford, T.W. Jr., Ajie, H.O. and Morris, D.P. (2002). Arlington Springs revisited. In *Proceedings of the Fifth California Islands Symposium*, ed.

D.R. Brown, K.C. Mitchell and H.W. Chaney, pp. 541–5. Santa Barbara, CA: Santa Barbara Museum of Natural History.

King, C.D. (1976). Chumash inter-village economic exchange. In *Native Californians: A Theoretical Retrospective*, ed. L.J. Bean and T.C. Blackburn, pp. 288–318. Ramona, CA: Ballena Press.

Kramer, D.G. and Brown, S.T. (1984). Sexually transmitted diseases and infertility. *International Journal of Gynaecology and Obstetrics* **22**,19–27.

Lambert, P.M. and Walker, P.L. (1991). Physical anthropological evidence for the evolution of social complexity in Southern California. *Antiquity* **65**, 963–73.

Lasuén, F.F. (1965). *Writings of Fermín Francisco de Lasuén*, transl. F. Kenneally. Washington: Academy of American Franciscan History.

Librado, F., Harrington, J.P. and Hudson, T. (1977). *The Eye of the Flute: Chumash Traditional History and Ritual*. Santa Barbara: Santa Barbara Museum of Natural History.

Librado, F., Harrington, J.P. and Hudson, T. (1979). *Breath of the Sun: Life in Early California, as Told by a Chumash Indian, Fernando Librado, to John P. Harrington*. Banning, CA: Malki Museum Press.

McLendon, S. and Johnson, J.R. (1999). *Cultural Affiliation and Lineal Descent of Chumash Peoples in the Channel Islands and the Santa Monica Mountains*. Final report. Washington, D.C.: National Park Service Archeology and Ethnography Program.

Milliken, R. (1995). *A Time of Little Choice: The Disintegration of Tribal Culture in the San Francisco Bay Area, 1769–1810*. Menlo Park, CA: Ballena Press.

Moore, S.E., Cole, T.J., Collinson, A.C., Poskitt, E.M.E., McGregor, I.A. and Prentice, A.M. (1999). Prenatal or early postnatal events predict infectious deaths in young adulthood in rural Africa. *International Journal of Epidemiology* **28**,1088–95.

Payeras, M. (1995). *The Writings of Mariano Payeras*, ed. D. Cutter. Santa Barbara: Bellerophon Books.

Perissinotto, G.S.A., Rudolph, C. and Miller, E. (1998). *Documenting Everyday Life in Early Spanish California: The Santa Barbara Presidio Memorias y Facturas, 1779–1810*. Santa Barbara: Santa Barbara Trust for Historic Preservation.

Portola, G., Smith, D.E. and Teggart, F.J. (1909). *Diary of Gaspar de Portola during the California Expedition of 1769–1770*. Berkeley: University of California Press.

Preston, W. (1996). Serpent in Eden: dispersal of foreign diseases into pre-mission California. *Journal of California and Great Basin Anthropology* **18**, 2–37.

Rigau-Perez, J.G. (1995). Surgery at the service of theology: Postmortem cesarean sections in Puerto Rico and the Royal Cedula of 1804. *Hispanic American Historical Review* **75**, 377–404.

Simpson, L.B. (1961). *Journal of José Longinos Martínez: 1791–1792*. San Francisco: John Howell Books.

Smith, D.E. and Teggart, F.J. (1909). Diary of Gaspar de Portola during the California Expedition of 1769–1770. *Publications of the Academy of Pacific Coast History*, vol. 1, no. 3. Berkeley: University of California Press.

Thornton, R. (1987). *American Indian Holocaust and Survival: A Population History since 1492*. Norman: University of Oklahoma Press.

Tibesar, A. (1955). *Writings of Junípero Serra*. Washington: Academy of American Franciscan History.

Valle, R.K. (1974). The cesarean operation in Alta California during the Franciscan Mission Period (1769–1833). *Bulletin of the History of Medicine* **48**, 265–75.

Wagner, R.H. (1929). *Spanish Voyages to the Northwest Coast of North America in the Sixteenth Century*. San Francisco: California Historical Society.

Walker, P.L. (2002). *A Spanish Borderlands Perspective on La Florida Bioarchaeology*. In *Bioarchaeology of La Florida: Human Biology in Northern Frontier New Spain*, ed. C.S. Larsen. Gainesville: University of Florida Press. (In press.)

Walker, P.L. and Hudson, T. (1993). *Chumash Healing: Changing Health and Medical Practices in an American Indian Society*. Banning, CA: Malki Museum Press.

Walker, P.L. and Johnson, J.R. (1992). The effects of European contact on the Chumash Indians. In *Disease and Demography in the Americas: Changing Patterns Before and After 1492*, ed. J. Verano and D. Ubelaker, pp. 127–39. Washington, D.C.: Smithsonian Institution Press.

Walker, P.L. and Johnson, J.R. (1994). The decline of the Chumash Indian population. In *In the Wake of Contact: Biological Responses to Conquest*, ed. C.S. Larsen and G.R. Milner, pp. 109–20. New York: Wiley-Liss.

Walker, P.L., Lambert, P.M. and DeNiro, M. (1989). The effects of European contact on the health of California Indians. In *Columbian Consequences: Archaeological and Historical Perspectives on the Spanish Borderlands West*, ed. D.H. Thomas, pp. 349–64. Washington, D.C.: Smithsonian Institution Press.

Walker, P.L., Lambert, P.M., Schultz, M. and Erlandson, J. (2003). The evolution of treponemal disease in the Santa Barbara Channel area of southern California. In *Treponemal Disease in the New World*, ed. M.L. Powell and D.C. Cook. Gainesville: University of Florida Press. (In press.)

Webb, E.B. (1983). *Indian Life at the Old Missions*. Reprint edition. Lincoln: University of Nebraska Press.

Westrom, L.V. (1994). Sexually transmitted diseases and infertility. *Sexually Transmitted Diseases* **21**, S32–7.

Wright, P. (1989). An examination of factors influencing black fertility decline in the Mississippi Delta, 1880–1930. *Social Biology* **36**, 213–39.

5 Children of the poor: infant mortality in the Erie County Almshouse during the mid nineteenth century

ROSANNE L. HIGGINS

Introduction

Katz (1995) suggests that the purpose of poor relief historically was to prevent death from starvation, disease, homelessness, and lack of clothing. Thus the growing number of destitute people in the United States at the beginning of the nineteenth century prompted the establishment of almshouses to provide relief. These institutions were intended to isolate individuals from the corrupting influences of the outside world that were alleged to lead to a life of laziness, alcohol abuse and, ultimately, pauperism (Katz 1983, 1986, 1995). The failure of the almshouse system to provide for its inmates is well documented (Katz 1983, 1986, 1995; Lawrence 1976; Leiby 1978; Rothman 1971). Almshouses were overcrowded with immigrants who spent their last pennies on passage to America, arriving destitute and often sick. Inmates frequently endured appalling living conditions, inadequate food and exposure to infectious diseases. The almshouses in New York State exemplify these deficiencies. Inspections in 1856 indicated that living conditions in many of the State's institutions were '. . . badly constructed, ill-arranged, ill-warmed and ill-ventilated' (New York State Senate Report of 1857, in Katz 1986). Mortality records from the Monroe County Almshouse in Rochester, New York, also suggest that these institutions were pesthouses, where many people who were suffering from infectious diseases came to die (Higgins et al. 2002; Higgins and Sirianni 1995; Lanphear 1986; Sirianni and Higgins 1995).

For the last two decades anthropologists have been investigating the biology of poverty during the nineteenth century through the analysis of almshouse cemetery collections (Grauer and McNamera 1995; Higgins and Sirianni 1995; Lanphear 1986; Phillips 2001; Sirianni and Higgins 1995; Steegmann 1991; Wesolowsky 1991). In general, these skeletal studies confirmed reports from economic historians (Katz 1983, 1986; Rothman 1971) that almshouse inmates lived in overcrowded, unsanitary conditions and were exposed to or carrying infectious diseases that contributed to their deaths. Other studies suggest that

78

poverty during the nineteenth century was particularly hard on children (Ball and Swedlund 1996; Goldin and Margo 1989; Grauer and McNamera 1995; Higgins *et al.* 2002; Higgins and Sirianni 1995; Sirianni and Higgins 1995) and that infant and child mortality were extremely high (Ball and Swedlund 1996; Goldin and Margo 1989; Grauer and McNamera 1995; Higgins *et al.* 2002; Swedlund 1990). Juvenile inmates (below the age of 15) often died from acute infectious diseases such as cholera, typhus fever and measles (Higgins and Sirianni 1995; Sirianni and Higgins 1995). Such diseases typically do not leave skeletal lesions; therefore institutional records can provide information on the health of juvenile inmates that cannot be determined through skeletal analysis.

This chapter adds to the previous research on the health and mortality of children living in nineteenth century almshouses by focusing on infant mortality. Records from the Erie County Almshouse (Buffalo, New York) are available from 1829, when the institution opened, through the turn of the twentieth century. These records provide insight into the demographic composition of juvenile inmates, as well as how successful the Erie County Almshouse was at buffering infants and expectant mothers from the consequences of poverty (i.e. poor nutrition, increased susceptibility to illness, and death).

Goldin and Margo (1989) used documentary evidence at the Philadelphia Almshouse to study the birth weights among infants born there in the early to mid nineteenth century. The mean birth weight for infants during this period was 3375 g, or 7.43 lb, comparing well with modern statistics. This suggests that this institution was reasonably successful in providing adequate prenatal care to poor women. However, when birth weight was held constant, the relative risk of fetal or first day mortality proved to be much higher at the almshouse. The authors suggest that the living conditions associated with urban poverty, such as poor nutrition, as well as nineteenth century obstetrical practices (for instance, the possible use of forceps in delivery) may have been important factors in raising the risks of mortality among almshouse infants.

There is no information on prenatal care available to expectant mothers, or on the care of infants at the Erie County Almshouse, in municipal documents such as the New York State Annual Reports of the Board of State Commissioners of Public Charities or the inmate records. However, the State Board of Charities Reports for 1868 and 1895 suggest that conditions at the Erie County Almshouse were better than the standard considered adequate at the time. The Board of State Commissioners of Public Charities was established in 1867 to oversee the charitable institutions (except prisons) that received state funding. According to these reports, the Erie County Almshouse provided clean living conditions, adequate food and the separation of sick and healthy inmates. Therefore, the possibility that conditions were favorable for the survival of infants at this institution can serve as a testable hypothesis for this chapter.

The records

The New York State Board of Charities Agency History Report indicates that in 1875 each almshouse was required by law to keep individual records on inmates and to send copies to the Board of State Commissioners of Public Charities. The regulations for record keeping at almshouses in New York prior to this date are not available insofar as we are able to determine. Summaries of the number of inmates relieved in almshouses, in addition to the expenditures and productivity of almshouses in New York, are published in the New York State Annual Reports of the Statistics of the Poor (1830–1863). Therefore, it is reasonable to assume that almshouse keepers were required to keep records prior to 1875.

Data for this study were collected from the Inmate Records (1830–1858) and the Children's Records (1863–1868) (Table 5.1). These records are housed in the archives at the Erie County Medical Center in Buffalo, New York, and were on loan for the duration of this project. They consist of bound volumes that were in very good condition. The handwritten documents were entered into a computer database and checked twice for transcription errors. All variables were collected from the inmate records, including name, age, sex, date of entry (or if born there, date of birth), nativity (place of birth), date of exit or death, and reason for admission into the almshouse. Cause of death was not listed.

As is often true of documentary sources, the records from the Erie County Almshouse are, in places, incomplete. The years 1850–1853 exist only on microfilm and are virtually illegible; they therefore were not used in this study. The records from 1859 are missing and no births were recorded from 1830 to 1849. It is impossible to ascertain whether no births occurred during this period, or whether the poorhouse keepers failed to record them, but the latter is most likely. Nativity (or place of birth) was not listed for approximately 21% of the infants in this study, mostly from 1854 to 1858 (see Table 5.4). Perhaps during this period the poorhouse keepers only listed nativity when inmates were foreign and thus many of the infants lacking a place of birth are actually native-born. Dates of exit from the almshouse were also often missing, making it impossible to calculate the total length of stay for some infants in this analysis (approximately 38%). The results of this study are therefore intended to provide only tentative answers to questions concerning infant mortality in New York State almshouses during the nineteenth century.

Table 5.1. *Sources of documentary data*

Source	Dates
Erie County Almshouse Inmate Records	January 1830–December 1839
	January 1840–December 1849
	January 1854–September 1858
Erie County Almshouse Children's Records	October 1863–April 1868

Table 5.2. *Infant records by sex and decade*

	Boys		Girls		Unknown		
	n	%	*n*	%	*n*	%	Total
1830–39	12	05.2	50	21.4	08	07.4	70
1840–49	62	26.8	08	03.4	03	02.8	73
1854–58	111	48.0	135	58.0	00	00.0	246
1863–68	46	20.0	40	17.2	97	89.8	183
Total	231	100.0	233	100.0	108	100.0	572

The Children's Record book was kept to document the number of children 'bound out' between October 1863 and April 1868. The practice of bounding out pauper children refers to sending children to work for those who would agree to care for them at the lowest rate of remuneration (Katz 1986). It is uncertain how long this practice persisted in Erie County, or if other records exist. The Children's Record book listed each child's name, age, sex, nativity (birthplace) and reason for entering the almshouse. Also included are the dates children were admitted to the almshouse (or dates of birth if they were born in the institution), the dates they were bound out, who they were bound out to, where they went, when they were returned to the almshouse and their dates of discharge or death. Again, cause of death was not listed. In most cases the bound out date is the same as the discharge date. Occasionally, a child was returned to the almshouse. Some infants were listed as having been bound out ($n = 24$), although it is more likely that they were placed in foster homes or adopted (although the records do not provide any further explanation). However, one infant was admitted with her mother on March 26, 1867, bound out on July 20, 1867 (also listed as discharge date) and returned to the almshouse on July 22, 1867. No reason was listed for the return of this infant.

A total of 572 infant records were used in this study (Table 5.2). The sex ratio is virtually equal, with 233 girls and 231 boys identified in the sample. Sex could not be determined for 108 infants either because they were not named or because their name (or first initial) was not indicative of their sex.

Infant mortality rates at the Erie County Almshouse were compared to those from the Monroe County Almshouse located in the nearby town of Brighton (Rochester, New York). Historical records are available from this institution in the Brighton Town Clerk's Record of Births, Deaths and Marriages only for the period of 1847–1850. These documents recorded vital statistics for alms-house residents as well as for the general population of the town of Brighton, New York.

Poor relief in the United States

While poverty was a growing problem as early as the eighteenth century, rapid industrialization, which began in the early decades of the nineteenth century, resulted in a shift from self-employed agricultural labor to a position of de-pendence on others for employment in manufacturing. Many immigrants also settled in urban areas, increasing the already high demand for steady employ-ment (Katz 1983, 1986). For those who fell on hard times, charitable contribu-tions of fuel, food and money (called 'outdoor relief') were often provided for through informal networks of friends, family or neighbors (Cray 1988; Hannon 1984; Katz 1983, 1986, 1995; Levine 1988; Rothman 1971). However, early in the nineteenth century outdoor relief was under attack by reformers in both the U.S. and Britain (whose influence can be seen in many aspects of poor relief in the United States). Many thought that this system perpetuated pauperism because it slowly deteriorated self-esteem, eroded the work ethic and promoted laziness (Katz 1983, 1986, 1995; Levine 1988; Rothman 1971).

Ultimately, as the numbers of people in need rapidly increased, the matter of poor relief was placed in the hands of the state governments. Many states established city and county almshouses as a form of 'indoor relief.' These institutions were established as a form of temporary assistance for those in need (Katz 1983, 1986; Rothman 1971). In the early years, the almshouse provided relief mainly for the urban poor. Many rural areas, where the image of the rural poor as 'neighbors' continued to persist, maintained the traditional forms of outdoor relief (Cray 1988).

An important goal of the almshouse system was to provide moral guidance to individuals. Reformers thought that if paupers could be provided with the proper skills (family values, good hygiene, a strong work ethic, and temperance) they could avoid dependency (Katz 1983, 1986; Rothman 1971). However, the underlying motive was deterrence. These institutions have been described as an attempt by the state governments to reduce the expenses of providing for the poor and at the same time deter those in need from asking for relief. Inmates were required to maintain the poorhouse in exchange for their keep. It was thought that the prohibition of alcohol and mandatory work requirements would keep all but the most needy from asking for relief (Katz 1983, 1986, 1995; Klabaner 1976; Rothman 1971). In San Francisco, for example, female almshouse inmates were given tasks such as sewing, washing, nursing the sick and cleaning. Those able-bodied women who refused to work were not permitted to stay (Smith 1896). In New York, male inmates were a source of agricultural labor for the almshouses (Katz 1986).

Local almshouses soon became crowded with the destitute, sick, pregnant, or mentally ill. Some paupers seeking temporary relief remained there for many years (Katz 1983, 1986; Rothman 1971). Katz (1986) characterizes many of the nation's poorhouses as overcrowded, dirty and dilapidated. Acute infectious diseases, a common hazard of nineteenth century life, were often a problem in almshouses. Epidemics of cholera and typhus that swept through major cities, for instance, had a particularly devastating impact on the poorhouses (Briggs 1988; Lanphear 1986; Lawrence 1976; Rosenberg-Naparsteck 1983).

Poverty was devastating for children during the nineteenth century. Epidemics, alcohol abuse and the American Civil War left many children without one or both parents. In the early decades of the century, private orphanages, most with religious affiliations, cared for some children. However, most children were treated as adult paupers, bound out into service, auctioned to the lowest bidder or admitted into almshouses (Bremner 1974; Katz 1986; McKelvey 1947). On the other hand, the idea that children were unique and precious, requiring special attention and care, was emerging during this time (Ashby 1984; Katz 1986; Rothman 1978; Tiffin 1982). As an extension of this sentiment, there was considerable controversy over the presence of children in almshouses during the last three decades of the nineteenth century. Reformers contended that this environment was insufficient to provide the proper education, as well as religious and moral guidance, that children needed. Additionally, there were strong concerns that children would learn the lazy and immoral habits ascribed to their parents and thereby continue the cycle of pauperism (Katz 1986). Between 1875 and 1883, laws were passed calling for the removal of children from almshouses in many states, including New York, Pennsylvania, Ohio, Indiana and Michigan. Other states, such as Minnesota, Rhode Island and Kansas, built state schools for pauper children (Folks 1974). In New York, the Children's Act of 1875 was intended to remove small children (ages 3–16) from these institutions and place them into orphanages. Katz (1986) argues that although the public rationale for removing children from almshouses was to break the cycle of dependence (which was thought to be inherited) and provide children with education and a positive example, the underlying theme was deterrence. Those who entered almshouses would be separated from their children.

The Erie County Almshouse

The Erie County Almshouse opened its doors in 1829 to provide relief to the growing numbers of those in need in the greater Buffalo area. Throughout its existence, the Almshouse was occupied by the destitute, sick, injured, disabled, or insane. The original Erie County Almshouse was often characterized as overcrowded, unsanitary, and in a bad state of repair before it was moved to Main Street in Buffalo in 1850 (Briggs 1988). After it was rebuilt (Fig. 5.1), the new facility was reported by the Secretary in the Second Annual Report of the Board of the State Commission of Public Charities (1868) to be a considerable improvement over the original facility. It had separate buildings to house the insane. Children as well as the aged were housed separately from the other paupers. There were adequate facilities for bathing and the inmates appeared suitably clothed and well fed.

In addition to offering shelter to the poor, the Erie County Almshouse provided a place for unwed mothers to give birth. Children were always in

Fig. 5.1. The Erie County Almshouse, *c.* 1920.

residence, even after the Children's Act of 1875. Inmates stayed for short periods of time, typically not more than a few weeks, but more often only for a few days (Briggs 1988; Higgins 1998; Katz 1983, 1986). Overcrowding was common, particularly during the winter months when occupancy was two or three times that of the summer months (Briggs 1988).

Katz (1983, 1986) identified four major demographic trends during the first 50 years of the operation at the Erie County Almshouse. First, from 1829 through the 1840s, the almshouse was occupied mainly by native-born families. Second, by the 1840s, the number of foreign-born inmates (particularly Irish-born immigrants) began to outnumber native-born inmates and the number of families declined. Higgins (1998) has shown moreover that foreign-born inmates accounted for 54% of the population at the Erie County Almshouse between the years 1829 and 1849 and 62% between the years 1850 and 1858. Third, during the American Civil War (1861–1865) the number of young men entering the almshouse declined, while the number of young, unmarried women increased. Finally, beginning in the decade after the American Civil War, the elderly population began to increase, such that when the institution moved from Main Street in Buffalo to Alden, New York, in the early years of the twentieth century, it was used primarily as a home for the aged. Today, the institution is still in operation as the Erie County Home and Infirmary, an elder care facility.

Birthplace of infants at the Erie County Almshouse

Researchers have reported higher mortality rates among foreign-born children, or children born to foreign-born mothers, in the United States (Haines 1977, 1991; Preston and Haines 1991). Therefore, the birthplace of children (nativity) was examined so that patterns of mortality in the Erie County Almshouse could be more clearly understood. Table 5.3 describes four categories of infant listed in the records (1830–1868) and Table 5.4 breaks down their distribution into four periods (1830–39; 1840–49; 1854–58; 1863–68). The first group consists of infants who were recorded as having been born in the almshouse ($n = 169$). Overall, they accounted for 30% of the infant sample. As Table 5.4 shows, however, no births were recorded in the Erie County Almshouse during the early periods (1830–39, 1840–49), but approximately 40% of the infant sample was born there during each of the later two periods (1854–58, 1863–68). The second and largest component of the sample consists of native-born infants who were admitted after birth ($n = 205$). They accounted for 36% of the total sample, but their relative proportions fluctuated dramatically from period to period (3% to 80%). The third category consists of infants listed in the records as foreign-born. These children were either born outside of the US (mostly

Table 5.3. *Birthplaces of infants at the*
Erie County Almshouse

	n	%
Born in Almshouse	169	30.0
Native-born	205	36.0
Foreign-born	76	13.0
Unknown	122	21.0
Total	572	100.0

Table 5.4. *Birthplaces of infants at the Erie County Almshouse*
through time

	1830–39		1840–49		1854–58		1863–68		
	n	%	n	%	n	%	n	%	Total
Born in Almshouse	0	0	0	0	98	39.8	72	39.3	169
Native-born	56	80.0	47	64.4	8	3.3	86	47.0	205
Foreign-born	11	15.7	20	27.4	31	12.6	18	9.8	76
Unknown	3	4.3	6	8.2	109	44.3	7	3.8	122
Total	70	100.0	73	100.0	246	100.0	183	100.0	572

in Ireland or Germany) or were children of foreign-born parents. This group
of infants ($n = 76$) made up 13% of the total sample. Information on nativity
was missing for 122 infants (21%), 89% of whom ($n = 109$) were listed in the
records between 1854 and 1858 (Table 5.4). It is uncertain why the birthplace
of so many infants is unknown during this interval, but the poorhouse keepers
may only have listed nativity when inmates were foreign-born and hence many
of these infants may actually be native-born.

It is clear from this breakdown that the majority of infants (66%) in the
Erie County Almshouse were native-born; if infants of unknown place of birth
are excluded, this percentage rises to 83% (374/450). It is likely that the true
proportion lies somewhere between these two values, especially since it is
probable that most of the infants whose place of birth is unknown are native-
born. Even when the data are partitioned by period, the foreign-born component
never exceeded 27% of the sample for each period.

Length of stay

The length of stay among infants in these three groups was also examined.
Total length of stay was calculated using the entry/birth dates and exit/death

Table 5.5. *Length of stay of infants at the Erie County Almshouse by period*

	<50 days		50–200 days		>200 days		
	n	%	*n*	%	*n*	%	Total
1830–39	46	66	19	27	5	7	70
1840–49	43	59	21	29	9	12	73
1854–58	114	61	59	32	13	7	186
1863–68	71	59	35	29	14	12	120
Total	274	61	134	30	41	9	449

dates. Records were complete with regard to entry/exit dates and nativity for 69 of 70 infants (99%) during 1830–39, 69 of 73 infants (95%) during 1840–49, 94 of 246 infants (38%) during 1854–58 and 120 of 183 infants (66%) during 1863–68. Exit dates were missing in the cases where length of stay could not be counted.

Table 5.5 shows that the majority of infants (61%) were recorded as residing in the Almshouse for less than 50 days, which is slightly less than two months. Another 30% had either left or died within 200 days, or about 7 months. Less than 10% were recorded with stays greater than 200 days. These patterns remained remarkably consistent through time. During all four periods significantly more infants were observed in the Erie County Almshouse records for fewer than 50 days.

The probabilities of infant death at the Erie County Almshouse

Period life table mortality rates were used to describe mortality among the infants who were born in the almshouse (see Table 5.7). However, it was inappropriate to apply this type of rate to infants who were admitted to the almshouse after birth (Table 5.6) because these groups were seldom in residence for an extended period of time. The risk of dying for inmates depended not only on age, but also on the length of time they spent in the almshouse. Therefore, using the total length of stay for each infant group ($_n s_x$) as the denominator produces a more meaningful death rate, $_n q_x = {_n d_x}/{_n s_x} \times 365/12$ (the monthly probability of dying within a year spent at risk at the almshouse).

Given the small sample size, these results are difficult to interpret. The probabilities of infant death fluctuated during all periods (Table 5.6). During the period of 1830–39 the probability of dying among foreign-born infants was 0.108, compared with 0.049 for the native-born sample. In other words, if a

Table 5.6. *Length of stay based probability of dying at the Erie County Almshouse*

	Infants (n)	Deaths (n)	Total length of stay (days)	$_nq_x$ per 1000 per month
(a) Native-born infants				
1830–39	35	8	4948	0.049
1840–49	47	10	3894	0.078
1863–68	53	21	5289	0.063
(b) Foreign-born infants				
1830–39	11	2	563	0.108
1840–49	20	2	1587	0.038
1854–58	39	4	1282	0.095
1863–68	14	3	972	0.094

Table 5.7. *Births, deaths, and standard period life table infant death rates per 1000 at the Erie and Monroe County Almshouses*

	Live births	Deaths	Death rate/1000
Erie County Almshouse 1854–58, 1863–68	167	16	0.095
Monroe County Almshouse 1847–50	69	17	0.246

hypothetical 1000 infants entered the almshouse in a given month, about 108 foreign-born infants would likely have died compared with about 49 native-born infants during that month. During the following period of 1840–49, about 78 of every 1000 native-born infants were likely to die each month, compared with about 38 foreign-born infants. During the period of 1854–58, native-born mortality was not calculated because birthplace was not recorded for 109 infants during this period, as discussed earlier. During the final period (1863–68), the risk of dying at the almshouse again appeared to be higher for foreign-born (94 per 1000) than native-born infants (63 per 1000).

Using period life table mortality rates, infant mortality was also compared between the Erie County Almshouse and the nearby Monroe County Almshouse (Town of Brighton, City of Rochester, New York) for children who were born at these institutions (Table 5.7). The Monroe County Almshouse data come from the Brighton Town Clerk's Record of Births, Deaths and Marriages (1847–1850). Unfortunately, there are no inmate records for this almshouse and thus no data on the length of time infants spent there. Sixteen of the 167 infants born at the Erie County Almshouse during the combined period of 1854–58 and 1863–68 died there. The probability of infant death at this institution was 0.095, indicating that 95 out of every 1000 infants born there were likely to die. At the Monroe County Almshouse a total of 17 deaths were recorded, hence, the probability of infant death there was 0.246, markedly higher than that observed for pauper infants in Erie County.

Discussion

Using archival evidence to answer questions about human biology can be difficult. Oftentimes one is faced with similar challenges as face those attempting to collect demographic data from contemporary indigenous communities; namely, sample sizes are small, records are incomplete and details of their origin and purpose are often difficult to ascertain. The present study of infant mortality exemplifies these challenges. However, although the records from the Erie County Almshouse are incomplete in many places and represent only a small group of infants in a limited time period, these documents still add to our understanding of the mortality experiences of almshouse infants. For example, although place of birth was not recorded for about 21% of the total sample, these records still provide important information about the nativity of many of the infants living and dying at the Erie County Almshouse. During the first two decades of operation, the majority of infants were native-born and recorded births were not evident in the inmate records. The high frequency of native-born infants seems unusual, given that more than half of the population of inmates during this time was listed as foreign-born (Higgins 1998). Also, native-born infants continue to outnumber foreign-born infants even as the adult foreign-born population increases (during the 1850s and 1860s). The documents indicate that the birthplace of the babies of immigrants is listed in the almshouse records as the nativity of their parents; thus this pattern may not be attributed in any straightforward fashion to the infants of foreign-born parents being recorded as native-born. Katz (1986) noted an increase in the number of unmarried women entering the Erie County Almshouse just prior to the Civil War. Higgins *et al.* (2002) found that female-headed, single-parent families were also common in the Monroe County Almshouse in the 1850s and 1860s. It is possible that many of the women were native-born and entered the Erie County Almshouse to provide food and shelter for recently born babies.

It seems unusual that there were no births recorded during this time. Perhaps there were so few births that it did not seem necessary to record them, or perhaps keepers of the poor simply were not required to record such events, or were simply negligent. Another possible reason for the lack of recorded births in the early decades may have been the social stigma associated with giving birth outside of wedlock. Perhaps many women came to the almshouse to give birth during this period knowing that there would be no record of the event, thereby allowing them to retain anonymity. During the period 1854–1858, 96 live births were recorded at the Erie County Almshouse, compared to 71 for the period 1863–1868. Similarly, there were 69 live births recorded in the Brighton Town Clerk's Records for the Monroe County Almshouse in Rochester, New York (1847–1850). The inmate records at the Erie County Almshouse do not indicate the legitimacy of the infants born there. However, the Brighton Town Clerk's Record of Births listed 30 of the 69 live births at the Monroe County Almshouse as illegitimate. An additional 30 were listed with fathers unknown, perhaps also suggesting illegitimate births. Together, they account for a remarkable 87% of all the births at the Monroe County Almshouse. These records also list 226 live births in the town of Brighton (outside of the almshouse). Two parents are listed for all but six infants and none are listed as illegitimate. Katz (1986) noted that the Erie County Almshouse served, perhaps unofficially, as a maternity hospital for unmarried women throughout its existence. It appears as though the Monroe County Almshouse served a similar function.

Another area in which these records were incomplete was the exit dates for inmates. For this reason, length of stay could not be recorded for all infants in the sample. However, approximately 62% (352/572) of the infant records were complete with regard to entry and exit dates and provide some insight on how long infants resided at the Erie County Almshouse. These records indicate that infants tended to stay in the almshouse for relatively longer periods of time compared to adults. Katz (1986) reported that approximately 60% of the inmates at the Erie County Almshouse stayed there for a maximum of six weeks, or 42 days. The minimum average length of stay for infants was 47 days, while the maximum was 88 days, over 12 weeks. This is also not surprising considering the special care required for a newborn infant. Women without husbands or homeless women would likely have no choice but to stay at the almshouse to be able to provide their babies with food and shelter. However, staying at the almshouse for any length of time likely exposed both mothers and infants to infectious diseases and increased the risk of dying while in residence there.

Given the small sample sizes analyzed here, the estimates of the probability of dying must also be interpreted with caution; nevertheless, they do suggest that the risk of dying was higher for infants living at the almshouse compared to other infants in the US at the time. Haines (1998) reported an infant mortality rate of 229 infant deaths per 1000 live births for the United States in 1850. The data for the Erie County Almshouse indicate that the monthly probabilities of infant death fluctuated throughout the time periods examined. For native-born infants admitted after birth, the monthly probabilities of dying ranged from 49 deaths per month per 1000 infants to 78 deaths per month per 1000 infants.

For foreign-born infants admitted after birth, the monthly probabilities of dying ranged from 38 deaths per month per 1000 to 108 deaths per month per 1000. If these probabilities are annualized, it is evident that infants admitted to the Erie County Almshouse, whether native- or foreign-born, were more likely to die than infants in the general population of the United States.

Given that these probabilities were estimated using length of stay as the denominator, the rates calculated for native-born infants during the 1830s and 1860s are probably inflated because length of stay could not be recorded for many infants (21 infants during the 1830s and 33 infants during the 1840s). However, the comparison of mortality between infants born at the Erie and Monroe County Almshouses suggests that neither of these institutions offered favorable accommodation to newborns. For infants born at the Erie County Almshouse, the mortality rate was 95 infant deaths per 1000 live births compared to 76 deaths per 1000 live births in Erie County for 1855 and 51 deaths per 1000 live births in 1865 (Higgins 1998). The infant mortality rate was 246 infant deaths per 1000 live births for infants born at the Monroe County Almshouse. For the city of Rochester, New York, the infant mortality rate was 132 per 1000 live births in 1840 and 145 per 1000 live births in 1850 (Higgins *et al.* 2002). Many of these deaths were likely attributed to infectious diseases, since during the nineteenth century these were often the leading causes of death, particularly in almshouses (Ball and Swedlund 1996; Higgins 1998; Higgins and Sirianni 1995; Lanphear 1986; Preston and Haines 1991; Sirianni and Higgins 1995; Swedlund 1990).

At the Monroe County Almshouse, respiratory infections, mainly tuberculosis, were the leading cause of death for infants. Unfortunately, cause of death was not recorded in either set of records for the Erie County Almshouse during the periods examined. However, during later decades (1880–1899) cause of death was listed for paupers in the Mortality Records for the Erie County Almshouse. Starvation and infectious diseases (particularly cholera and respiratory infections) were listed as leading causes of neonatal and post-neonatal mortality during both decades. Swedlund (1990) suggests that both nutrition and feeding practices in the early months of life were highly influential correlates of infant mortality during this period. There is no information available on the care provided to infants or expectant mothers during this period at the Erie County Almshouse. A newspaper article of uncertain date from the Buffalo Courier (other articles on the same page suggest the winter of 1900) does describe the care given to infants at the Erie County Almshouse around the turn of the century. The article indicates that infants were bottle-fed a mixture of cow's milk, cream, barley water, limewater and milk sugar every two hours. This report goes on to point out that the bottles were sterilized before they were used and that each bottle was labeled with an infant's name. This article presents a 'squeaky-clean' image of the poorhouse, including pictures of a well-organized and spotless facility. However, the article was written at the turn of the century and thus may reflect significant improvement over the care received by the infants that were examined in this study. The New York State Annual Reports of the Board of State Commissioners of Public Charities indicate that the almshouse fell in and out of periods of disrepair and neglect. Prior to 1850, the Erie County Almshouse was described as overcrowded, unsanitary, and in disrepair. Also, a new facility was built in 1850 after a fire destroyed the building (Briggs 1988). While the new facility was described in the 1860s as having separate quarters for children and the aged, there is no mention of the separation of sick from healthy inmates (Second Annual Report of the Board of the State Commission of Public Charities 1868). Infants would have been particularly vulnerable to the spread of infectious diseases here, as is indicated by mortality due to tuberculosis and other infectious diseases.

Conclusion

Although the records used in this study clearly have limitations, they do provide some useful insight into the lives and deaths of infants who spent time at the Erie County Almshouse, and therefore contribute important information to anthropological literature regarding conditions of life in the nineteenth century and particularly in almshouses. The records from this institution indicate that most of the infants in the Erie County Almshouse were native-born, compared to the increasingly foreign-born population of adults described by Katz (1986) and Higgins (1998). Also, on average, this group resided at the almshouse for longer periods of time than did adults. High infant mortality rates compared to those of the general populations of Buffalo and Rochester, combined with infectious disease mortality, suggests that both prenatal and postnatal care at these institutions were inadequate. However, without more information on the details of prenatal care and feeding practices at almshouses, the estimated probabilities of infant death presented here can only provide indirect inferences on this issue.

Katz (1983, 1986) has suggested that policymakers did not establish almshouses to meet the true needs of the poor, nor did they truly understand the underlying causes of poverty. Ball and Swedlund (1996) have echoed similar sentiments regarding the tendency of policymakers to blame high infant mortality on the ignorance of poor women during the nineteenth century. The probabilities of infant death observed at the Erie and Monroe County Almshouses support the contention that these institutions failed to buffer infants from the consequences of poverty because they were not designed to provide relief, but rather to be a place of last resort for the most desperate members of society. This study also lends support to the idea suggested by Katz (1983, 1986) that almshouses were established with an underlying motive of deterrence. Overcrowded, unsanitary living conditions may have deterred all but the most desperate (such as poor, unwed mothers-to-be) from seeking relief from the almshouse.

References

Annual Report of the Board of the State Commission of Public Charities (1868). Albany: New York State Board of Charities, The Argus Company Printers.

Annual Report of the Board of the State Commission of Public Charities (1895). Albany: New York State Board of Charities, Wynkoop Hallenbeck Crawford Company.

Annual Reports of the Statistics of the Poor, 1830–1863. New York Legislative Documents, New York State Senate.

Ashby, L. (1984). *Saving the Waifs: Reformers and Dependent Children, 1890–1917.* Philadelphia: Temple University Press.

Ball H.H. and Swedlund, A.C. (1996). Poor women, bad mothers: Placing the blame for turn of the century infant mortality. *Northeast Anthropology* **52**, 31–52.

Bremner, R.H. (1974). Introduction. In *Care of Dependent Children in the Late Nineteenth and Early Twentieth Century,* ed. R.H. Bremner. New York: Arno Press.

Briggs, J. (1988). *History of the Erie County Home and Infirmary.* Unpublished MS available at the Erie County Home and Infirmary, Alden, New York.

Brighton Town Clerk's Record of Births, Deaths and Marriages for the Years 1847–1850. Local History Division, Rochester Public Library.

Cray, R.E., Jr. (1988). *Paupers and Poor Relief in New York City and Its Rural Environs, 1700–1830.* Philadelphia: Temple University Press.

Folks, H. (1974). The removal of children from almshouses. In *Care of Dependent Children in the Late Nineteenth and Early Twentieth Century,* ed. R.H. Bremner, pp. 119–32. New York: Arno Press.

Goldin, C. and Margo, R. (1989). The poor at birth: Birth weights and infant mortality at Philadelphia's Almshouse Hospital, 1848–1873. *Explorations in Economic History* **26**, 360–79.

Grauer, A. and McNamera, E.M. (1995). *Bodies of Evidence: Reconstructing History Through Skeletal Analysis,* ed. A. Grauer, pp. 91–103. New York: Wiley Liss.

Haines, M.R. (1977). Mortality in nineteenth century America: Estimates from New York and Pennsylvania census data, 1865 and 1900. *Demography* **14**, 311–31.

Haines, M.R. (1991). *The use of historical census data for mortality and fertility research.* Working Paper Series on Historical Factors in Long Run Growth. Cambridge, Massachusetts: National Bureau of Economic Research.

Haines, M.R. (1998). Estimated life tables for the United States, 1850–1910. *Historical Methods* **28**, 149–69.

Hannon, J.U. (1984). Poverty in the antebellum Northeast: The view from New York State's poor relief rolls. *Journal of Economic History* **44**, 1007–32.

Higgins, R.L. (1998). *The biology of poverty: Epidemiological transition in western New York.* Ph.D. Dissertation, State University of New York at Buffalo.

Higgins, R.L. and Sirianni, J.E. (1995). An assessment of health and mortality of nineteenth century Rochester, New York using historic records and the Highland Park skeletal collection. In *Bodies of Evidence: Reconstructing History Through Skeletal Analysis,* ed. A. Grauer, pp. 121–36. New York: Wiley-Liss.

Higgins R.L., Walsh, L., Haines, M. and Sirianni, J.E. (2002). The biology of poverty: Evidence from the Monroe County Poorhouse. In *The Backbone of History: Health and Nutrition in the Western Hemisphere,* vol.1, ed. R.H. Steckel and J.C. Rose, pp. 121–36.

Katz, M.B. (1983). *Poverty and Policy in American History.* New York: Academic Press.

Katz, M.B. (1986). *In the Shadow of the Poorhouse: A Social History of Welfare in America.* New York: Basic Books.

Katz, M.B. (1995). *Improving Poor People.* Princeton: Princeton University Press.

Klabaner, B.J. (1976). *Public Poor Relief in America, 1790–1860.* New York: Arno Press.

Lanphear, K.M. (1986). *Health and mortality in a nineteenth century poorhouse skeletal sample.* Ph.D. Dissertation, SUNY Albany. Albany: University Microfilms.

Lawrence, C. (1976). *History of the Philadelphia Almshouses and Hospitals.* New York: Arno Press.

Leiby, J. (1978). *A History of Social Welfare and Social Work in the United States.* New York: Columbia University Press.

Levine, D. (1988). *Poverty and Society: The Growth of the American Welfare State in International Comparison.* New Brunswick: Rutgers University Press.

McKelvey, B. (1947). Historic origins of Rochester's social welfare agencies. *Rochester History* **9**, 6–11.

Phillips, S. (2001). Social stigma, disease and death in two county institutions. *Northeast Anthropology* **61**, 27–47.

Preston, S.H. and Haines, M.R. (1991). *The Fatal Years: Child Mortality in Late Nineteenth Century America.* Princeton: Princeton University Press.

Rosenberg-Naparsteck, R. (1983). Life and death in nineteenth century Rochester. *Rochester History* **45**, 2–24.

Rothman, D.J. (1971). *The Discovery of the Asylum: Social Order and Disorder in the New Republic.* Boston: Little, Brown and Company.

Sirianni, J.E. and Higgins, R.L. (1995). A comparison of the death records from the Monroe County Almshouse with the skeletal remains from the Associated Highland Park Cemetery. In *Grave Reflections: Portraying the Past through Cemetery Studies,* ed. S.R. Saunders and A. Herring, pp. 71–92. Toronto: Canadian Scholars Press.

Smith, M.R. (1896). *Almshouse Women: A Study of the Two Hundred and Twenty-eight Women in the City and County Almshouse of San Francisco.* Stanford: Stanford University Press.

Steegmann, A.T. (1991). Stature in an early mid-nineteenth century poorhouse population: Highland Park, Rochester, New York. *American Journal of Physical Anthropology* **85**, 261–68.

Swedlund, A. (1990). Infant mortality in Massachusetts and in the United States in the nineteenth century. In *Diseases in Populations in Transition,* ed. G. Armelagos and A.C. Swedlund, pp. 161–82. New York: Greenwood, Bergin and Garvey.

Tiffin, S. (1982). *In Whose Best Interest? Child Welfare Reform in the Progressive Era.* Westport, CT: Greenwood Press.

Wesolowsky, A.B. (1991). The osteology of the Uxbridge paupers. In *Archaeological Excavations at the Uxbridge Almshouse Burial Ground in Uxbridge, Massachusetts,* ed. R.J. Elia and A.B. Wesolowsky, pp. 230–53. Oxford: BAR International Series 564.

6 Worked to the bone: the biomechanical consequences of 'labor therapy' at a nineteenth century asylum

SHAWN M. PHILLIPS

Introduction

During the nineteenth century in North America, the erection of custodial institutions, such as prisons and asylums for the mentally ill, initiated a legacy of long-term inmate care, structured around social and medical philosophy, which has persisted to the present day (Dwyer 1987; Foucault 1965; Grob 1994; Phillips 2001; Rothman 1995). The historiography of those institutions demonstrates that inmates were viewed as social deviants and potentially threatening to the fabric of society. As such, the purpose of the emergent nineteenth-century institutions, from prisons to asylums, was to attempt to reform the inmates for possible return to mainstream society. In this effort, inmates' lives were structured around contemporary medical and sociological concepts of normalcy (Dwyer 1987; Ferguson 1994; Rothman 1995). In the attempt to institute normalcy, asylum superintendents hoped to shepherd inmates' habits away from deviant behaviors and within the sphere of normalcy. In this atmosphere, inmates of long-term custodial institutions were exposed to regimens established to bring about normalcy, in many cases, for their entire adult lives. The aim of this study is to examine some of the biological consequences, in the skeletal record, to inmates of long-term custodial institutions who were exposed to socially and medically prescribed regimens. Some basic questions this study poses include: How can the skeletal record be utilized to test hypotheses generated from the historical record? Since institutional inmates were virtually a voiceless group, how can we recover their day-to-day experiences? What treatments were practiced to bring about reform and how did that affect human biology? And, how do we reconcile discrepancies between the historical and skeletal records in order to reach a more accurate picture of human biological variation in historic period contexts? This study addresses those questions in a case study of the historical documents and bioarcheological remains of a nineteenth-century asylum for the mentally ill.

Bioarcheological study of populations in nineteenth-century North America has made great strides during the past decade in the refinement of research questions. Though not enough data are yet available for broad regional comparisons, scholars investigate health patterns in terms of urban health (Grauer 1995; Murry and Perzigian 1995; Rankin-Hill 1997), rural health (Larsen *et al.* 1995; Rose 1985; Winchell *et al.* 1995), class formation and the biology of poverty (Higgins and Sirianni 2000; Phillips 1998; Wood 1996), the health consequences of institutional contexts like the military (Liston and Baker 1996; Pfeiffer and Williamson 1991), and slavery (Rathbun 1987). This analysis extends the base of nineteenth-century North American bioarcheology to examine the biomechanical consequences of institutionalization in a long-term asylum for the mentally ill.

The challenge of this study is to demonstrate methods to link qualitative documentary sources with quantifiable measures of human biology. In order to test for the consequences of the treatment long-term institutional inmates received, biomechanical observations of skeletal samples from differing contexts are investigated. Since the philosophy of the treatment was to have inmates mirror the habits of those in the general population, observations of skeletal robusticity, cortical maintenance, and vertebral burst fractures are taken in this study to examine the effects of instituted normalcy. Skeletal robusticity and cortical maintenance are measures of skeletal biology that reflect the degree of activity during life. In addition, vertebral burst fractures, also known as Schmorl's nodes, result from biomechanical loading during discrete moments of overexertion (Larsen 1997; Lovell 1997; Pfeiffer and Williamson 1991; Rankin-Hill 1997). Though a consistent pattern of high frequencies of vertebral burst fractures among populations engaged in heavy work loads (enslaved populations, soldiers, homestead pioneers) is present in the bioarcheological literature, some scholars remain more cautious and suggest the etiology of vertebral burst fractures may be idiopathic (Roberts and Manchester 1995). Patterns for these skeletal markers of biomechanical stress are compared across samples and with age.

The biocultural perspective is taken in this study to develop a holistic interpretation of biomechanics in institutional contexts. The documentary record is utilized to develop the social, historical, and environmental components of the model, and biomechanical observations comprise the biological outcome data. Measures of 'daily life' are recovered from the skeletal sample in the form of robusticity indices, measures of cortical maintenance, and vertebral burst fractures. Of significance to the research question is that long-term, nineteenth-century institutions functioned to recreate the ideal of family and social life for those deemed social deviants. Thus, a goal of this study is to investigate the extent to which long-term institutions were successful in recreating a 'normal' environment. Though skeletal activity markers only serve as proxy measures for activity patterns during life, the activity measures examined here should distinguish general similarities or differences between skeletal samples representative of institutional inmates and the general population. Comparison of the biomechanical markers seen in the skeletons of those in the general population and those in long-term institutions can indicate differences in activity between these populations; any such differences can in turn be correlated with archival descriptions to obtain a picture of just how 'normal' the activities of long-term inmates really were.

Oneida County Asylum for the mentally ill

The Oneida County Asylum, Rome, central upstate New York, serves as a case study for understanding the day-to-day experiences of inmates in long-term institutions in nineteenth-century North America (see Fig. 6.1). Since the Oneida Asylum was an institution for the mentally ill that operated during the second half of the nineteenth century, the following provides a brief review on how care for the mentally ill developed in nineteenth-century America, and definitions of mental illness during that period. The skeletal sample associated with the Oneida Asylum is well suited for this study since the documentary

Fig. 6.1. Oneida County Asylum, circa the late 1880s. Photograph courtesy of the New York State Archives.

records indicate that most inmates entered the institution as young adults and, since the recovery rate was zero, were exposed to the asylum environment for the duration of their adult lives. This chapter examines historical details on the Oneida Asylum that include the practice and implementation of therapies, medical attention by physicians, attendant care, and the basic accommodations provided at the institution, as part of the effort to situate the analysis of the skeletal remains. The picture that emerges reveals that despite the effort to recreate normal society within the institution, each skeletal measure from the Oneida Asylum sample differed significantly from samples representative of the general population.

Definitions of mental illness during the nineteenth century

Throughout the nineteenth century, in Europe and North America, those afflicted with mental illness were thought to have suffered a loss of humanity and degenerated into an animalistic state (Gilman 1988). Although little was known about the causes of mental illness, it was possible to distinguish mental disabilities from certain forms of mental illness. For example, if an individual grew up in a community, participated in its commerce as an adult, and then began to exhibit aberrant behaviors to an extent to be labeled 'insane,' that sequence was noticeably different from someone born with a mental disability that never permitted them access to community activities (Ferguson 1994; Trent 1994). During the nineteenth century, the fall from a sound mind was considered the most dehumanizing condition since it could, and did, happen to anyone, including pillars of the community (Grob 1994). Despite the theme of a fall from humanity, late nineteenth-century physicians, psychiatrists, and sociologists developed a number of competing theories to categorize and explain the causes of mental illness in order to help treat the condition.

By the last quarter of the nineteenth century, admission to an asylum for the mentally ill in New York was contingent upon legal documentation that certified an individual was mentally ill (New York Legislature [NYL], New York Statute 446.1, 1874). Commitment to an asylum for the mentally ill required the certification of two physicians, who testified under oath, and the signature of a local district judge. The behaviors that resulted in one being certified insane during the nineteenth century were vague. For example, behaviors deemed unusual, dangerous, antisocial, fixatious, or well beyond the realm of normal activities could qualify as symptoms of mental illness (Dunglison 1872).

Accommodations at the Oneida County Asylum

Between the mid-1850s and the 1890s, the character of the Oneida County Asylum changed immensely. In 1856, the county institution functioned as a typical almshouse that provided general, and admittedly poor, care to a variety of unfortunates, 28 of whom were classed as mentally ill (NYL, New York Statute 71.0, 1870). Just ten years later, the Oneida institution was transformed into an asylum that specialized in the care of the chronically insane and had the charge of nearly 400 inmates at its close in 1892 (State Board of Charities [SBC] Annual Report 1893). The transformation at the Oneida institution was largely precipitated by investigations of the appalling conditions of the mentally ill in county almshouses, the cause championed by Dorothea Dix, and the passage of New York's Willard Act of 1865, which ordered state-level care of the mentally ill and, consequently, carried the largest state public charities tax in New York history (Dwyer 1987; Grob 1994). The direct implication of the Willard Act was that the abuses suffered by the mentally ill under county care were too egregious to ignore and the purpose of the statute was to strip local governments, the counties, of their charge. Many county authorities, however, believed the interests of the mentally ill could best, and less expensively, be served at the county level. Since the implementation of the Willard Act was contingent on the state constructing new asylums to house the mentally ill, counties that opposed Willard had time to improve their local facilities with the hopes of avoiding the looming Willard tax.

New York's effort to gain control over the care of the mentally ill within the state climaxed with failure at the opening of the Willard Asylum in Ovid, New York, in December 1869. Within a year of the Willard's opening, it filled to capacity and it was clear that New York had failed in the effort to accommodate the needs of all the mentally ill. In 1869, for example, 1500 mentally ill inmates were under county care. By 1875, 900 of the mentally ill had been transferred from various counties to the Willard Asylum, yet 1300 still remained under county care (NYL, New York Statute 15.0, 1875). The problem of what to do with New York's mentally ill remained, leaving the SBC, the body governing the care of the mentally ill in the state, confronted with difficult choices. Coincidentally, Oneida County officials anticipated the limits of the Willard Asylum and, in 1869, constructed a new county asylum with a capacity for 135 inmates when the county only had the charge of 117 inmates (SBC Annual Report 1870). During the 1870 SBC inspection, Oneida County officials informed the inspectors of their wish to retain local control of the mentally ill. It was evident to the SBC inspectors that Oneida County had taken serious steps to retain care of their chronically mentally ill, considering funds from the tight county budget were risked to build a local asylum at a time when state law (the Willard Act) intended to prohibit county care of the mentally ill by 1870. Despite the risk, Oneida County's new asylum reflected a state of the art asylum format in that it kept strict separation of the sexes, provided modern heating and ventilation designs, and made accommodations only for chronically mentally ill inmates. When the Willard Asylum filled to capacity in 1870, the SBC was hamstrung and several counties applied for, and obtained, exemption from the Willard Act, thereby dodging the new state tax within the county while accepting responsibility for the care of their mentally ill. The expense Oneida County put into building and improving their asylum facilities thus paid off when it received exemption from the Willard Act, on September 5, 1871 (SBC Annual Report 1872). And, when other counties were denied exemption from Willard due to inadequate facilities, the SBC granted Oneida County the unusual benefit of not restricting the eventual size of the asylum. This clause was added because Oneida was slated to be an 'overflow' asylum, which therefore retained not only its own chronically insane but also inmates from all over the state, and received the tax funds to care for inmates from outside Oneida County.

Throughout the history of the Oneida Asylum, it continually strove to keep pace with the growing needs of its inmates and to promote its image as self-sufficient. For example, the asylum acquired nearly 100 acres of farm land each decade of operation, until it had 356 acres in 1889, so there would be enough land for inmates to work and to keep the growing asylum supplied with food (Oneida County Asylum Annual Report 1870, 1880, 1889; SBC Annual Report 1870, 1889). Likewise, Oneida County also added to the asylum structure three times to accommodate their ever-increasing inmate population. And, by the end of the 1880s, the efforts at the Oneida Asylum paid off with its comparison to other exempted county asylums. For example, the value of the Oneida Asylum (including the asylum and grounds) was US$247 400 whereas the average value of the other county asylums was US$80 457.65, only a third of the Oneida Asylum's value (SBC Annual Report 1889). The difference in value represents Oneida County's commitment to maintaining the standards set by the SBC as the Willard Asylum's overflow inmates from around New York State were funneled into the continually expanding Oneida Asylum.

The accomplishments of the Oneida Asylum were recognized statewide. For example, the SBC noted such a distinction between the Oneida Asylum and the other exempted county asylums when, as early as 1882, it recommended that county asylums be no smaller than 250 inmates (SBC Annual Report 1882). Most of the county asylums retained fewer than 100 inmates, in comparison to Oneida's nearly 300 during the early 1880s. In comparison to other counties, Oneida County officials were willing to risk funds from the local budget to increase the size of the Oneida Asylum as well as maintain the low attendant to inmate ratio (circa 20 : 1) in lieu of paying the state tax. The SBC's recommendation was based on the assumption that only asylums willing to make such commitments were qualified to continue to care for the mentally ill. In addition to other accolades from the SBC (SBC Annual Report 1880, 1885, 1889), the State Charities Aid Association (SCAA) (a reformist group that continued Dix's campaign to end county care of the mentally ill) could not resist praising the Oneida Asylum (SCAA Annual Report 1880, 1885). Despite the fact that the SCAA desperately wanted to put an end to county care of the mentally ill, in promotion of sole state-level care, as Dix had advocated, the association praised the Oneida Asylum as a model institution and the best appointed asylum in the state. Likewise, the SCAA inspectors praised the Oneida superintendent and matron for constantly working to anticipate and improve the asylum in advance of the problems that plagued other county asylums.

Praise from the SBC and the SCAA, however, did little for Oneida County when New York's state government passed the State Care Act of 1890, which had the same purpose as the Willard Act of 1865. This time, however, New York made certain it was ready to accommodate the state's mentally ill, and thereby nullified the prior exemptions awarded to all counties that permitted them to operate local asylums. Oneida Asylum officials pleaded in open hearings for an explanation for why they were to lose control over their mentally ill (SBC Annual Report 1890). The Oneida officials questioned those present if they could recall any infractions ever reported during Oneida's tenure as an exempted county asylum; no one could recall any criticisms of the Oneida Asylum. In the end, the state purchased the Oneida Asylum buildings and grounds, for just over US\$200 000 (20% less than its estimated value), from the county in 1892 to create a new state asylum for mentally disabled children (Ferguson 1994). In an ironic footnote, approximately 200 of the Oneida Asylum inmates were retained at the new state facility to perform labor to keep the institution running since no funds were available to hire paid workers. By 1896, the last of the held-over Oneida inmates were relocated to state asylums for the mentally ill.

Therapeutics at the Oneida Asylum

Perhaps the greatest reform in the care of the mentally ill during the nineteenth century was the widespread adoption of labor therapy. Dorothea Dix's campaign to put the plight of the mentally ill on the agenda for state governments during the first half of the century (Dwyer 1987; Grob 1994) resulted in the abolition of the common practice of placing the mentally ill in physical restraints (for extended periods) in favor of the new practice of implementing regimented labor schedules for the mentally ill. Labor therapy, though implemented with differing justifications, was new and popular in other American nineteenth-century long-term institutions (Ferguson 1994; Morris and Rothman 1995). In New York's penitentiaries, for example, the 'Auburn System' was the model of prison operations followed by the other states beginning in the ante-bellum era (Rothman 1995). The Auburn System inmates began their day at 5:00 a.m. to labor in large work groups with meals as the only break in the orderly, monotonous schedule. Justification for inmate labor at prisons was that it served punitive, reformatory, and economic ends. In the most positive justification, it was hoped that a routinized work schedule would literally (and forcibly) change criminals' habits, thereby reforming them for their return to society as industrious citizens (Rothman 1995). Unlike prison reform, regimented labor for the mentally ill was meant to be therapeutic, to provide a normal routine and an opportunity for the lost mind to find reality once more. For example, a New York Senate committee charged with investigating the conditions of the mentally ill in the state's county institutions, largely in response to Dix's findings, suggested asylums should be built on fertile soil with labor therapy (primarily agriculture) in mind because 'fertility of the soil is necessary, in order to give interest to the cultivation of the grounds, and that the insane may find the harvest an ample recompense for their labor' (NYL, New York Statute 71.0, 1856:21). Basically, if the mentally ill followed a schedule similar to those in the sane world, they might become like them and recover their sanity.

The reasoning behind labor for the mentally ill paralleled the notion of re-forming prison inmates by forcibly changing their habits. Yet labor therapy was implemented differently in asylums for the mentally ill than in prisons. Rather than the militaristic and punitive agenda in prisons, asylums for the mentally ill adopted a 'family system of home influences' (County Superin-tendents of the Poor of New York, Annual Report 1875, 1880, 1885, 1889). In this framework, the asylum superintendent and the matron (typically a married couple) were symbolic parents of the asylum family who guided the inmates, symbolically viewed as children, to the plane of the sound mind through moral care and labor therapy. Moral care for the mentally ill was a new develop-ment compared to their treatment during the first half of the nineteenth century. For example, in terms of morality, it was significant to recognize the men-tally ill as deserving of basic humane care, like adequate nutrition, clothing, heat, lighting, ventilation, and sanitary conditions (Dwyer 1987). Moreover, it became of the utmost importance to segregate male and female inmates to prevent pregnancies from occurring at asylums. Prior to the Willard Act (1865), New York had not mandated basic standards for the care of the men-tally ill and many, like Dix, had documented such offenses as commonplace in New York's county houses and deemed the conditions lacked a basic 'morality.'

By many measures, the Oneida Asylum stretched the tenets of labor therapy to their most economically efficient end. Year after year, SBC and SCAA inspectors complimented the Oneida Asylum for its effective use of labor therapy (SBC Annual Report 1867–1890; SCAA Annual Report 1869–1890). Female inmates made the asylum clothes and linens, washed the laundry, mended tattered clothes, knitted the asylum hats, mittens, and scarves, worked in the kitchen and dining halls, cleaned the asylum from top to bottom, and made soap. Male inmates cleared fields of stones and tree stumps, did farm labor, gardened, chopped wood, tended livestock (the asylum had 60 cows, milk and butter from which was consumed at the asylum), and performed an assortment of other tasks. (See Figs. 6.2 and 6.3 for photographs of the asylum female and male inmates engaged in labor therapy in large work groups.) The Oneida Asylum gained economic benefits from the inmates' labor in that it was not necessary to hire laborers to perform the work or purchase the materials the inmates manufactured. The average value, for example, of inmate labor per year at New York's exempted county asylums during the 1880s was US$1155.91, while the average for Oneida was over six times greater at US$7000.00. The effort of the inmate labor is also reflected in the value of the institutional farm's products. The yearly average estimated value of farm products, for example, at county asylums overall was US$3833.95 while the Oneida Asylum average was nearly four times greater at US$13 309.22 (SBC Annual Report 1889). Though the Oneida Asylum was two to three times larger than the comparable county asylums, the estimated values for its inmate labor and farm products were astronomical in comparison. The SBC, SCAA, and Oneida County officials continually turned to such figures as markers of the therapeutic success at the Oneida Asylum. And, to keep the inmates engaged in labor therapy, the Oneida Asylum staff found that a reward system helped to promote industrious habits (SBC Annual Report 1882). Likewise, Dwyer (1987) reports that the Willard Asylum, the state facility the Oneida Asylum modeled itself after, offered food rewards and greater freedom to move about the institution to the hardest working inmates.

The therapeutics practiced at the Oneida Asylum are significant in understanding the biocultural context since it offers details on the activities the majority of the inmates were engaged in for the duration of their incarceration (i.e. their entire adult lives) at the asylum. Furthermore, since the goal of the new asylum format was to mimic the family and the goal of labor therapy was designed to mimic a 'normal' industrious life, it is possible to test what the biomechanical effects of institutionalization were, in comparison to skeletal samples representative of the general population.

A

B

Fig. 6.2. Male asylum inmates at work chopping wood in the winter (A) and clearing stones from fields (B). Photographs taken circa the late 1880s. Courtesy of the New York State Archives.

A

B

Fig. 6.3. Female asylum inmates at work washing (A) and ironing (B) the institution's laundry. Photographs taken circa the late 1880s. Courtesy of the New York State Archives.

Attendant interaction with inmates at the Oneida Asylum

Attendants at the Oneida Asylum spent the most time with the inmates, worked closest with them, and, by many measures, defined how well asylums fared in inspection evaluations by the SBC and the SCAA. The records of attendants' responsibilities for inmates and their notations in inmate case files provide the closest information available on what institutional inmates' daily lives entailed during the last quarter of the nineteenth century. Documentary data drawn from state legislation, inspection reports (state and charity associations), inmate case files, the asylum's annual reports, and medical reports are detailed below to develop the restrictions on attendants' authority, their general duties and responsibilities, and the extent of personal care attendants provided to inmates. These issues not only offer an image of the quality of treatment the inmates experienced, but also offer a direct record for a voiceless group.

The Oneida Asylum established a set of 'Rules and Regulations' (Letchworth and Carpenter 1882) that included the duties and responsibilities of the asylum attendants. Of the seventeen rules, thirteen defined the parameters and limits of the attendants' responsibilities. The rules generally followed moralistic ideals concerning the separation of sexes. For example, a primary rule was that no male attendants could enter female wards without the direct permission of the matron. Moreover, attendants were not to engage in disagreements, fights (verbal or physical), or games of chance with inmates. By and large, the rules concerning attendant conduct acknowledge the inmates' susceptibility to their attendants' authority and, thus, served to guard against physical, sexual, economic, and mental abuses from their caretakers. All these abuses were rampant in New York in the pre-Willard legislation era in county facilities, as documented by Dix and other investigators, which had helped to provide fodder for the need of the Willard legislation (Dix 1844).

Furthermore, attendant duties at the Oneida Asylum included monitoring inmates at all times (a ratio of 20 inmates per attendant was maintained at the Oneida Asylum [Letchworth and Carpenter 1882]), shepherding inmates through the daily schedule, evaluating situations to keep inmates from getting too 'excited,' and administering restraint or punishments. Each day at the Oneida Asylum began at 5:00 a.m. Attendants rang the morning bell and coordinated the meals, distributed labor assignments, and organized recreational activities (singing, reading, crocheting, etc.) as required by the asylum and inmates' needs. Through this process, attendants were the first line to assess, and curtail, disruptive inmate behaviors like general disobedience, fighting, or masturbating. Attendants determined when restraint of disruptive inmates was necessary (muffs, cribs, strong chair, etc.) or when punishment was required. It was only after a disruptive situation was resolved with restraint or punishment that an attendant was required to notify the superintendent that the punitive measure had been taken. Punishments other than restraint included grueling, monotonous tasks. The 'Polisher,' for example, was a push broom with a 100-pound cement block on the end (covered with a soft cloth), assigned to unruly inmates to push in the asylum halls for a week at a time. A similar practice was used at the Willard Asylum (Dwyer 1987). The Polisher punishment served to wear down disruptive inmates and illustrates the close interaction and latitude attendants maintained over inmates' experiences at the Oneida Asylum.

In addition to their general duties, the Oneida Asylum attendants also provided a great deal of personal care to inmates. For example, attendants were responsible for bathing (inmate bathing occurred once a week or 'as often as required'), dressing, and feeding inmates unable to perform those personal tasks (West 1882). Since each attendant had charge of about 20 inmates, and was responsible for adhering to the Asylum schedule, there must have been some degree of pressure to complete the expected routine each day. In fact, attendants' most common comments in inmate case files (see Oneida County Asylum, Inmate Case Files 4–199, 1880–94) were observations of whether inmates could feed and dress themselves and if they had 'filthy habits' (filthy habits included incontinence, uncontrolled bowels, and masturbation). Inmates with filthy habits who could not feed or dress themselves tended to be described in the most derisive manner by attendants in their case file notations. Undoubtedly, those inmates not only created the most work for attendants, and probably the most unsavory tasks, but also must have slowed attendants in keeping up with the daily agenda. Despite an apparent friction between attendants and inmates, the SCAA complimented the Oneida Asylum attendants by saying 'the insane [were] treated very tenderly, everything [was] done for their comfort and distraction.'

Evidence: historical documents, cemetery artifacts, and skeletal remains

The Oneida County Asylum skeletal sample ($n =$ ca. 100) serves as the case study to test for the biomechanical consequences of exposure to long-term institutions. Several criteria make the Oneida Asylum an adequate test case for this analysis of biological consequences of exposure to a long-term asylum environment. First, the asylum was a long-term institution from the 1860s to the 1890s in Rome, New York, a rural town in the central region of the state. Second, the Asylum was always described as a 'model' long-term institution by inspectors (both state and public charities agents) in its administration and implementation of mainstream social and medical philosophy. Finally, sufficient materials, in the form of historical documents and a skeletal sample, are available to fulfill the requirements for biocultural analysis.

The Public Archeology Facility at SUNY-Binghamton recovered the Oneida Asylum skeletal sample during the summer of 1988 (Nawrocki 1989; Santangelo 1989). The institution now serves as a state prison and the sample was recovered from the prison grounds as part of a salvage archeology project. Initially, the sample was thought to be part of an almshouse cemetery, although several factors dispelled that notion. First, nineteenth-century almshouse skeletal samples tend to be composed of large percentages of children and infants in addition to adults, whereas the Oneida sample contains all adults and only one neonate. Second, during the period of interments, the 1880s, the institution functioned as an asylum for adult mentally ill inmates. Oneida County still provided alms to the itinerant poor for brief periods, one to three days, and through 'outdoor' relief. Outdoor relief means the county supplied the needy with food and necessities in their homes rather than providing space in an institution. Hence, during the period of interments, the institution's primary purpose was as an asylum for the mentally ill. It is possible that individuals with other forms of dependency, like mental retardation or paralysis, may have been present in the asylum. Still, the institutional records and New York State Legislature records indicate the asylum primarily functioned to care for the mentally ill (Phillips 1997, 2001).

The material culture associated with the inmate graves is sparse. Typical grave goods include buttons, rings (gold and ebony), hairpins, clothing and shroud fragments, shroud pins, Christian medallions, and rosaries. The dearth of material culture is common for graves associated with institutional cemeteries. What is unusual, however, is the presence of jewelry made with precious metals and individuals with gold fillings in their teeth. Since the Oneida skeletal sample is associated with upper-working- and middle-class families (Phillips 1998, 2001), it is possible that their precious possessions were permitted to remain with them.

To date, no other nineteenth-century northeastern American skeletal sample has contained individuals with dental work (Elia and Wesolowsky 1991; Lanphear 1988; LoRusso 1990; Pfeiffer and Williamson 1991; Rankin-Hill 1997). The absence of dental work in the bioarcheological record does not indicate that dental work did not exist in nineteenth-century America. Rather, it reflects the low social and economic status of those disinterred and their financial inability to acquire dental work. The historiography of the field of dentistry suggests dental work was a thriving industry in the United States throughout the nineteenth century (see Charles 1982; Glenner 1990, 1994; Huff 1985; Meinke 1983; Phillips 2001). Also, Linebaugh and Phillips (2001) report 16% of the adults from a prosperous, rural, nineteenth-century Kentucky family cemetery exhibited some form of dental work (fillings, gold caps, dental appliances).

Skeletal samples representative of the general population in nineteenth-century North America were selected for comparison with the Oneida Asylum skeletal sample. The Albany County Almshouse skeletal sample ($n =$ ca. 40) comprises the original comparison data presented in this analysis. Previously published comparison skeletal data include the Uxbridge Almshouse ($n = 29$) (Elia and Wesolowsky 1991) and the First African Baptist Church ($n = 75$) (Rankin-Hill 1997). All three comparison samples represent poorer segments of the general population of nineteenth-century northeastern North America. To date, samples representative of the other income strata have not been available. Together, these data sets offer comparison between the poor, working-class segment of nineteenth-century North America and an institutionalized group. Since the reality of labor therapy was a routine of continual low-skill heavy work, it is likely that the life created in the institution was most similar to the poor working class. If so, then the data sets utilized in this study should provide a good comparison to test the hypothesis.

The Albany Almshouse skeletal sample serves as the primary comparison sample in this study. The Albany sample was selected because it fit a number of criteria that made it ideal for comparative purposes. First, the Albany sample is associated with late nineteenth-century northeastern North America; individuals in the sample were interred circa 1880 to 1900. Second, the context of the sample is sufficient to link it with the general population in the greater Albany area. The Albany Almshouse skeletal sample was recovered in the fall of 1990 by the Cultural Resources Survey Program of the New York State Museum as part of an emergency mitigation project (LoRusso 1990). Thirty-four discrete burials and approximately 20 disturbed burials were recovered. Forty individuals in the sample are adult. Sample preservation is excellent. The material culture associated with the Albany Almshouse cemetery was very sparse. Typical of almshouse burials, less than half the burials contained material goods. Burials with grave goods were limited to milk glass or hard rubber buttons. The only adornment recovered was a medallion, one of many given to the homeless at late nineteenth-century Albany charities missions. LoRusso (1990) suggests that the complete lack of material culture in more than half of the burials indicates the individuals were interred nude. Given the excellent state of preservation in this sample, that conclusion is likely to be accurate. Coffin shape in this sample alternated between six- and four-sided frames. Between the late 1880s and 1890s, coffin styles shifted from the traditional six sides to four. Thus, the archeologists linked the material culture, historic, and coffin shape data to determine the date of interments to have occurred between 1880 and 1900.

Though the Oneida and Albany institutions operated during the same time period in upstate New York, they served very different purposes in their respective counties. The Oneida Asylum maintained a long-term custodial population. The SBC and the SCAA monitored the Oneida Asylum's daily operations. Asylum inmates entered as young adults and remained for the duration of their lives. In contrast, the Albany Almshouse was typical of nineteenth-century poorhouses. The almshouse served as the county's social welfare safety net. Individuals in financial strain turned to the almshouse as a last resort and did not stay long. The almshouse was a place to get food, water, and shelter while securing a means to support oneself. In fact, the almshouse offered Spartan accommodations and a poor diet as a deterrent to those angling to take advantage of the tax-supported system. In the emergent Industrial era, wage labor kept many at the poverty mark in a tenuous position within view of the almshouse (Katz 1996). For the poor, mortal illness could also land those without family in the almshouse during their final days.

Methods

Standard skeletal markers of age and sex were used to develop the demographic structure of the Oneida and Albany skeletal samples. Cranial and pelvic indicators of age and sex are considered the most reliable in skeletal analysis. Sexual dimorphism of cranial and pelvic features was examined in concert for sex determination. Morphology of the mental eminence, brow ridge, mastoid process, nuchal lines, palate breadth, and mandibular gonial angle comprised some of the cranial features considered in sex determination. Pelvic morphology markers of sex examined included depth and width of the sciatic notch, angle of the subpubic concavity, thickness/sharpness of the medial aspect of the ischio-pubic ramus, presence/absence of the ventral arc on the pubis, and the dimensions of the sacrum. Other postcranial markers of sex included general skeletal robustness and femoral head diameter. Ageing indicators focused on the morphology of the auricular surface, pubic symphysis, sternal rib ends, cranial suture closure, and epiphyseal fusion (Buikstra and Ubelaker 1994; Ubelaker 1989). Since the Oneida sample is composed of adults, epiphyseal fusion indicators centered on the transverse sacral lines and the medial clavicle since those ossification centers fuse during early adulthood. The age and sex indicators were assessed according to established standards (Buikstra and Ubelaker 1994) and seriated to determine the range of variation present in these idiosyncratic samples. Seriation as a methodology helps to control for inter-observer error and to identify sample biases.

Table 6.1. *Mortality distribution of the Oneida and Albany skeletal samples*

Age	Oneida	Albany
18–29	12.66%	13.33%
30–39	20.25%	26.67%
40–49	25.32%	40%
50–59	22.78%	16.67%
60+	18.98%	3.33%

Table 6.1 describes the mortality distribution of the Oneida and Albany skeletal samples. The Albany Almshouse skeletal sample best mirrors the mortality pattern of the general population for late nineteenth-century America, when life expectancy hovered just below 50 years (Leavitt and Numbers 1996). In contrast, the Oneida sample demonstrates a greater survival rate to older age groups. This may be an artifact of sample recovery or an indication that institutional life sheltered inmates from some mortal risks present in the general population. There were no significant age differences between males and females in either sample (Student's t-test, $p = 0.05$). Since this study focuses on the consequences of adult long-term institutionalization, juveniles in the Albany sample were excluded from this analysis.

Basic statistical methods were used to test for significant differences in comparisons within and between samples. Where appropriate, Student's t-test, analysis of variance, cluster analysis, and regression analyses were applied to identify statistical patterns (Haskins and Jeffrey 1991; Thomas 1986). Prevalence fractions (ratio, percentage, and index) are used to present vertebral burst fracture data (Waldron 1994). In this study, the p value was set at 0.05 to constitute a significant difference.

Activity patterns can be discerned from skeletal remains via numerous methods. Long bone robusticity (based on skeletal muscular marker hypertrophy/atrophy), cortical maintenance, and the frequency of vertebral burst fracture observations have been used to reconstruct behavior patterns that reflect activity levels during life (Kennedy 1989; Pfeiffer and Williamson 1991). A prominent theory in the study of skeletal biology is Wolff's Law (Larsen 1997), the concept that bones respond, as a living tissue in a constant state of renewal, and adapt to the mechanical forces and stresses exerted upon them during life. Throughout life bone tissue is renewed through osteoblastic action with the 'building' of new bone cells, while osteoclastic action removes dead or dying bone cells. With age, osteoclastic cell removal outpaces osteoblastic building, generally

Table 6.2. *Long bone measurements*[a]

Upper long bone measures		Lower long bone measures	
Humerus	Maximum length	Femur	Maximum length
	Minimum circumference		Bicondylar length
	of the shaft		Maximum subtrochanteric A/P diameter
	Midshaft circumference		Maximum subtrochanteric M/L diameter
	Maximum midshaft diameter		Midshaft circumference
	Minimum midshaft diameter		Maximum A/P midshaft diameter
Ulna	Maximum length		Maximum M/L midshaft diameter
	Physiological length	Tibia	Maximum length
	Minimum shaft circumference		A/P diameter at the nutrient foramen
	Transverse diameter of shaft		M/L diameter at the nutrient foramen

[a] Definition of measurements in Schwartz (1995).

around the fifth decade, and can result in osteoporosis. High activity levels can help the osteoblastic process keep pace with osteoclastic action whereas lower activity levels result in thinner bones and advance the onset of osteoporosis. By and large, in order for behaviors to have a lasting impression on the skeletal system, the activities must be fairly strenuous and performed habitually for a number of years (Kennedy 1989).

Skeletal robusticity indices quantify the development of skeletal muscular markers. Robusticity indices, in general, are proxy measures for muscular skeletal markers that provide information on activity levels by demonstrating the relationship between the length and girth of a long bone. The higher the index value, expressed as a ratio, the more robust the bone. Greater skeletal robusticity, following the tenets of Wolff's Law, suggests greater activity level during life. Since the skeletal system is continually renewed throughout life and the skeletal tissues respond to muscular pull and torsion forces, skeletal robusticity is an excellent, quantifiable measure of activity during life. Table 6.2 lists the osteometrics used to calculate skeletal robusticity indices in this study. Table 6.3 shows the skeletal robusticity indices calculated for each long bone and Table 6.4 shows the major muscle groups represented in the robusticity indices. These muscles are responsible for flexion, extension, abduction, adduction, pronation, and supination of the arms and legs. Furthermore, these muscle groups also cross the major joints of the shoulder, elbow, hip, and knee in addition to the finer manipulation of the hands and feet to function with other muscle groups in the movement of the body. Guidelines for bone measurement can be found in Buikstra and Ubelaker (1994) and Schwartz (1995). Long bone measurement materials included an osteometric board, sliding calipers, and a

Table 6.3. *Robusticity indexes*

Femur		
Robusticity Index	=	(A/P midshaft + M/L midshaft) × 100 / bicondylar length
Pilastric Index	=	A/P midshaft × 100/M/L midshaft
Subtrochanteric Robusticity Index	=	(A/P subtroch + M/L subtroch) × 100 / bicondylar length
Midshaft Circumferential Index	=	Midshaft circumference × 100 / bicondylar length
Tibia		
Robusticity Index	=	(A/P @ nutrient foramen + M/L @ nutrient foramen) × 100 / maximum length
Humerus		
Robusticity Index	=	Minimum shaft circumference × 100 / maximum length
Diaphyseal Index	=	Minimum midshaft diameter × 100 / maximum length
Circumferential Index	=	Midshaft circumference × 100 / maximum length
Deltoid Index	=	Maximum midshaft diameter × 100 / maximum length
Ulna		
Caliber Index	=	Minimum shaft circumference × 100 / physiological length
Interosseous Crest Robusticity	=	Maximum diameter of crest × 100 / maximum length

Table 6.4. *Muscle groups represented in robusticity indexes*

Long bone	Muscle region	Muscle
Humerus	Shoulder joint	Deltoid, supraspinatus, pectoralis major, latissimus dorsi, teres major, triceps, biceps
	Elbow joint	Brachialis anticus, biceps, anconeus, triceps
Ulna	Wrist	Flexor profundus digitorum
	Elbow joint	Pronator teres, biceps
Femur	Hip	Gluteus maximus, adductor magnus, adductor longus, hamstrings, iliopsoas, pectineus, quadratus femoris, biceps
	Knee	Quadriceps femoris, semimembranosus, biceps femoris, sartorius, popliteus, gastrocnemius
Tibia	Knee	Biceps femoris, tensor fasciae latae, popliteus, gracilis
	Ankle	Tibialis posterior, peroneus longus, gastrocnemius, soleus

flexible metric tape measure. Long bone measurements from the left side were utilized where possible; bones from the right side were used in absence of the left.

Cortical maintenance refers to the degree to which cortical bone is retained during life. Since bone is a living tissue, it remains in a process of renewal during life. Longitudinal studies on osteoporosis in living populations have shown that higher activity levels can improve cortical maintenance (Purtilo and Purtilo 1999). To study the patterns of cortical maintenance, basic geometric principles are applied to the elliptically shaped cross section of long bones. For example, area measurements are taken to differentiate the medullary cavity space, cortical area, and the total area of the cross section. From these measures, the percent cortical area can be calculated. Cross sections from the Oneida sample were taken at the mid-section of the left femur; the right side was sectioned when the left was unavailable. Femurs were cut during the late 1980s while under the supervision of personnel at Syracuse University. The femurs from the Albany skeletal collection were not sectioned. Cortical maintenance data generated in this study are compared with previously published First African Baptist Church (FABC) data.

To assess cortical maintenance, values for periosteal and medullary area of the bone cross sections were determined. To find those values the femur cross sections were photographed under magnification with a graduated metric scale placed at the same plane as the cross section. The images were then scanned, projected onto a digitizer screen, and traced with a digitizer stylus at one-millimeter increments. **Imagetool**, the digitizer used in this study, performs computerized area measurements as one of its functions and was designed for use with biomedical specimens. The metric scale in the image served as the reference point to calculate the area measurements. Percent cortical area (PCA) is the value reported to represent cortical maintenance and is calculated with the formula:

Periosteal Area − Medullary Area / Periosteal Area × 100 = PCA.

Vertebral burst fractures, also known as Schmorl's nodes, represent a measure of biomechanical loading on the spinal column. During heavy exertion, like lifting, the intervertebral discs can rupture, causing the herniated disc to fracture the vertebra (Aufderheide and Rodriguez-Martin 1998). Bioarcheological studies report higher frequencies of spinal burst fractures in populations that undertake heavy workloads in comparison to populations with lighter loads. Moreover, populations under the strain of very heavy workloads (enslaved people and wartime soldiers) exhibit nearly double the frequency of burst fractures in comparison to samples that represent the general population. Given the utility of spinal burst fractures to roughly gauge spinal loading pressure, their

Table 6.5. *Robusticity indexes: whole sample comparisons*

	Oneida	Albany	Uxbridge
Femur			
Robusticity Index	12.87	12.8	12.44
Circumferential Robusticity Index	20.12	20.0	na[a]
Subtrochanteric Robusticity Index	13.42	13.3	12.8
Pilastric Index	107.1	104.1	105.9
Tibia			
Robusticity Index	16.52	16.17	15.52
Humerus			
Robusticity Index	20.2	20.18	na
Circumferential Robusticity Index	21.9	21.4	na
Diaphyseal Index	6.0	5.7	na
Deltoid Robusticity Index	7.53	7.3	na
Ulna			
Caliber Index	17.3	16.5	na
Interosseous Crest Index	7.1	6.8	6.7

[a] na, Not available.

frequency provides another objective means to test the effects of labor therapy in institutional contexts. The identification of vertebral burst fractures in this study follows Lovell's (1997) diagnostic protocol. Since vertebral burst fractures are the result of violent herniations that fracture the disc, they are easily identified macroscopically as discrete sharp-edged depressions on the inferior and superior aspects of the vertebral body. The frequency of burst fractures is reported here as percentages per individual and per the number of vertebrae in the whole sample to ameliorate biases in sample representation due to variable preservation.

Skeletal robusticity

Table 6.5 shows the robusticity index sample means for four long bones for the Oneida, Albany, and Uxbridge skeletal samples. Tables 6.6 and 6.7 show the female and male subsample means, respectively. These three tables demonstrate that the Oneida long bone robusticity values are greater than those in the other two samples, and the Uxbridge sample is the most gracile of the three groups. In Table 6.6, the Oneida females are more robust than the other groups in all robusticity comparisons save two (the Uxbridge Pilastric Index and the Albany Femur Robusticity index). With these exceptions, the Oneida females' robusticity indices are significantly greater ($p < 0.05$; Student's t-tests) than

Table 6.6. *Robusticity indices: Female subsample comparisons*

	Oneida	Albany	Uxbridge
Femur			
Robusticity Index	12.67	12.74	12.32
Circumferential Robusticity Index	19.8	19.78	na[a]
Subtrochanteric Robusticity Index	13.27	13.15	12.58
Pilastric Index	104.21	101.6	107.2
Tibia			
Robusticity Index	16.38	15.87	14.97
Humerus			
Robusticity Index	19.81	19.65	na
Circumferential Robusticity Index	21.5	20.88	na
Diaphyseal Index	5.84	5.6	na
Deltoid Robusticity Index	7.35	7.17	na
Ulna			
Caliber Index	16.98	15.76	na
Interosseous Crest Index	6.98	6.48	6.44

[a]na, Not available.

Table 6.7. *Robusticity indices: Male subsample comparisons*

	Oneida	Albany	Uxbridge
Femur			
Robusticity Index	13.07	12.85	12.59
Circumferential Robusticity Index	20.43	20.12	na[a]
Subtrochanteric Robusticity Index	13.53	13.43	13.01
Pilastric Index	109.91	105.69	104.15
Tibia			
Robusticity Index	16.64	16.42	16.46
Humerus			
Robusticity Index	20.59	20.51	na
Circumferential Robusticity Index	22.28	21.71	na
Diaphyseal Index	6.06	5.77	na
Deltoid Robusticity Index	7.67	7.41	na
Ulna			
Caliber Index	17.56	16.81	na
Interosseous Crest Index	7.15	7.0	6.99

[a]na, Not available.

those of the Uxbridge females in the ulna, tibia, and femur indices. Table 6.7 demonstrates that the Oneida males have greater robusticity in all limbs in comparison to the Albany and Uxbridge males with significant differences from the Uxbridge males ($p < 0.05$).

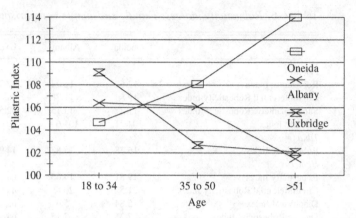

Fig. 6.4. Pilastric Index: changes in males with age.

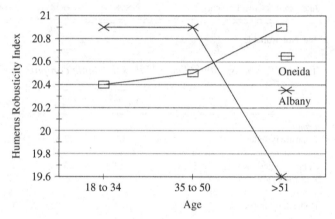

Fig. 6.5. Humerus Robusticity Index: changes in males with age.

Figures 6.4–6.9 show age and sex patterns in robusticity between the Oneida, Albany, and Uxbridge skeletal samples. In all the robusticity figures, the Oneida sample shows an increase in robusticity with age that diverts from the trends exhibited in the Albany and Uxbridge data. Figures 6.4 and 6.5 show changes in male femur and humerus indices, respectively. In Fig. 6.4, male Pilastric Index declines with age in the Albany and Uxbridge skeletal samples whereas it increases with age in the Oneida sample. In Fig. 6.5, the Humerus Robusticity Index is greater in the Albany sample until the third age group (>51) when it declines with age. In the Oneida sample the Humerus Robusticity Index stays the same until the third age group (>51), when it increases sharply. Figures 6.6 and 6.7 show changes by age in the Tibia Robusticity Index and the Ulna

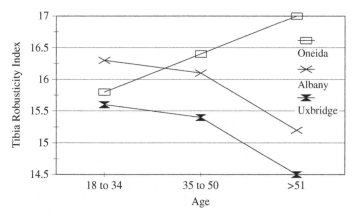

Fig. 6.6. Tibia Robusticity Index: changes in females with age.

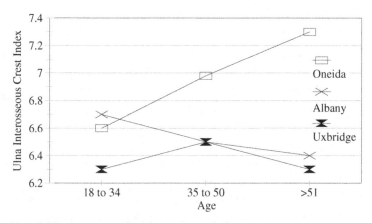

Fig. 6.7. Ulna Interosseous Crest Index: changes in females with age.

Interosseous Crest Index (UICI) for the Oneida, Albany, and Uxbridge females. In Fig. 6.6, the Tibia Robusticity Index values for the Albany and Uxbridge skeletal samples remain the same until the oldest age group (>51) when they decline. In the Oneida sample, the female Tibia Robusticity Index increases with each age group. Finally, in Fig. 6.7, the UICI remains the same and decreases slightly in the Albany and Uxbridge females while the UICI increases with age among Oneida females.

In Fig. 6.8, percent cortical area (PCA) data for the Oneida and FABC males are plotted as they change with age. The patterns for the two groups are similar except in the oldest age group (>51) where the FABC males have a sharp decline in PCA. Figure 6.9 shows PCA changes in females from the Oneida and FABC skeletal samples. The two groups of females show a similar pattern

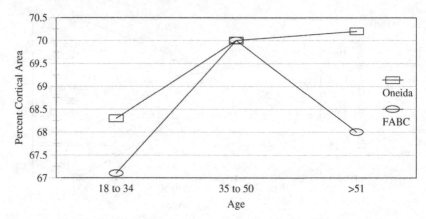

Fig. 6.8. Percent Cortical Area: changes in males with age.

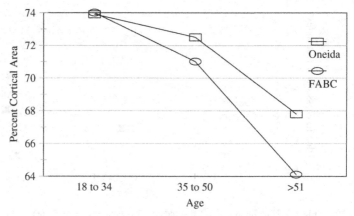

Fig. 6.9. Percent Cortical Area: changes in females with age.

in PCA, though the Oneida females have slightly better cortical maintenance, and do not show a sharp decline in PCA until after age fifty.

Finally, vertebral burst fractures are ten times more common in the Oneida sample (9.6) in comparison to the Albany sample (0.9) when the frequency is compared per vertebral element. When the number of vertebral burst fractures is compared per individual, the Oneida sample has nearly five times the rate of the fractures (45.5) in comparison to the Albany sample (9.5). Another difference in vertebral burst fractures between the samples is that all the fractures in the Albany sample take place in the lower half of the thoracic and lumbar vertebrae. However, over 20% of the vertebral burst fractures in the Oneida sample occurred in the upper thoracic vertebrae (Fig. 6.10).

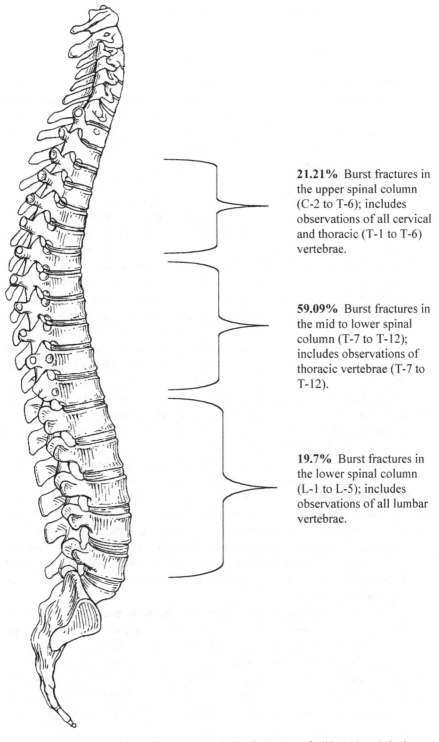

21.21% Burst fractures in the upper spinal column (C-2 to T-6); includes observations of all cervical and thoracic (T-1 to T-6) vertebrae.

59.09% Burst fractures in the mid to lower spinal column (T-7 to T-12); includes observations of thoracic vertebrae (T-7 to T-12).

19.7% Burst fractures in the lower spinal column (L-1 to L-5); includes observations of all lumbar vertebrae.

Fig. 6.10. Burst fracture (Schmorl's nodes): distribution in the Oneida Asylum skeletal sample.

Discussion

As mentioned, this chapter aims to recover evidence of the impact of the day-to-day experiences of a voiceless group exposed to medically informed therapeutics in a long-term custodial institution. The historical data show that long-term institutions, from prisons to asylums for the mentally ill, implemented labor therapy as part of the effort to reform or rehabilitate deviant inmates. In the Oneida County Asylum, as with other asylums for the mentally ill, the purpose of labor therapy was to impose a normal routine. The philosophy behind the practice was that, if mentally ill inmates went through the motions of a normal, productive day, then the opportunity was presented for them to regain their sanity. An intriguing aspect to this concept is that for the asylum to practice that form of therapy, the semblance of normalcy had to be recreated and imposed within the asylum. How well the asylum, and other long-term institutions, recreated a normal environment is what is at issue here.

Nearly all of the sample comparisons of skeletal robusticity indices indicate that the Oneida Asylum inmates were the more robust group in terms of skeletal muscle markers. This finding is notable since four major long bones from the upper and lower body halves (humerus, ulna, femur, and tibia) that represent major muscle groups demonstrate similar patterns of robusticity. Moreover, the pattern of greater robusticity persists when the samples are broken down into comparisons of male and female groupings. Although the Oneida Asylum skeletal robusticity means are not significantly greater ($p = 0.05$) than the Albany sample, they are consistently more robust in each of the four long bones presented. Based on observations of greater skeletal robusticity, it is valid to conclude that the Oneida Asylum inmates had more developed muscular skeletal markers, and therefore participated in more strenuous activities than their counterparts from poorer segments of the general population. Based on what we know of labor therapy, inmates repetitively performed heavy manual labor which, one would expect, would develop muscular skeletal attachment sites. However, since the comparison skeletal samples are of segments of society that probably also participated in similar work, it seems that the demands of labor therapy may have exceeded the level of work of those in the general population.

When the long bone skeletal robusticity indices are traced with age, a very striking pattern emerges. In the samples that represent the general population, skeletal robusticity decreases with age. In the Oneida asylum sample, on the other hand, we see the opposite pattern in each long bone and for both sexes; skeletal robusticity increases with age and the Oneida robusticity indices are significantly greater ($p = 0.05$) in comparison to the Albany and Uxbridge samples in the oldest age category. The decrease in skeletal robusticity in the general population likely reflects what is observed in most modern societies: as people age the heavier work is done by the younger generation of adults. Thus, even in poorer classes in industrial America, people beyond age 40 or 50 took on lighter work than they did in their 20s and 30s. However, the Oneida inmates do not seem to have shared in the transition to a lighter workload once they passed young adulthood. Despite its ostensible intention to create 'normalcy,' the application of labor therapy resulted in institutional inmates' skeletal robusticity deviating from that of the general population.

The increase in skeletal robusticity with age in the Oneida inmates reveals a dynamic aspect of the long-term institutional experience; since the inmates did not recover the sole therapy practiced was continued. It is important to note that greater robusticity with age does not indicate that inmates became 'stronger' with age, only that, as outlined in Wolff's Law, their skeletal systems continued to respond to an active and strenuous lifestyle. In addition to recreating a normal life, institutional superintendents had other motivations for keeping inmates actively engaged in labor. First, heavy labor wore inmates down and made them more tractable. Second, inmate labor was free, so not only was the institution spared the expense of hiring laborers, it also benefited from the products of inmate labor (mostly crops and garments). Thus, in addition to its usefulness as a *bona fide* medical treatment, institutions also had a financial interest in keeping inmates busy with labor therapy. It is possible that these combined factors kept inmates engaged in heavy labor even at ages that their cohorts in the general population were able to take on lighter work. In any case, exposure to the medically prescribed therapy led to a skeletal robusticity pattern that was the opposite of the general population after which it was modeled.

The data on cortical maintenance, presented as percent cortical area (PCA), are not as consistent as the skeletal robusticity data. To date, the FABC sample is the only nineteenth-century data available to compare with the Oneida asylum PCA data. Overall, the Oneida sample has greater PCA than the FABC sample, though the difference is not significant. And, unlike the skeletal robusticity data, males and females from both samples exhibit varying patterns of cortical bone maintenance with age. The Oneida males show an increase in PCA until the middle-age group and then a leveling off, whereas the FABC males show a sharp decline in PCA after age fifty. The Oneida and FABC females show a similar pattern in cortical bone maintenance in that PCA declines with age. The major difference between the two samples is that the Oneida females do not demonstrate a sharp decline in PCA until after age fifty.

The varying PCA patterns in these historic period samples indicate that cortical bone loss is highly dependent on external factors. The increased cortical maintenance of the Oneida inmates, in comparison to the FABC sample, possibly reflects the effects of the high activity demanded in labor therapy in conjunction with a substantial diet. Asylum records, which were echoed in inspection reports, state the inmates were supplied with an ample diet of meat, vegetables, and fatty foods. In addition, the asylum permitted inmates to eat as much as they chose and attendants stated they offered food rewards to hard-working inmates. The absence of an adequate food supply would have depleted cortical bone maintenance if inmates were pushed to participate in activities that demanded high energy levels. Though these findings are consistent with what is found in contemporary research on how activity levels affect cortical bone maintenance and the causes of osteoporosis, more PCA data is required from nineteenth-century contexts to more fully extrapolate the interpretation of these data.

The prevalence of spinal burst fractures per individual is nearly five times greater in the Oneida sample (45.5%) in comparison to the Albany sample (9.5%). The distribution of vertebral burst fractures in the Oneida skeletal sample extend from the second thoracic vertebrae to the first sacral vertebrae (Fig. 6.10). In contrast, spinal burst fractures are limited to the lower thoracic vertebrae in the Albany sample. The wider distribution of burst fractures in the Oneida sample suggests greater strain and more variability in body position during activities, which resulted in the fractures. In the military sample from the War of 1812, Pfeiffer and Williamson (1991) report vertebral burst fractures in the upper thoracic vertebrae, though most occurred in the lower thoracic region of the spine. Perhaps the wide distribution and high frequency of vertebral burst fractures represent, along with increased skeletal robusticity with age, a form of 'signature' in the bones of the biological effects of long-term exposure to labor therapy. Bioarcheological research on historic period North American samples report the highest frequencies (reported as prevalence per individual) of burst fractures in military (48% (Pfeiffer and Williamson 1991)) and African-American slave skeletal samples (39% (Rathbun 1987)). The high frequency of vertebral burst fractures in those skeletal samples is not surprising given the context in which they were associated. Rose (1985) reports vertebral burst fractures in 25% of an African-American, Reconstruction era, skeletal sample from Arkansas, and Rankin-Hill (1997) reports burst fractures in 18.7% of the FABC sample. Larsen et al. (1995) report vertebral burst fractures in 18.2% of a mid-nineteenth-century rural Illinois farmstead skeletal sample. These studies indicate a fairly large difference in vertebral burst fracture frequency in nineteenth-century North America between institutional contexts (military and slavery) and the general population. In institutional contexts, burst fractures tend to occur in nearly 40% or more of the individuals whereas in the general population, either urban or rural contexts, the prevalence of burst fractures occurs in less than 25% of the individuals in skeletal samples (see Table 6.8 for sample comparisons of vertebral burst fractures).

Table 6.8. *Comparison of spinal burst fractures in 'institutional'
and general population skeletal samples*

Note that spinal burst fractures occur in nearly 40% or greater of the institutional
skeletal samples compared with less than 25% of the general population samples.

Sample	Occurrence of spinal burst fracture (%)
Institutional samples	
Oneida Asylum inmates	45.5
South Carolina plantation slaves	48
Snake Hill soldiers	39
General population samples	
Albany Almshouse	9.5
Cedar Creek	25
First African Baptist Church	18.7
Illinois pioneer homestead	18.2

Both the skeletal robusticity and vertebral burst fracture data support the suggestion that exposure to the institutional environment incurred biomechanical consequences. Both biomechanical measures suggest a life of continual and strenuous labor. The cortical bone maintenance data are not as clear, though they also suggest improved cortical maintenance in comparison to a contemporary skeletal sample. From these data we can confirm that life for inmates was composed of continual toil. The hope that labor therapy would result in a cure was ill-founded; in reality, the asylum functioned more as a custodial institution where inmates were kept busy with the labor it took to keep the edifice going. The reverse pattern of skeletal robusticity (increase with age) in comparison to the general population and the high frequency of vertebral burst fractures in the Oneida Asylum sample may, in future research, be shown to be comparable to other skeletal samples from institutional contexts. Since other nineteenth-century institutions, like prisons, implemented versions of labor therapy, as in the Auburn penal system, it is possible that the divergent pattern found in the Oneida sample may be present in those contexts as well.

In conclusion, this study demonstrates a means to address research questions generated from the historical record through methods applied to the skeletal record. An interesting irony developed as a result of this analysis. Reformers, politicians, and physicians lauded the efforts of institutions like the Oneida Asylum for their application of labor therapy. A large part of that praise was due to the comparison of the deplorable conditions mentally ill people suffered in the United States in the pre-1865 era. Despite that praise, the skeletal remains reveal that the inmates' labor served needs that were likely to have benefited others more than it served their own interests. Moreover, the skeletal robusticity indices, cortical maintenance, and spinal burst fracture data all indicate that the Oneida inmates were engaged in heavier work loads in comparison to their cohorts in the general population. And, the spinal burst fracture data place the Oneida inmates in a category of tractable 'institutional' groups (soldiers and enslaved people) who also had very little agency or control over the activities in which they participated. The merger of historical data and human biological data will refine the picture of the past that is currently available and it is likely to supply access to more subtle and complex research questions concerning the not-so-distant past.

Acknowledgements

I am very grateful to Ann Herring and Alan Swedlund for their helpful suggestions for this chapter and to Kathryn Denning for her efforts as Managing Editor. I thank Lisa Anderson (New York State Museum) and Brenda Baker (Arizona State University) for arranging my Research Associate appointment with the New York State Museum, which permitted access to the skeletal collections, laboratory space, and other support necessary to complete the data collection. Linda Braun (New York State Library) greatly enhanced my Research Residency tenure with the New York State Library by helping to track down idiosyncratic documents, setting up a workspace in the stacks, and processing document photocopies. This project was supported with a Doctoral Fellowship from the State University of New York at Albany, small grants from the Graduate Student Organization and Benevolent Association (both of SUNY-Albany), and professional development support from the State Historical Society of Wisconsin. Finally, I thank Lisa W. Phillips (Edgewood College) for her critiques of the manuscript and her continual support.

References

Aufderheide, A.C. and Rodriguez-Martin, C. (1998). *Cambridge Encyclopedia of Human Paleopathology*. New York: Cambridge University Press.
Buikstra, J.E. and Ubelaker, D.H. (eds) (1994). *Standards for Data Collection from Human Skeletal Remains*. Arkansas Archeological Survey Research Series No. 44.
Charles, A.D. (1982). The story of dental amalgam. *Bulletin of the History of Dentistry* **30**, 2–7.
County Superintendents of the Poor of New York. (1877–96). *Proceedings of the County Superintendents of the Poor.* MS at the New York State Archives.
Dix, D. (1844). *Petition to the New York State Legislature: On Behalf of the Insane.* Albany: New York State Legislature.

Dunglison, R. (1872). *Dictionary of Medical Science*. Philadelphia: Henry C. Lea.

Dwyer, E. (1987). *Homes for the Mad: Life Inside Two Nineteenth-Century Asylums*. New Brunswick: Rutgers University Press.

Elia, R.J. and Wesolowsky, A.B. (eds) (1991). *Archaeological Excavations at the Uxbridge Almshouse Burial Ground in Uxbridge Massachusetts*. BAR International Series 564.

Ferguson, P.M. (1994). *Abandoned to Their Fate: Social Policy and Practice Toward Severely Retarded People in America, 1820–1920*. Philadelphia: Temple University Press.

Foucault, M. (1965). *Madness and Civilization: A History of Insanity in the Age of Reason*. New York: Random House.

Gilman, S.L. (1988). *Disease and Representation: Images of Illness from Madness to AIDS*. Ithaca: Cornell University Press.

Glenner, R.A. (1990). The scaler. *Bulletin of the History of Dentistry* **38**, 31–3.

Glenner, R.A. (1994). Gold fillings. *Bulletin of the History of Dentistry* **42**, 129–30.

Grauer, A. (ed.) (1995). *Bodies of Evidence: Reconstructing History Through Skeletal Analysis*. New York: Wiley-Liss.

Grob, G.N. (1994). *The Mad Among Us: A History of the Care of America's Mentally Ill*. New York: Free Press.

Haskins, L. and Jeffrey, K. (1991). *Understanding Quantitative History*. New York: McGraw-Hill Publishing.

Higgins, R. and Sirianni, J. (2000). Poverty in the nineteenth century: Investigating almshouse life through skeletal and archival analysis. *American Journal of Physical Anthropology* **30** (Suppl.), 17–19.

Huff, D.E. (1985). A comparison of American and German dentistry. *Bulletin of the History of Dentistry* **33**, 47–53.

Katz, M.B. (1996). *In the Shadow of the Poorhouse: A Social History of Welfare in America*. New York: Basic Books.

Kennedy, K. (1989). Skeletal markers of occupational stress. In *Reconstruction of Life from the Skeleton*, ed. M. Iscan and K. Kennedy, pp. 129–60. New York: Alan R. Liss.

Lanphear, K.M. (1988). *Health and mortality in a nineteenth-century poorhouse skeletal sample*. Ph.D. dissertation, Department of Anthropology, State University of New York at Albany. Ann Arbor: University Microforms.

Larsen, C.S. (1997). *Bioarchaeology: Interpreting Behavior from the Human Skeleton*. Cambridge: Cambridge University Press.

Larsen, C.S., Craig, J., Sering, L.E., Schoeninger, M.J., Russell, K.F., Hutchinson, D.L. and Williamson, M.A. (1995). Cross homestead: Life and death on the Midwestern frontier. In *Bodies of Evidence: Reconstructing History through Skeletal Analysis*, ed. A.L. Grauer, pp. 139–60. New York: Wiley-Liss.

Leavitt, J.W. and Numbers, R.L. (eds) (1996). *Sickness and Health in America: Readings in the History of Medicine and Public Health*. Madison: University of Wisconsin Press.

Letchworth, W. and Carpenter, S. (1882). *Report on the Chronic Insane in Certain Counties, Exempted by the State Board Charities, From the Operation of the Willard Asylum Act*. Albany: Weed, Parsons, and Co.

Linebaugh, D.W. and Phillips, S.M. (2001). Exploring a silent city: Excavation and analysis of the Holmes-Vardeman-Stephenson Cemetery, Lincoln County, Kentucky. Paper presented at the Kentucky Heritage Council Meetings, Franfort, KY.

Liston, M.A. and Baker, B.J. (1996). Reconstructing the massacre at Fort William Henry, New York. *International Journal of Osteoarchaeology* **6**, 28–41.

LoRusso, M. (1990). *Archaeological Excavations at the Wadsworth Complex*. MS on file at the New York State Museum, Albany.

Lovell, N.C. (1997). Trauma analysis in paleopathology. *Yearbook of Physical Anthropology* **40**, 139–70.

Meinke, H. (1983). A short history of dental advertising. *Bulletin of the History of Dentistry* **31**, 36–42.

Morris, N. and Rothman, D. (eds) (1995). *The Oxford History of the Prison: The Practice of Punishment in Western Society*. New York: Oxford University Press.

Murry, E.A. and Perzigian, A.J. (1995). A glimpse of early nineteenth century Cincinnati as viewed from Potter's Field: An exercise in problem solving. In *Bodies of Evidence: Reconstructing History Through Skeletal Analysis*, ed. A.L. Grauer, pp. 173–84. New York: Wiley-Liss.

Nawrocki, S.P. (1989). *Developing Archaeological Methodology: The Oneida Burial Project*. MS on file at the New York State Museum, Albany.

New York Legislature (1860–96). *New York Statutes*. Albany: New York Legislature.

Oneida County Asylum (1877–94). *Annual Reports of the Oneida County Asylum*. Albany: New York State Legislature.

Oneida County Asylum (1880–94). *Inmate Case Files 4–199*. MS at New York State Archives.

Pfeiffer, S. and Williamson, R.F. (eds) (1991). *Snake Hill: An Investigation of a Military Cemetery from the War of 1812*. Toronto: Dundurn Press.

Phillips, S.M. (1997). Recovering lost minds: Evidence of insanity in a late nineteenth-century almshouse skeletal sample. In *In Remembrance: Archaeology and Death*, ed. D. Poirier and N. Bellantoni, pp. 79–92. Westport, CT: Bergin and Garvey.

Phillips, S.M. (1998). Stature and stress in a nineteenth-century middle class skeletal series. *American Journal of Physical Anthropology* (Suppl. 26).

Phillips, S.M. (2001). *Inmate life in the Oneida County Asylum, 1860–1895: A biocultural study of the skeletal and documentary records*. Ph.D. dissertation, Department of Anthropology, State University of New York at Albany. Albany: University Microfilms.

Purtilo, D.T. and Purtilo, R.B. (1999). *A Survey of Human Diseases*. Toronto: Little, Brown, and Co.

Rankin-Hill, L.M. (1997). *A Biohistory of 19th-Century Afro-Americans: The Burial Remains of a Philadelphia Cemetery*. Westport, CT: Bergin and Garvey.

Rathbun, T.A. (1987). Health and disease at a South Carolina plantation: 1840–1870. *American Journal of Physical Anthropology* **74**, 239–53.

Roberts, C. and Manchester, K. (1995). *The Archaeology of Disease*. Ithaca: Cornell University Press.

Rose, J.C. (ed.) (1985). *Gone to a Better Land: A Biohistory of a Rural Black Cemetery in the Post-Reconstruction South*. Arkansas Archeological Survey Research Series No. 25.

Rothman, D. (1995). Perfecting the prison: United States, 1789 to 1865. In *The Oxford History of the Prison: The Practice of Punishment in Western Society*, ed. N. Morris and D. Rothman, pp. 111–30. New York: Oxford University Press.

Santangelo, M. (1989). *DOCS Oneida Supplement*. MS on file at the New York State Museum, Albany.

Schwartz, J. (1995). *Skeleton Keys: An Introduction to Human Skeletal Morphology, Development, and Analysis*. New York: Oxford University Press.

State Board of Charities (1860–96). *Annual Report of the State Board of Charities*. Albany: New York Legislature.

State Charities Aid Association (1865–96). *Annual Reports of the State Charities Aid Association*. Albany: State Charities Aid Association.

Thomas, D. (1986). *Refiguring Anthropology: First Principles of Probability and Statistics*. Prospect Heights, IL: Waveland Press.

Trent, J.W. (1994). *Inventing the Feeble Mind: A History of Mental Retardation in the United States*. Berkeley: University of California Press.

Ubelaker, D. (1989). *Human Skeletal Remains*, 2nd edn. Washington, DC: Taraxacum Press.

Waldron, T. (1994). *Counting the Dead: The Epidemiology of Skeletal Populations*. New York: Wiley Press.

West, W. (1882). *Questionnaires on Hospital Accommodations of County Poor-Houses: Oneida County Asylum, 1881*. Albany: New York State Legislature.

Winchell, F., Rose, J. and Moir, R. (1995). Health and hard times: A case study from the middle to late nineteenth-century in eastern Texas. In *Bodies of Evidence: Reconstructing History through Skeletal Analysis*, ed. A.L. Grauer, pp. 161–72. New York: Wiley-Liss.

Wood, L. (1996). Frequency and chronological distribution of linear enamel hypoplasia in a North American colonial skeletal sample. *American Journal of Physical Anthropology* **100**, 247–60.

7 Monitored growth: anthropometrics and health history records at a private New England middle school, 1935–1960

LYNETTE LEIDY SIEVERT

The more complicated the human organism appeared and the more intricate the causation of its numerous ills, the more determinedly the really enlightened physician yearned for a science of human biology ... In this extremity a few medical scientists turned to physical anthropology, the illegitimate offspring of medical and zoological indiscretion.

E.A. Hooton (1940: 198)

Introduction

In 1936, a young Harvard-trained doctor accepted the position of Resident Physician at a private New England middle school (name of school withheld at the Headmaster's request). On the heels of a Surgical Residency at the Children's Hospital in Boston, this meticulous man was kept busy with the mundane injuries that adolescent boys inflict upon themselves and each other. His records include such notations as 'bean removed from left ear,' 'skiing accident,' 'pencil point in roof of mouth removed,' 'ruptured spleen from jumping off roof,' 'kicked by horse,' 'fell out of window,' 'football accident' (notes in the health records of individual boys). The Resident Physician taught a few health-related courses, supervised the cafeteria, and was a key decision maker in the lineup of sports teams (Fig. 7.1). Each year he wrote hundreds of letters and dozens of telegrams to parents advising them of their sons' injuries and bouts of infectious disease. In addition, he monitored the growth and maturation of every boy under his care. Today, the health records of these boys are stored in a dormitory basement. These archives provide a superb opportunity to examine physical growth and frequency of infectious disease among wealthy, white, and generally well-nourished young adolescents from 1935 to 1960.

The purpose of this study was to examine whether there was a secular trend[1] in height among boys born from 1918 to 1952 and to see how the height of

130

Fig. 7.1. Skiers.

these private school boys compared to the height of boys measured during the same period of time in Iowa City, Iowa. In addition, this study considers the rate and timing of infectious disease experience, height in relation to the number of diseases experienced, and growth in relation to history of chickenpox, measles, mumps, rubella, and whooping cough. A final, overarching goal of this study was to understand the health records of this private middle school within their historical context.

The sample

Records were available for 955 boys who attended the school from 1935 to 1960. The sample includes 188 boys who first enrolled during the years 1935–1939, 191 boys who enrolled during 1940–1944, 193 boys who enrolled during 1945–1949, 228 boys who enrolled during 1950–1954, and 153 boys who enrolled during 1955–1960. If boys stayed beyond 1960, those records were not made available for this study. The records were made available as an act of trust on the part of the Headmaster – trust that the school would not be identified and trust that the confidentiality of individual boys would be maintained.

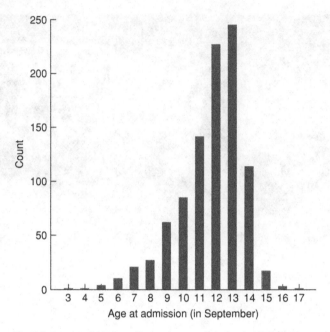

Fig. 7.2. Ages at admission of boys who attended school, 1935–1960.

Age at admission ranged from 3 to 17 years with a mean of 11.7 years (s.d. 1.9). As Fig. 7.2 shows, most boys were 12 or 13 years of age at admission. Most boys attended the school for 2 years, although as Fig. 7.3 shows, 45 were enrolled for 5 years, 20 for 6 years, and 5 for 7 years. Mean length of attendance was 2.5 years (s.d. 1.2). Because the archived data can be connected to living people, the information is both stimulating and constraining. For example, is there a relationship between body mass index (wt/ht^2) or waist/hip ratio at puberty and risk of diabetes in later life? These data present a unique opportunity to follow up on the chronic diseases experienced by these boys (now men aged 49 to 83); however, such a follow-up study is not yet approved. The study presented here will only deal with the years of attendance and health histories as recorded on standardized forms.

Many of the boys ($n = 371$) were from New England, but more than a quarter of the population came from either upstate New York ($n = 144$) or New York City ($n = 117$). A substantial portion ($n = 124$) hailed from New Jersey, Ohio, and Pennsylvania (particularly the steel towns of Pittsburgh and Bethlehem). Thirty-six states are represented, including Michigan ($n = 17$), Illinois ($n = 16$), Texas ($n = 15$), and California ($n = 14$). Diplomats sent their sons from Washington, D.C. Thirty-eight boys came from outside of the United States. For example, oil executives sent their sons from Venezuela.

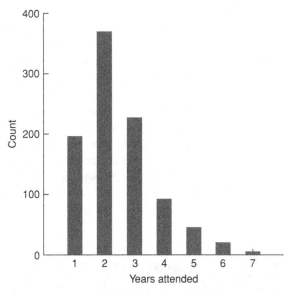

Fig. 7.3. Number of years boys attended the private middle school, 1935–1960.

Fig. 7.4. Mealtime.

This is by no means a cross section of the general American population, but a sample of privileged sons of wealthy or well-connected families at a school that promised, in a 1930s brochure: '. . . to bring about the fullest development of each boy's character and abilities. Our inspiration is afforded by the fact that some of the most vital years of a boy's life are, to a great extent, in our hands.' The middle school provided a structured environment in which boys rose together in dormitories, ate breakfast together in a dining hall, went to classes, lunched together, ended the afternoon with competitive sports, ate supper together (Fig. 7.4), and were supervised in the dormitory until bedtime (Fig. 7.5). The structure and togetherness may have built character, but it also may have facilitated the spread of infectious disease. For example, in 1943, the Resident Physician diagnosed 19 cases of chickenpox and 23 cases of measles during the school year.

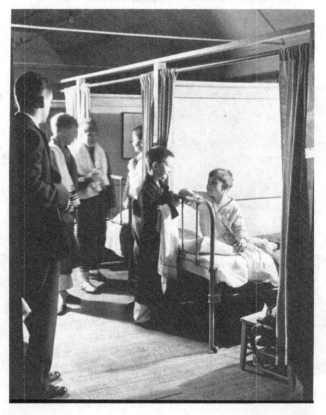

Fig. 7.5. Dormitory.

Monitoring growth

With time, the Resident Physician realized the potential of his young charges as study subjects. Copies of letters sent home to parents indicate that the doctor was aware of ongoing studies of young men in New England. For example, a letter of September 24, 1945, thanked a father 'for the Hooton book which came to my office a few days ago. I am very interested in comparing similar work in our age group here . . .' Later, on December 14, 1945, he added, 'I have Heath's book at last and am just starting it now.' Both of these books describe the Grant Study of the Department of Hygiene of Harvard University, in which 268 Harvard students were studied over a period of four years to determine the characteristics of healthy, normal young men (Heath 1945; Hooton 1945). Of particular resonance would have been A.V. Bock's introduction, which stated that the investigators' concern 'lies primarily in the interest of young men upon whom the future burdens of the world will fall' (Heath 1945: ix). Similarly, the private middle school sought to recruit 'able boys, those with a capacity for later leadership.' The school sought 'to send them on . . . determined to render useful service in a world that needs new vision and courage' (school booklet, written during the War, 1940s).

At nearby Phillips Academy in Andover, Dr J. Roswell Gallagher also believed that there were benefits to studying the 'normal adolescent who has the hereditary and nutritional background of the upper and middle class social level' (from Gallagher's grant proposal to the Carnegie Corporation, cited in Prescott 1992), and his work was touted to be of promise 'not only for education, but for an understanding of human biology and sociology' (Hooton 1940: 226). Gallagher carried out 725 physical examinations per year (Gallagher 1943) and published results specific to color vision testing (Gallagher *et al.* 1943), dental surveys (Brown and Gallagher 1941), sensitivity to coccidioidin (Aronson and Gallagher 1942), and how to test physical fitness (Gallagher and Brouha 1943). In 1940, Gallagher held a conference for physicians serving private schools in the Northeast to share research findings from his Adolescent Study Unit. According to Prescott (1992), Gallagher probably felt that by conducting research at Phillips Academy he could enhance his standing in the medical community. It is quite likely that the Harvard-trained Resident Physician at the private middle school felt the same.

It is also likely that the Resident Physician was aware of Hooton's complaint that medical practitioners lacked 'a comprehension of the basic principles, methods, and tools of the biometry which deals with groups and with multiple variables' (1940: 206). According to Hooton, the physician needed to learn to collect data that were 'precise, orderly, and clearly categorized. . . . (His art) may cure the patient, but it contributes nothing to the general corpus of knowledge we seek – that is knowledge derived from the analysis of large masses of comparable data' (1940: 207–8). Like Gallagher at Andover, the Resident Physician at the private middle school took up the challenge to contribute to the general corpus of knowledge about childhood growth. In contrast to Gallagher, the Resident Physician collected anthropometric measures beyond height and weight.

In the early years of his tenure (1936, 1937), the Resident Physician collected heights and weights just once, at admission. In 1938, precise, orderly, and clearly categorized forms were introduced (Fig. 7.6). Stature, the 'best single measurement for registering gross linear size and rate of linear growth during the school years' (Stuart and Meredith 1946: 1368), was recorded four times per year. Body weights, the best measure of 'general mass and overall rate of growth in mass' (Stuart and Meredith 1946: 1368), were recorded at least five times per year. Circumferences (head, chest, shoulders, abdomen, hips, umbilicus) and arm span were measured once per year as 'useful indicators of the degree of skeletal size and build' (Stuart and Meredith 1946: 1368). Genitals were characterized as 'not developed,' 'development beginning,' and 'well developed.'

In 1946, one year after reading Harvard's Grant Study (Heath 1945; Hooton 1945), the Resident Physician introduced a new form that systematized the collection of stature to four times per year and the collection of body weights to nine times per year. Circumferences (chest, shoulders, abdomen, and hips) and genital maturation scores (1–5) were assessed once per year (Fig. 7.7).

Inter- and intra-observer error in measurements cannot be known. The Resident Physician remained the same from 1935 to 1960; however, distinctive handwriting suggests a change in the nurses who recorded body measurements. Also, it is not known whether or not the same instruments (e.g. platform scales) were used throughout the study period.

According to teachers who remember these years, getting weighed and measured was a rainy day activity. Dormitory 'parents' would call the nurse, or the nurse would call the dormitory, and the boys were trooped over to the infirmary, stripped to their underwear and socks, lined up, and measured. Sometimes boys missed heights and weights because they were wearing heavy casts for broken bones (noted on records).

Secular trend

The Resident Physician's collection of stature and body weights was carried out at a time when secular increases in body size were documented in the general population of the United States (Bock and Sykes 1989; Meredith 1941; Stoudt

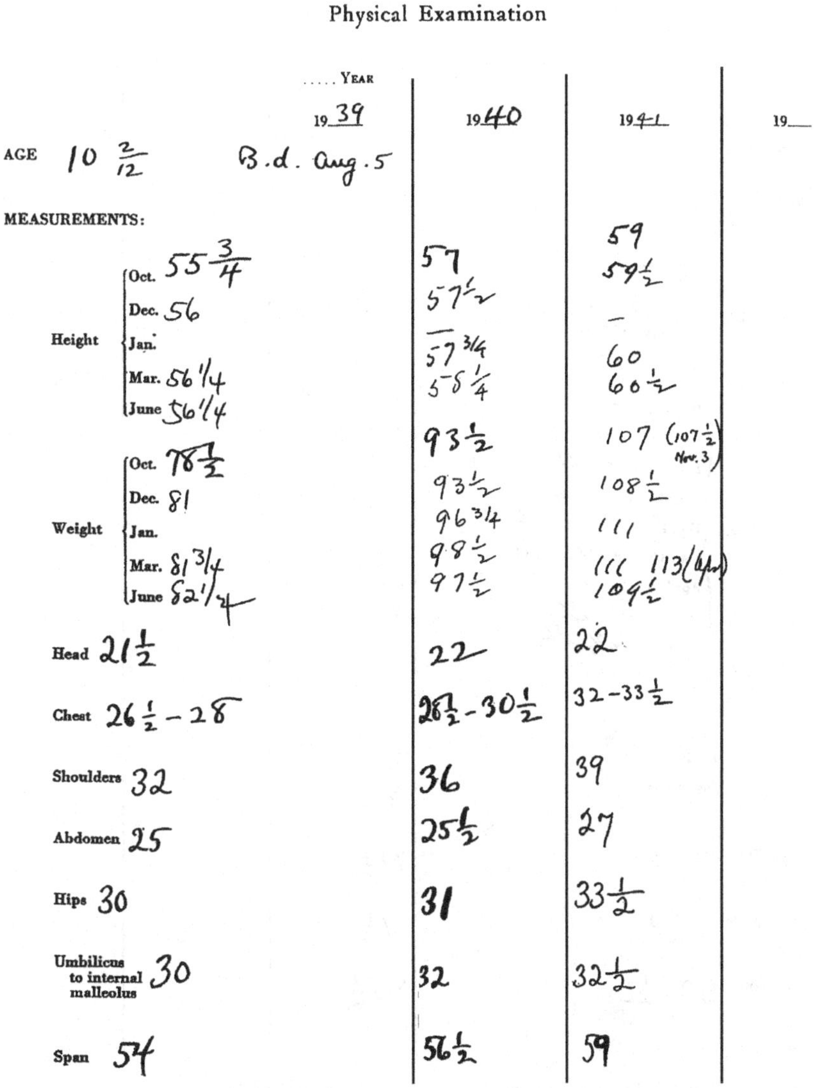

Physical Examination

	YEAR 19 39	19 40	19 41	19__
AGE	10 2/12 B.d. Aug.5			

MEASUREMENTS:

Height	Oct. 55 3/4 Dec. 56 Jan. Mar. 56 1/4 June 56 1/4	57 57 1/2 —57 3/4 58 1/4	59 59 1/2 — 60 60 1/2	
Weight	Oct. 78 1/2 Dec. 81 Jan. Mar. 81 3/4 June 82 1/4	93 1/2 93 1/2 96 3/4 98 1/2 97 1/2	107 (107 1/2 Nov. 3) 108 1/2 111 111 113(4m.) 109 1/2	
Head	21 1/2	22	22	
Chest	26 1/2 - 28	28 1/2 - 30 1/2	32 - 33 1/2	
Shoulders	32	36	39	
Abdomen	25	25 1/2	27	
Hips	30	31	33 1/2	
Umbilicus to internal malleolus	30	32	32 1/2	
Span	54	56 1/2	59	

Fig. 7.6. Physical Examination Form used from 1938 to 1945.

et al. 1965), Western Europe, Australia, and Japan (Meredith 1976). This trend of increasing body size across successive generations was attributed to many variables, including improved nutrition and sanitation, increased availability of preventative and medical care, reduced family size, urbanization, upward social mobility, and the genetic possibilities of selection or assortative mating

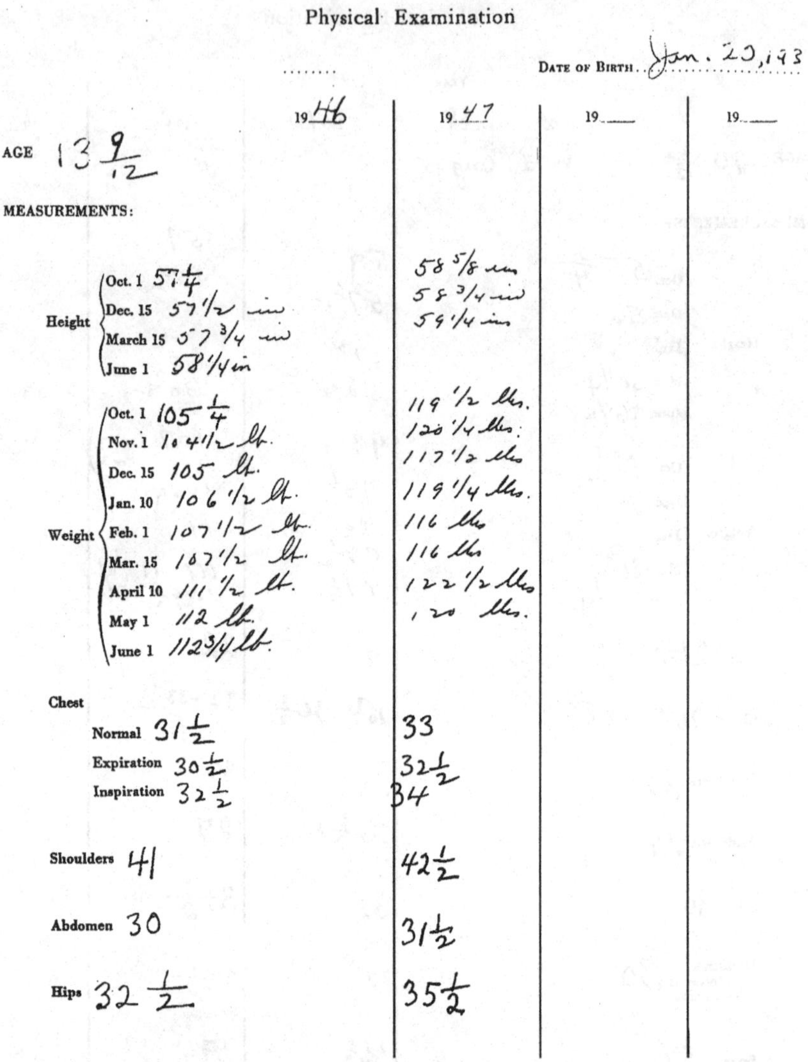

Physical Examination

Date of Birth. *Jan. 23, 193*

	19 *46*	19 *47*	19 __	19 __

AGE *13 9/12*

MEASUREMENTS:

Height
- Oct. 1 *57 1/4*
- Dec. 15 *57 1/2 in*
- March 15 *5 7 3/4 in*
- June 1 *58 1/4 in*

58 5/8 in
5 8 3/4 in
59 1/4 in

Weight
- Oct. 1 *105 1/4*
- Nov. 1 *104 1/2 lt.*
- Dec. 15 *105 lt.*
- Jan. 10 *106 1/2 lt.*
- Feb. 1 *107 1/2 lt.*
- Mar. 15 *107 1/2 lt.*
- April 10 *111 1/2 lt.*
- May 1 *112 lt.*
- June 1 *112 3/4 lt.*

119 1/2 lbs.
120 1/4 lbs.
117 1/2 lbs
119 1/4 lbs.
116 lbs
116 lbs
122 1/2 lbs
, 120 lbs.

Chest
- Normal *31 1/2*   *33*
- Expiration *30 1/2*   *32 1/2*
- Inspiration *32 1/2* *34*

Shoulders *41*   *42 1/2*

Abdomen *30*   *31 1/2*

Hips *32 1/2*   *35 1/2*

Fig. 7.7. Physical Examination Form used from 1946 to 1960.

for stature (Damon 1968; Eveleth and Tanner 1990; Meredith 1976). Of interest to this study, the greatest secular increase in stature occurred at the age of 14. For example, from 1920 to 1960, mean height at age 14 increased by 13.2 cm in Oslo, Norway. In Japan, from 1904 to 1970, mean height at age 14 gained an additional 13.4 cm. In the United States, from 1875–76 to 1963–70, boys at age 14 were 14.8 cm taller (reviewed by Meredith 1976).

Table 7.1. *Mean stature in September of boys in selected age categories by birth year cohort*

Age at admission in September	Birth cohort				
	1918–1925	1926–1930	1931–1935	1936–1940	1941–1945
	cm[a] (n)[b]	cm (n)	cm (n)	cm (n)	cm (n)
8	—	133.4 (6)	134.1 (6)	135.4 (7)	126.5 (3)
10	141.0 (3)	141.5 (18)	143.5 (17)	145.3 (18)	145.0 (15)
12	155.7 (27)	156.0 (28)	154.2 (49)	157.0 (47)	153.4 (49)
14	163.3 (20)	164.6 (13)	164.1 (23)	165.9 (25)	167.1 (28)

[a] Original measurements were made in inches.
[b] Number of boys.

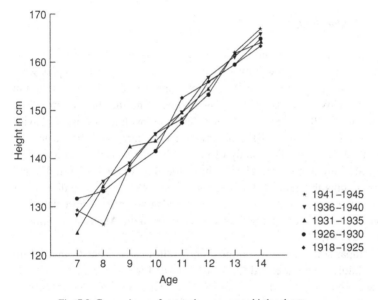

Fig. 7.8. Comparisons of stature by age across birth cohorts.

The boys in this sample were born from 1918 to 1952. Figure 7.8 shows the height of boys born 1918 to 1925 ($n = 117$), compared to the height of boys born 1926 to 1930 ($n = 161$), 1931 to 1935 ($n = 204$), 1936 to 1940 ($n = 228$), and 1941 to 1945 ($n = 231$). The only ages at which a trend may be interpreted from the heights are at ages 10 and 14 (Table 7.1), although the difference of 1.5 inches (3.8 cm) at age 14 is nowhere near the difference documented in Meredith's (1976) review.

In general, these boys had every opportunity to be well nourished across every decade; therefore, given the standard assumption that diet lies behind the secular trend, it is not surprising that no clear secular trend in height emerges from the data. These finding are consistent with a study carried out in Denver, Colorado, where it appeared that 'in a stable, economically- and educationally-privileged population such as that of the Child Research Council study series, secular changes in size or rate of maturation have not occurred in the last 30 years' (Maresh 1972). Similarly, a comparison of Harvard freshmen, from the 1930s to 1957–58, showed that public school graduates experienced an increase in height, whereas private school graduates (i.e. boys like those studied here) did not (Bakwin and McLaughlin 1964). Further research with Harvard freshmen also suggested that early in the twentieth century, 'for men at the very top of the ladder, the end [to the secular trend] may have come' (Damon 1968: 49).

Comparability

Resident physicians at private schools were able to take advantage of their position to study child growth over time; however, their study samples were limited to wealthy children. During the same time period, children residing in or near Iowa City were also enrolled in studies of normal human growth at the Iowa Child Welfare Research Station, later the Institute of Child Behavior and Development (Tanner 1981). One investigation, carried out between 1946 and 1960, was a longitudinal study of normal human growth from ages 5 to 10 years. Measurements were taken within a few days of the fifth birthday and every six months thereafter (Meredith and Knott 1962: 147). The director of the project, Howard Meredith, was a professional auxologist[2] whose data 'had the deserved reputation of being the most accurate on the American continent' (Tanner 1981: 307).

The comparison data below come from studies carried out from 1930 to 1945 on children who attended the University of Iowa experimental schools (Stuart and Meredith 1946). The boys of Iowa City were almost entirely of northwest European descent, predominantly from professional and managerial classes, and physically normal. They are, therefore, a good 'norm' for comparison with the white, upper-class boys attending the private middle school in New England.

How do the boys from the private middle school compare to the well-nourished children of Iowa City? Most studies of the time concluded that families of upper socioeconomic status tended to produce children who were taller than the general population (Meredith 1941; NCHS 1972). In addition to good nutrition, the private school boys had other advantages associated with high

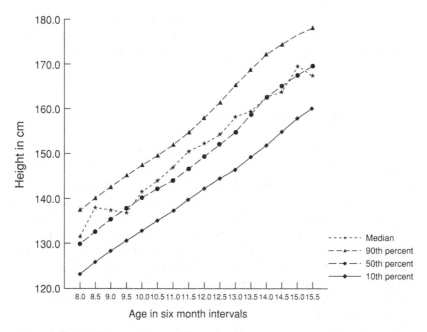

Fig. 7.9. Comparison of stature between private middle school boys (star) and Iowa City boys at 10th (square), 50th (filled dot), and 90th (arrow) percentiles.

family income, including access to the best health care of the time – as evidenced by a 97% vaccination rate for smallpox, an 85% immunization rate for diphtheria, and a 76% tonsillectomy rate.[3]

Figure 7.9 shows median heights per six month age intervals[4] for the New England private school boys (stars) measured from 1935 to 1945 in relation to the Iowa boys measured from 1930 to 1945 (10th, 50th, and 90th percentiles). Compared to the median height of boys from Iowa, privately educated boys were taller at most ages. It may be that the boys drawn from the private New England middle school had more access to health care, although there is no reason to believe that they experienced lower rates of infectious disease (see below). Alternatively, there is some support for the idea that taller men have greater upward social mobility (Bielicki and Charzewski 1983; Bielicki and Waliszko 1992). The positive association between height and social mobility was stronger in the past among men than among women (reviewed in Bielicki and Charzewski 1983). It may be that high-quality parental investment was associated with better nutrition and health care as well as with encouragement and assistance for boys to make the most of educational opportunities (Bielicki and Waliszko 1992). Owing to social selection, financially or politically successful fathers (i.e. fathers able to afford a private middle school) may have been taller than average, hence have sons with the potential to be taller than average. Eventually, genotypes could become unequally distributed along the social scale.

Infectious disease

While infectious disease has been called 'the most important of the adjuvent causes of poor growth' in developing countries, most investigators agree that 'in well-nourished children the effects on growth of minor diseases such as measles . . . are minimal' (Tanner 1990: 133, 147). For example, in Iowa City, a study of 90 children ages 5 and 6 demonstrated no evidence of any significant relationship between amount of illness and size of body or rate of physical growth (Martens and Meredith 1942). A second, prospective study of 135 Iowa City children, from ages 5 to 10, concluded that among children with good home and community support, 'body size and form at age 5 years is not systematically related to amount of illness during the succeeding years of childhood' (Meredith and Knott 1962: 150).

In well-nourished children, even if growth slows during an illness there follows an immediate period of catch-up that restores the child to his normal growth curve (Eveleth and Tanner 1990). Although there is some evidence that boys are at greater risk of experiencing symptomatic disease (Green 1992) and are more stressed by illness during the early years compared to girls (Hummert and Goodman 1986), in well-nourished boys there is little evidence for a lasting effect on long-term growth potential. Nonetheless, it is still of interest to use the data made available by the private middle school to investigate the frequency of infectious disease among boys of this age, and to examine whether or not a slowing of growth is apparent in relation to particular diseases. The data also encourage broader questions about infectious disease, such as whether or not the prevalence of individual diseases increased or declined across birth cohorts prior to the introduction of immunizations.

Monitoring infectious disease

At admission to the private middle school, parents and family doctors filled out standardized health information forms. Doctors were asked to record the year during which a boy experienced common infectious diseases. Recent bouts of infectious disease were also recorded on the interim health forms that parents filled out after their boys were home for the summer. In addition, school health records documented if boys contracted a disease while school was in session and, given the close quarters of the first dormitory (Fig. 7.5), a good number of boys did.

When boys skipped a year of school, a second admissions form was filled out by parents and/or the family physician. This allowed an assessment of recall accuracy. Happily, there was a 100% concordance in history (yes/no) of disease; however, there was far less agreement on dates of infection. In addition, many parents or family physicians wrote '4 years ago,' 'at age 6 or 7,' or simply 'yes' rather than the actual year of infection. For this reason, in the following analyses, timing of infection was treated as a rough dichotomy (before age 10, at age 10 or later) to differentiate between those who experienced an infection prior to the onset of puberty and those who may have been in the beginning of pubertal development when they contracted the disease.

In addition to disease history, under a section called 'Prophylaxis' on the admission forms, parents were asked: 'Has boy been vaccinated against smallpox?' 'Has he ever had preventative inoculation against diphtheria?' 'Has he had any other protective vaccinations or inoculations?' The most common vaccinations or inoculations besides smallpox (96.9%) and diphtheria (85.2%) were tetanus (56.5%), polio (27.9%), pertussis (26.2%), and typhoid fever (20.8%). Some boys (10.5%) received additional immunizations against scarlet fever, measles, yellow fever and other diseases (depending on a boy's summer travels).

Stature in relation to number of infectious diseases

Most boys did not just get one infectious disease. As Fig. 7.10 shows, they reported from 0 to 7 diseases, with a mean of number of 2.9 (s.d. 1.06, $n = 933$). Diseases included in the count were: chickenpox, measles, mumps, whooping cough, rubella, scarlet fever, polio, diphtheria, and other (e.g. rarely reported diseases such as typhoid and yellow fevers).

Thirteen boys had none, 39 boys had five diseases, seven boys had six, and one unfortunate boy reported seven different infectious diseases. It is important to note that this count does not include the numerous cases of ear infections and 'grippe' which occurred at the school. Nor does it reflect the rate of complications from some of the diseases (for example, encephalitis from measles was noted on a few charts). The overall number of infectious diseases experienced by individual boys did not change across the birth cohorts.

Although researchers generally agree that, among well-nourished children, communicable diseases have a minimal effect on growth (Eveleth and Tanner

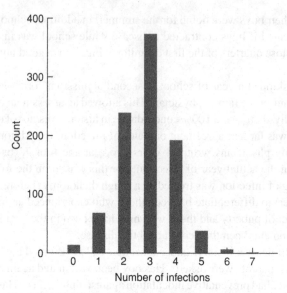

Fig. 7.10. Number of infections experienced by individual boys before or during middle school years.

1990; Hummert and Goodman 1986; Martens and Meredith 1942; Meredith and Knott 1962; Tanner 1990), a few studies have shown measureable short-term effects related to minor illnesses (Hummert and Goodman 1986) and respiratory symptoms (Rona and du V Florey 1980). It has also been suggested that vulnerability to infection might be greater among well-nourished children, at least for tuberculosis (Rona *et al.* 1983). In this private middle school, is stature positively or negatively associated with number of infectious diseases?

Table 7.2 shows that there is no clear pattern of heights in relation to the number of infectious diseases. At some ages, boys with zero to two infections are taller (e.g. ages 11.5–12) compared to boys with more infections. At other ages, boys are taller if they have experienced more than two infections (e.g. ages 12.5–14.5). When sexual development is held constant, developing (G2–3) boys aged 12.5 years with a history of one or two infectious diseases are shorter (one disease 59.8 inches, $n = 6$; two diseases 59.8 inches, $n = 15$), compared to those with three (61.1 inches, $n = 28$), four (61.4 inches, $n = 8$), or five infections (62.5 inches, $n = 3$). There are similar findings at other ages; however, in agreement with the findings of Meredith and Knott (1962), stature did not differ significantly at any age between boys reporting few (0–2) compared with many (3 or more) episodes of infectious disease.

Table 7.2. *Stature (cm)^a in relation to number of infections recorded in health records*

Age	Number of infections $(n)^b$					
	0	1	2	3	4	5+
8	—	132.1	133.6 (6)	127 (7)	132.1	—
8.5	—	139.7	132.1	129.5	—	—
9	—	135.2	135.9 (11)	136.7 (11)	141.2	—
9.5	—	140.5	137.2 (9)	138.2 (6)	137.1 (8)	—
10	—	141.7	143.0 (14)	141.0 (14)	142.2 (5)	—
10.5	148.1	149.6	140.0 (9)	144.0 (20)	144.0 (8)	—
11	—	146.8 (8)	146.6 (14)	147.3 (23)	147.3 (8)	148.8
11.5	152.6	144.0	150.6 (18)	148.3 (23)	150.4 (14)	150.9(5)
12	159.5	155.2 (6)	154.2 (24)	152.2 (40)	152.7 (15)	153.2(7)
12.5	153.7	152.7 (8)	152.9 (22)	154.4 (46)	155.4 (13)	155.7(5)
13	169.7	152.9 (6)	158.5 (18)	158.0 (41)	161.0 (28)	158.0
13.5	162.8	161.5 (5)	158.5 (27)	160.3 (47)	157.7 (32)	159.8
14	165.1	168.7	163.1 (24)	164.6 (28)	163.6 (26)	169.7(6)
14.5	—	162.3	162.3 (15)	165.6 (28)	168.7 (17)	162.1
15	—	167.1	170.4	165.1 (5)	185.4	176.5
15.5	—	—	172.5	170.7	162.8	—

^{a}Original measurements were made in inches.
^{b}Number of boys less than 5 if not noted.

Stature in relation to particular diseases

The following is an examination of stature at age of admission in September in association with history of chickenpox, mumps, measles, rubella, and whooping cough. Frequency and characteristics of the disease are described, in part to give historical context. For example, there was a low prevalence of diphtheria in the medical records. This reflects the high immunization rate. Diphtheria antiserum had been used as a vaccine since the end of the nineteenth century (Stratton *et al.* 1994). As another example, there was a low prevalence of polio in the medical records; however, four of the 15 cases of polio reported by all boys occurred while they were at the school. The Salk polio vaccine was given to 225 boys at the school in the mid-1950s by the Residing Physician (93% of all polio vaccines received). Most other vaccines were not introduced until the 1960s, hence the greater prevalence of infection.

In addition to examining stature in relation to the history of a particular disease, rates of change in stature[5] were examined in relation to particular infections. Then stature and rate of change in stature were re-examined with level of maturity held constant (undeveloped, G1; developing and mid-puberty,

Table 7.3. *Percentage of boys with history of infectious disease* (n = *940*) *with years when cases occurred at the school*

Disease	Frequency	Years of cases at school (*n*)
Chickenpox	83.9%	1935(1), 1936(1), 1938(9), 1942(1), 1943(19), 1950(1), 1952(5), 1953(2)
Measles	82.3%	1936(1), 1937(1), 1940(1), 1942(4), 1943(23), 1944(1), 1949(2), 1952(1)
Mumps	55.5%	1936(1), 1941(3), 1942(4), 1944(1), 1959(1)
Rubella	21.0%	1935(2), 1941(1), 1942(14), 1943(1), 1949(6), 1952(1), 1953(4), 1956(2)
Whooping cough	33.0%	1935(1), 1939(5), 1943(3), 1944(3), 1950(1), 1955(1)
Scarlet fever	6.9%	1936(1), 1944(1), 1950(1), 1952(1), 1953(1)
Polio	2.8%	1944(1), 1949(2), 1950(1)
Diphtheria	0.4%	—
Other	1.1%	—

G2–3; and well developed, G4–5). Finally, stature and rate of change in stature were examined in relation to early and late ages at onset of disease (infancy to age 9, age 10 and later).

Chickenpox

Chickenpox, or varicella, is an acute, generalized, viral disease of sudden onset with slight fever and mild systemic symptoms, including a rash that dries and crusts. The disease, caused by a herpes virus, is spread by direct contact, droplet, or airborne respiratory secretions (Wieczorek and Natapoff 1981). Table 7.3 shows that 84% of the boys who attended the private middle school had a history of chickenpox. Ages at which boys had chickenpox ranged from infancy to 16, with a mean of 6.9 years (s.d. 3.1) – prior to entry into the school. The peak number of cases experienced at the school was in 1943 (19 cases). Table 7.4 shows that the frequency of chickenpox increased slowly across birth cohorts.

Figure 7.11 shows that boys with a history of chickenpox were shorter at ages 12 and taller at age 14 ($p < 0.05$). Holding maturity constant, 12-year-old boys in mid-puberty (G2–3) are shorter if they had chickenpox (60.3 inches vs. 62.6 inches, $p < 0.05$). In contrast, 14-year-old boys in mid-puberty (G2–3) are taller if they had chickenpox (63.3 vs. 66.0 inches, $p = 0.08$). Also, 14-year-old boys in late puberty (G4–5) are taller if they had chickenpox (66.6 vs. 63.7 inches, $p < 0.05$). There is no difference at any age in rates

Table 7.4. *Percentage of boys with history of infectious disease by birth cohort*

| | Birth cohort | | | | |
Disease	1918–1925 (*n* = 115)	1926–1930 (*n* = 158)	1931–1935 (*n* = 204)	1936–1940 (*n* = 220)	1941–1945 (*n* = 227)
Chickenpox	77.4%	81.7%	82.4%	85.9%	89.0%
Measles	80.9	76.0	84.3	84.6	85.0
Mumps	47.0	50.6	54.7	55.3	64.0
Rubella	6.1	18.9	12.3	29.1	29.8
Whooping cough	61.7	57.2	37.8	20.5	11.0
Scarlet fever	8.7	10.1	6.4	7.3	4.4
Polio	1.7	1.9	2.9	3.2	3.5
Diphtheria	1.7	0	0.5	0	0.4

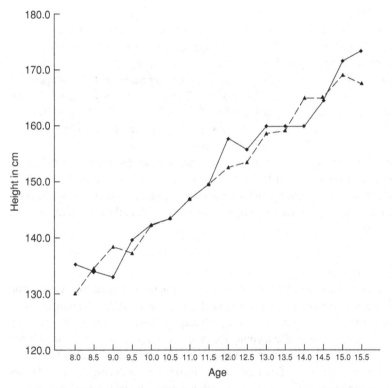

Fig. 7.11. Comparison of stature between boys with (square) and without (triangle) a history of chickenpox.

of growth between boys with or without a history of chickenpox. There is no difference at any age in height in relation to timing of chickenpox (disease before age 10, disease at age 10 or later).

Measles

Measles, or rubeola, is a highly communicable viral disease that is characterized by a reddish brown rash that lasts for five to six days. Although a rare side effect, postinfection encephalitis can lead to permanent brain damage (Wieczorek and Natapoff 1981). Almost every susceptible child exposed to another child in the early stages of measles will contract the disease (Burnet and White 1982). In the private middle school, measles was experienced by 82% of the boys and occurred from infancy to 16 years, with a mean of 7.7 years (s.d. 3.3). As with chickenpox, the peak number of cases of measles at the school ($n = 23$) occurred in 1943. The frequency of measles also increased slowly across birth cohorts. The vaccine for measles did not become available until 1963 (Stratton *et al.* 1994).

Boys with a history of measles were shorter at ages 8.5 ($p = 0.06$) and 10.5 ($p = 0.07$), as shown in Fig. 7.12. At age 8.5, boys with a history of measles were growing more slowly (1.0 inches/six months vs. 1.4 inches/six months, $p = 0.05$). Boys with a history of measles were significantly taller at ages 11.5 ($p < 0.05$) and 14.5 ($p < 0.01$). At age 11.5, undeveloped boys (stage G1) with a history of measles were taller (59.3 vs. 57.2 inches, $p < 0.05$). The difference in height by history of measles was not significant among boys in early (G2–3) or late (G4–5) puberty. Similarly, when maturity levels were held constant, there were no significant differences in height in relation to history of measles among 14- and 14.5-year-old boys. Among 14-year-old boys, those who had measles before age 10 were shorter (64 inches, s.d. 3.0, $n = 46$) compared to those with measles at age 10 or later (65.6 inches, s.d. 2.9, $n = 20$, $p < 0.05$).

Mumps

Mumps, or infectious parotitis, is a disease characterized by fever, swelling and tenderness in one or more salivary glands. There is no rash. The mumps virus is spread by droplets (Wieczorek and Natapoff 1981). At the private middle school, mumps occurred at a young mean age of 8.0 (s.d. 3.0), therefore it is not surprising that the peak number of mumps in any one year was only four cases in 1942. The frequency of mumps experienced before or during the school years increased from 47% to 64% across birth cohorts. The vaccine for mumps did not become available until 1968 (Sherris 1990).

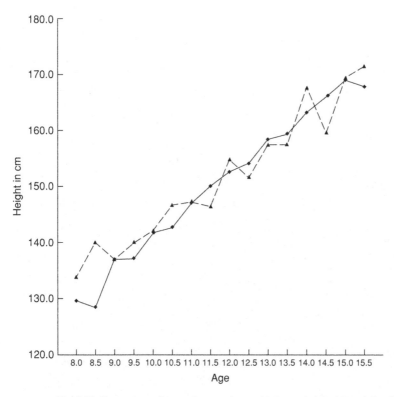

Fig. 7.12. Comparison of stature between boys with (square) and without (triangle) a history of measles.

There are no significant differences in stature in Fig. 7.13 between boys with and without a history of mumps. As for the timing of mumps, 11-year-old boys who experienced mumps before the age of 10 are slightly taller (56.8 inches) than are those who had mumps at age 10 or later (56.3 inches, $p < 0.05$). Eleven-year-old boys who had mumps at age 10 or later are also significantly shorter than boys who never had mumps (58.4 inches, $p < 0.01$). At age 11.5, boys with mumps before the age of 10 are growing faster (0.76 inches/ six months) compared to those with mumps the year or so before (0.43 inches/ six months, $p < 0.05$). However, by age 13.5, boys with mumps before the age of 10 are still shorter (62.2 inches, s.d. 2.8, $n = 27$) than boys with mumps at the age of 10 or later (64.0 inches, s.d. 3.5, $n = 26$, $p < 0.05$). Boys with later mumps are also significantly taller than boys who never had mumps at age 13 (62.1 vs. 63.8, $p < 0.01$) and age 13.5 (62.3 vs. 64.2, $p = 0.01$).

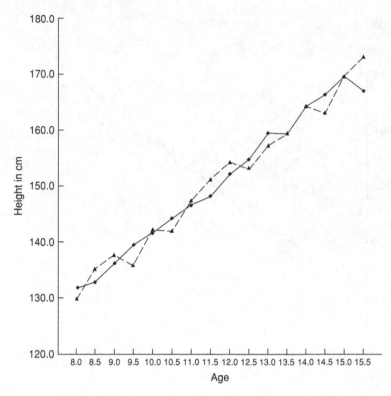

Fig. 7.13. Comparison of stature between boys with (square) and without (triangle) a history of mumps.

Rubella

Rubella, or German measles, is a viral disease characterized by fever, malaise, and a pinkish-red rash that lasts for 2–3 days (Wieczorek and Natapoff 1981). In the private middle school, rubella occurred at a mean age of 9.0 (s.d. 4.3) and peaked at the school with 14 cases in 1942. The prevalence of rubella increased among all boys admitted to the school from 6% among boys born before 1926 to 30% among boys born 1941–1945. The vaccine for rubella did not become available until 1969 (Sherris 1990).

Figure 7.14 shows that boys with a history of rubella were shorter at age 9.5 ($p = 0.07$). Results were significant when maturity was specified as undeveloped (G1) (52.2 inches with rubella, 55.2 inches with no history of rubella, $p < 0.05$). Boys with a history of rubella were significantly taller at ages 12.5 ($p < 0.05$) and 15.5 ($p < 0.05$). When maturity rates were held constant at early puberty (G2–3), boys aged 12.5 with a history of rubella were taller (62.0 inches) compared to boys with no rubella (60.3 inches, $p < 0.05$). Results

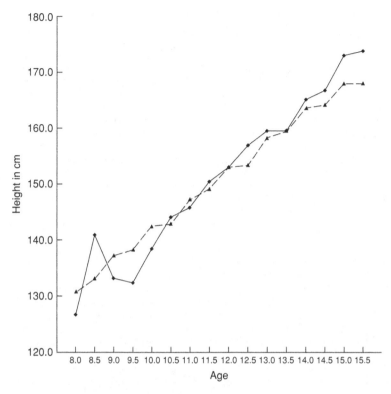

Fig. 7.14. Comparison of stature between boys with (square) and without (triangle) a history of rubella.

at age 15.5 were no longer significant when maturity rates were held constant. At ages 12.5 and 13, rate of growth was faster for boys who had experienced rubella (0.9 inches/six months vs. 1.2, $p = 0.06$; 0.9 vs. 1.3, $p < 0.05$, respectively). When age at rubella was examined, 13.5-year-old boys who experienced rubella early (prior to age 10) were significantly shorter (61.7 inches) than boys who experienced rubella later (at age 10 or later, 64.1 inches, $p < 0.05$). Boys at age 13 showed very similar results (62.0 inches for those with early rubella, 63.9 inches for those with later rubella, $p = 0.09$). Rates of growth did not differ in relation to timing of rubella.

Whooping cough

Whooping cough, or pertussis, is a bacterial disease characterized by recurrent spasms of coughing ending in a whooping inspiration (Taber's 1981). Among these boys, 33% reported whooping cough, from infancy to 14 years of age, with a mean age of 5.6 years (3.0). The finding that few boys experienced whooping

Table 7.5. *Percentage of boys reporting immunizations by birth cohort*

Disease	Birth cohort				
	1918–1925 (*n* = 115)	1926–1930 (*n* = 158)	1931–1935 (*n* = 200)	1936–1940 (*n* = 215)	1941–1945 (*n* = 225)
Smallpox	96.5%	96.2%	97.5%	97.7%	96.1%
Diphtheria	65.2	81.0	86.0	94.0	88.9
Whooping cough	7.0	14.6	39.3	35.3	23.5
Polio	0.9	1.9	0	26.4	83.8

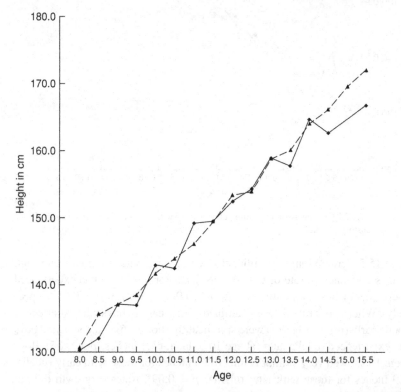

Fig. 7.15. Comparison of stature between boys with (square) and without (triangle) a history of whooping cough.

cough at the school – a peak of four in 1939 – fits the epidemiology of the disease, as it generally occurs in children under 5 (Jawetz *et al.* 1980). The frequency of whooping cough declined dramatically, from 62% to 11% across birth cohorts. During the 1940s, more effective whooping cough (pertussis) immunizations became available (Sherris 1990). According to the school's health records, only 7.0% of boys born prior to 1926 were immunized, compared with 14.6%, 39.3%, 35.3%, and 23.5% for each succeeding cohort (Table 7.5). The drop in immunization rates among boys born from 1941 to 1945 is unexplained, but may be related to the shortages of the war years. There was no significant difference in height in relation to history of whooping cough (Fig. 7.15).

Conclusion

It is of interest to ask why the Resident Physician did not publish results of his work, since Dr Gallagher, at nearby Phillips Academy, did (Gallagher 1943). Perhaps the Resident Physician was simply too busy with the accidents of growing boys. Later, he was kept busy by an independent pediatric practice in a nearby town. Alternatively, perhaps he never intended to publish the data, but instead used the measurements to assess the child's size, muscular development, nutritional status, progress in growth, and relationship to the normal range of variations encountered (Stuart and Meredith 1946). The difficulty with this explanation is that the work of the Resident Physician was profiled in a 1947 Alumni Newsletter, and an intent to publish the results was indicated by the final awkward sentence: 'We are equally certain that (the study) is the very stuff by which men live or we could not devote ourselves to a privileged group of wonderfully potential children without planning to turn to public use that which we are learning with them.' Perhaps, against the Physician's wishes, the school administration discouraged the 'public use' of the gathered information. In the same article, the Alumni Newsletter author notes that, 'it is certainly unwise to make our students feel they are in any sense "guinea pigs".' Even today, decades after the data were collected, the school does not want to be identified by name in order to protect students' identities.

The boys came to the school from different parts of United States, but this is probably not a problem for the comparisons of stature presented here (Meredith 1941). The difficulty lies in comparing mean heights among groups of boys that differ in pubertal development. For example, among 112 11-year-old boys, 85 are undeveloped (G1), 26 are developing (G2), and one is mid-puberty (G3). Just one year later, among 174 12-year-old boys, 69 are undeveloped (G1), 72 are developing (G2), 22 are mid-puberty (G3), 7 are more developed (G4), and 4 are considered fully developed (G5). By age 13, of 190 boys, 23 are still undeveloped (G1), but 70 are developing (G2), 41 are mid-puberty (G3), 42 are more developed (G4), and 14 are fully developed (G5). In other words, the data stop short of fully monitoring the adolescent growth spurt, and the variation in development makes comparisons of growth more complicated. This may lie at the heart of why the Resident Physician did not publish from the data so carefully collected.

There are, however, some conclusions that can be drawn from the collected data. First, there is no clear secular trend of increasing height among boys born from 1918 to 1952. In general, these boys should have had good nutrition and the best of medical care, and therefore a secular trend would be unlikely. Paralleling the lack of secular trend is evidence that preventative health care did not change appreciably across cohorts. Indeed, the frequencies of chickenpox, measles, mumps, rubella and polio rose across birth cohorts, while only the frequency of whooping cough and scarlet fever declined.

In general, the median heights of these boys exceeded the median heights of a comparable group of boys drawn from Iowa City. In this comparison, both groups of boys were well nourished and of similar (northern European) background. It may be that the boys drawn from the private New England middle school had more access to health care. Alternatively, the boys at the private school may have been expressing a genetic potential to be taller.

Rates of infectious disease were consistent with published rates of contagion. For example, 84% of the boys reported a history of chickenpox, which has a 75% attack rate among susceptible contacts. Eighty-three percent of the boys reported measles, a disease for which 85–95% of those infected will become ill. Mumps, whooping cough, and rubella were less contagious. Only 45–55% of infected persons experience mumps and only 30–60% of infected persons develop rubella (Sherris 1990). Whooping cough differed from mumps and rubella in that whooping cough was experienced at a much younger age.

Height did not vary in relation to number of infectious diseases, although some interesting trends were demonstrated (e.g. at age 12.5). Boys who experienced chickenpox at 12 and measles at 8.5 and 10.5 were shorter, whereas boys with a history of chickenpox at 14 and measles at 11.5 and 14.5 were taller. When maturity was held constant, the results were the same for chickenpox, but the differences in height in relation to measles at age 14.5 were no longer significant. There is a weak suggestion that chickenpox and measles slowed growth at earlier ages, but were related to increased heights at later ages.

With regard to mumps, at age 11, boys with a recent bout of mumps (at age 10) were shorter than boys who never had mumps, and their growth rate was still slowed (in comparison to boys with early mumps) during the first six months after age 11.5. By ages 13 and 13.5, boys with later mumps (at age 10 or later) were taller than boys who never had mumps at all and taller than boys with mumps at earlier ages. Similarly, rubella at early ages was associated with shorter stature (at age 9.5) and the effects persisted in boys aged 13.5 who had rubella at earlier ages, despite apparent catch-up growth at ages 12.5 and 13. Boys with rubella at later ages were taller than boys who never had the disease: at age 12.5, boys who never had the disease were significantly shorter (60.4 inches, s.d. 2.8, $n = 77$) than boys who had the disease at age 10 years or later (61.9 inches, s.d. 2.0, $n = 10$, $p < 0.05$). The pattern continued at age 13 (62.4 vs. 63.9, $p = 0.07$), age 13.5 (62.7 vs. 64.1, n.s.), age 14.5 (64.7 v.s 65.9, n.s.), age 15 (66.1 vs. 70.25, n.s.), and age 15.5 (66.2 vs. 68.5, $p < 0.05$). These findings would suggest that taller boys, approaching or at mid-puberty, put more energy into growth and were more susceptible to disease, particularly to diseases that were more difficult to contract (i.e. mumps and rubella). The same has been suggested in relation to tuberculous infection (Rona *et al.* 1983).

Stature and rate of growth were unrelated to history of whooping cough, suggesting that the early timing of whooping cough (mean 5.6 years) may have resulted in no effect on later growth.

In summary, these privately schooled boys were generally well-nourished and the lack of secular trend indicates that they had most likely achieved their potential height in the context of good nutrition and access to medical care. The wealth of these boys and assumed absence of variation in opportunity for good nutrition allows for a detailed examination of disease frequency and growth in relation to particular diseases. There was a suggestion that boys with more diseases were taller (although no significant differences were demonstrated), as well as a suggestion that an increase in height was associated with the less contagious of the communicable diseases. These are preliminary results from a data set yet to be fully mined. The Resident Physician has, indeed, contributed to Hooton's 'general corpus of knowledge.'

Acknowledgements

I am indebted to Jenny Foster for some key citations, and to Kathryn Denning, Ann Herring, and Alan Swedlund for helping to improve the manuscript. Finally, I am most grateful for having had the opportunity to access the records of these boys. It was a privilege to watch them grow – incrementally, progressively, generally slowly, sometimes by great leaps. It was also a privilege to read the medical records in search of measles and chickenpox. For brief moments I became familiar with the boys, their parents, and their doctor. Archival data became more than just numbers. When a particularly sick boy died (at home), I was heartbroken. It was an opportunity and a privilege to study a few of the years in the lives of these boys.

Notes

1 That is, a trend of increasing size across successive generations generally attributed to improved nutrition and better social conditions.

2 Auxology is the study of biological growth (Bogin 1999).
3 Gallagher (1943) reported a comparable rate of 'about 80%' for tonsillectomies
   among boys at Phillips Academy, Andover, Massachusetts. In describing the
   Harvard Grant Study, Hooton noted that 'only persons unconversant with the
   surgical fashions of the day will be surprised to hear that but 22.6 per cent of
   the Grant boys and 20.2 per cent of all of the Harvard freshmen have retained their
   tonsils' (1945: 21).
4 Age of boys was determined as year of age (e.g. 8) at time of measurement in
   September. Boys with birthdays from March to August were grouped (e.g. 8), as
   were boys with birthdays from September to February (e.g. 8.5).
5 Three rates of change in height were computed (a) from September to March during
   the first year of enrollment (mean increase 0.95 inches/six months, s.d. 0.48);
   (b) from the March measurement of the first year to September of second year
   (includes growth during summer, mean increase 1.34 inches/six months, s.d. 0.58);
   and (c) from September to March of the second year of enrollment (mean increase
   1.05 inches/six months, s.d. 0.47). All three measurements were available for 578
   boys. Maximum rate of change was 2.9 inches/six months (during first school year),
   3.5 inches/six months (across summer), and 2.8 inches/six months (during second
   school year).

References

Aronson, J.D. and Gallagher, J.R. (1942). Sensitivity to coccidioidin among boys in
    eastern preparatory school. *American Journal of Public Health* **32**, 636–9.
Bakwin, H. and McLaughlin, S.M. (1964). Secular increase in height. Is the end in sight?
    Lancet ii, 1195–6.
Bielicki, T. and Charzewski, J. (1983). Body height and upward social mobility. *Annals
    of Human Biology* **10**, 403–8.
Bielicki, T. and Waliszko, H. (1992). Stature, upward social mobility and the nature
    of statural differences between social classes. *Annals of Human Biology* **19**, 589–
    93.
Bock, R.D. and Sykes, R.C. (1989). Evidence for continuing secular increase in height
    within families in the United States. *American Journal of Human Biology* **1**, 143–8.
Bogin, B. (1999). *Patterns of Human Growth*, 2nd edn. Cambridge: Cambridge Univer-
    sity Press.
Brown, J.C. and Gallagher, J.R. (1941). Findings in dental survey of 354 preparatory
    school boys. *Journal of Dental Research* **20**, 511–20.
Burnet, M. and White, D.O. (1982). *Natural History of Infectious Disease*, 4th edn.
    New York: Cambridge University Press.
Damon, A. (1968). Secular trend in height and weight within old American fam-
    ilies at Harvard, 1870–1965. *American Journal of Physical Anthropology* **29**,
    45–50.
Eveleth, P.B. and Tanner, J.M. (1990). *Worldwide Variation in Human Growth*, 2nd edn.
    Cambridge: Cambridge University Press.

Gallagher, J.R. (1943). The health examination of adolescents. *New England Journal of Medicine* **229**, 315–18.

Gallagher, J.R. and Brouha, L. (1943). Simple method of testing physical fitness of boys. *Research Quarterly* **14**, 23–30.

Gallagher, J.R., Gallagher, C.D. and Sloane, A.E. (1943). Brief method of testing color vision with pseudo-isochromatic plates. *American Journal of Ophthalmology* **26**, 178–81.

Green, M.S. (1992). The male predominance in the incidence of infectious diseases in children: a postulated explanation for disparities in the literature. *International Journal of Epidemiology* **21**, 381–6.

Heath, C.W. (1945). *What People Are: A Study of Normal Young Men.* Cambridge, MA: Harvard University Press.

Hooton, E.A. (1940). *Why Men Behave Like Apes and Vice Versa or Body and Behavior.* Princeton: Princeton University Press.

Hooton, E.A. (1945). *"Young Man, You Are Normal" Findings from a Study of Students.* New York: G.P. Putnam's Sons.

Hummert, J.R. and Goodman, A.H. (1986). An assessment of the effects of minor childhood illness on growth in height and weight. *Growth* **50**, 371–7.

Jawetz, E., Melnick, J.L. and Adelberg, E.A. (1980). *Review of Medical Microbiology,* 14th edn. Los Altos:, Lange.

Maresh, M.M. (1972). A forty-five year investigation for secular changes in physical maturation. *American Journal of Physical Anthropology* **36**, 103–9.

Martens, E.J. and Meredith, H.V. (1942). Illness history and physical growth: Correlation in junior primary children followed from fall to spring. *American Journal of Diseases of Children* **64**, 618–30.

Meredith, H.V. (1941). Stature and weight of children of the United States with reference to the influence of racial, regional, socioeconomic, and secular factors. *American Journal of Diseases of Children* **62**, 909–32.

Meredith, H.V. (1976). Findings from Asia, Australia, Europe, and North America on secular change in mean height of children, youths and young adults. *American Journal of Physical Anthropology* **44**, 315–26.

Meredith, H.V. and Knott, V.B. (1962). Illness history and physical growth: Comparative anatomic status and rate of change for school-children in different long-term health categories. *American Journal of Diseases of Children* **103**, 146–51.

National Center for Health Statistics (1972). *Height and Weight of Children: Socio-economic Status United States.* Series 11, Number 119, DHEW Publ. No. 73-1601. Health Services and Mental Health Administration, Oct.

Prescott, H.M. (1992). *"A doctor of their own": The emergence of adolescent medicine as a clinical sub-specialty,* 1904–1980. Ph.D. thesis, Cornell University.

Rona, R.J. and du V Florey, C. (1980). National study of health and growth: respiratory symptoms and height in primary schoolchildren. *International Journal of Epidemiology* **9**, 35–43.

Rona, R.J., Chinn, S., Marshall, B.S.M. and Eames, M. (1983). Growth status and the risk of contracting primary tuberculosis. *Archives of Disease in Childhood* **58**, 359–61.

Sherris, J.C. (1990). *Medical Microbiology: An Introduction to Infectious Diseases,* 2nd edn. New York: Elsevier.

Stoudt, H.W., Damon, A., McFarland, R.A. and Roberts, J. (1965). *Weight, height, and selected body measurements of adults. United States, 1960–1962.* Publ. No. 1000-Series 11-No.8, Vital and Health Statistics, US Public Health Service, Government Printing Office, Washington, D.C.

Stratton, K.R., Howe, C.J. and Johnston, R.B. (1994). *Adverse Effects Associated with Childhood Vaccines: Evidence Bearing on Causality.* Washington, D.C.: National Academy Press.

Stuart, H.C. and Meredith, H.V. (1946). Use of body measurements in the school health program, Parts I and II. *American Journal of Public Health* **36**, 1365–86.

Taber's Cyclopedic Medical Dictionary (1981). Philadelphia, PA: F.A. Davis Company.

Tanner, J.M. (1981). *A History of the Study of Human Growth.* Cambridge: Cambridge University Press.

Tanner, J.M. (1990). *Fetus into Man.* Cambridge, MA: Harvard University Press.

Wieczorek, R.R. and Natapoff, J.N. (1981). *A Conceptual Approach to the Nursing of Children.* Philadelphia: JB Lippincott.

8 Scarlet fever epidemics of the nineteenth century: a case of evolved pathogenic virulence?

ALAN C. SWEDLUND AND ALISON K. DONTA

Introduction

Between approximately 1820 and 1880 there was a world pandemic of scarlet fever and several severe epidemics occurred in Europe and North America. It was also during this time that most physicians and those attending the sick were becoming well attuned to the diagnosis of scarlet fever, or scarlatina. They could differentiate the disease from diphtheria by the presence of the characteristic rash, or 'exanthem,' that accompanied the sore throat, fever, inflammation of lymph nodes and abscessing of the throat and tonsils.

Streptococci bacteria were probably first isolated by the Viennese surgeon Theodor Billroth in 1874, but the association of hemolytic streptococci with scarlet fever was not demonstrated until 1884, and the specifics not outlined until 1924 by George and Gladys Dick (Dowling 1977). Once there is the onset of symptoms such as sore throat and fever the course of the infection can progress very quickly and children in the nineteenth century epidemics were known to succumb within as few as 48 hours in some cases. Streptococci may be contracted through human contact, airborne droplets, or ingestion. A common source of the bacteria in historical outbreaks was unpasteurized milk handled by infected dairy workers.

When an outbreak occurs, symptoms can vary widely within the population, or even within the same family – ranging from asymptomatic individual carriers to acute rheumatic fever and severe tissue infections. For common strains the highest risk is for young children. Infants often are at lower risk owing to the presence of maternal antibodies, and by age 10 it is estimated that upwards of 80% of children have developed lifelong protective antibodies against the exotoxins (Darmstadt 1998). When President Abraham Lincoln's son Tad contracted scarlet fever during Lincoln's first presidential election it was a severe case, whereas Lincoln himself had symptoms of sore throat and headache, and assumed he also had the disease (Dowling 1977: 60).

159

Severe outbreaks of Group A *Streptococcus pyogenes* have occurred over the past two decades in North America, England, Europe and elsewhere. These outbreaks have occurred primarily among otherwise healthy populations, raising concern about the rise of new virulent strains and accompanying antibiotic resistance (see, for example, Stevens *et al.* 1989). The return of this nineteenth century scourge prompts consideration of its contemporary manifestations and historical correlates. In this chapter we briefly outline present-day observations and then turn to the historical evidence, using a microepidemiological case study from the Connecticut River Valley of Massachusetts.

We wish to test the limits of historical epidemiology for providing various levels of explanation of disease processes. Secondly, we wish to contrast the pathogenic/virulence evidence with the socioeconomic data for our particular case study. After a preliminary analysis of the available evidence, we conclude that the virulence hypothesis is consistent with our findings, but also suggest that explanations of these events are complex and should encompass socioeconomic and nutritional variables to varying degrees. By starting with an international pandemic, focusing in on a regional epidemic, and then 'microscopically' focusing in on four communities, our purpose is to ultimately examine the details about some of the specific individuals who succumbed to scarlet fever. Using the example of the town of Deerfield, we can attempt, literally, to map the course of events in this town during an outbreak. Stylistically and methodologically, we aim to validate the particular and the qualitative knowledge of these events as well as the epidemiological.

The return of scarlet fever

Strep Outbreak kills 26 in Texas

El Paso, March 7 (AP) – An outbreak of streptococcus has killed 26 Texans in the past three months, including 9 children, a new report says... The most recent deaths include a 5-year-old Channelview boy who died of toxic shock syndrome on Sunday and a 5-year-old El Paso boy who died on the way to the hospital last week... Health officials have no explanation for the apparent surge, but a Health Department spokesman, Doug McBride, said it may be partly attributable to recent changes in reporting procedures.

(*New York Times*, March 8, 1998)[1]

Beginning in the mid-1980s, increasing numbers of invasive streptococcal infections (including scarlet fever) were reported throughout the United States, Canada, Europe, and Australia (Stevens 1992). Outbreaks occurred between July 1989 and June 1991 in four of seven US Army basic training installations (Gunzenhauser *et al.* 1995) and Canadian researchers have reported an increasing frequency of all types of *Streptococcus pyogenes* infections among children during the period 1984–1990 (Cimolai *et al.* 1992). Streptococcal infections have now become such a public health problem that they have been the focus of increasing research by the World Health Organization (WHO) (Efstratiou 2000). In addition, invasive streptococcal disease, Type A, is a nationally reportable disease in the United States (CDC 1999). Scarlet fever and the associated invasive streptococcal infections can no longer be considered historical oddities, or the outcome of new reporting procedures by health officials, but are instead very contemporary clinical diseases with serious effects.

Symptoms in present-day cases range from classic scarlet fever to acute streptococcal toxic shock syndrome. Numerous deaths have occurred. Case fatality rates as high as 50% have been reported (Spencer 1995). In the United States during the period 1980–1998, the number of deaths per year from scarlet fever ranged from zero to three. This represented an average of 1.1 deaths per year during the 1980s and 0.6 deaths per year during the 1990s (Centers for Disease Control and Prevention data from the Compressed Mortality data set). For all invasive Type A streptococcal infections, however, there were approximately 10 200 cases (1300 of them fatal) in the United States in 1998 (CDC 1999).

This increase in the reported number of streptococcal infections in recent decades has been compared to the pandemic of fatal scarlet fever in 1825–1885 (Katz and Morens 1992) and suggests the possibility of varying virulence factors throughout time and across geographic areas. Several researchers are reporting the presence of new variants of the *speA* gene, which encodes the pyogenic exotoxin causing hemolytic damage and producing the characteristic red rash. New variants have been suggested as markers for monitoring future potential outbreaks (see, for example, Musser *et al.* 1993).

Because the decline of severe symptomatic scarlet fever in the US began sometime after 1880, but well before the antibiotic era, many have suggested that attenuation of the streptococcal organism may have occurred, only to arise again in the 1980s (see, for example, Stollerman 1988). These arguments center on hypotheses regarding host–pathogen coevolution, or on the absence of compelling nutritional evidence or on the lack of effective medical or public health intervention. Others have argued for a significant socioeconomic component mediated through nutritional status and health (see, for example, McKeown 1976; Szreter 1988; Hardy 1993; Duncan *et al.* 1996).

Historical epidemiology in bioanthropological perspective

A comprehensive view of an historical epidemic from a bioanthropological perspective involves at least four domains of interest.

1. Historicity: situating the epidemic in time and place.
2. Medical history: understanding how the epidemic was viewed 'then.'

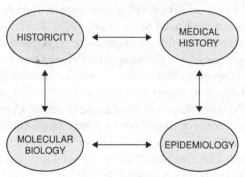

HISTORICAL EPIDEMIOLOGY AS
BIOANTHROPOLOGY

Fig. 8.1. Four domains of historical epidemiology from a bioanthropological
perspective.

3. Epidemiology: reevaluating the available data using contemporary models.
4. Molecular biology: understanding the pathogenicity and genetics of the
    disease.

Molecular biology

Taking the last topic first, we already have described briefly what is known
about Type A streptococcal infections and their potential for mutating into
new clones with high toxicity. The most certain method of demonstrating the
periodicity of virulent variations of streptococcus is to examine samples of the
bacterium over time, and this has, in fact, been done recently using multi-
locus enzyme electrophoresis on blood samples of individuals infected in
Canada in the1920s and 1930s (Musser *et al.* 1993). The results of this study
support the theory of cyclical virulence as an explanation for disease fluctuations both temporally and geographically. Examination of the serotypes
of Type A streptococcus associated with various outbreaks throughout the
world revealed an association between the M1 serotype and the most invasive,
epidemic-like outbreaks (Cleary *et al.* 1992; Colman *et al.* 1993; Efstratiou
2000; Spencer 1995).[2] This further supports the theory of fluctuating disease
virulence.

It is less likely (though not impossible, however) that we will be able to
isolate streptococci or the *speA* gene from the remains of individuals living
in the nineteenth century. Meanwhile, if we are going to suggest evidence for

virulence in the nineteenth century it would most likely be on the basis of the *absence or lack of evidence* for alternative explanations, with virulence being the residual explanation. Retrospective interpretation of past events is a hallmark of anthropology and historical epidemiology – for better or worse – and strong inference usually involves the rejection of some variables, with a lack of absolute certainty for those remaining.

Historicity

The demographic patterns of this historical period in the Connecticut River Valley of Massachusetts have been described in detail elsewhere (see, for example, Swedlund *et al.* 1976; Swedlund 1990). In brief, it was a rich agricultural valley with emerging industrialization between 1850 and 1900. The four towns selected for this microepidemiological study are Deerfield, Greenfield, Montague and Shelburne, Massachusetts. They vary in their dependence on agriculture, commercial, and industrial economies, but all are diversified with levels of each. Our data are drawn from the sum of deaths occurring in each community between the years 1850 and 1910, extending the time so that we could follow post-1880 transitions. In this chapter we concentrate on the 1858–59 epidemic, using the 1867–68 epidemic simply for the purposes of examining periodicity.

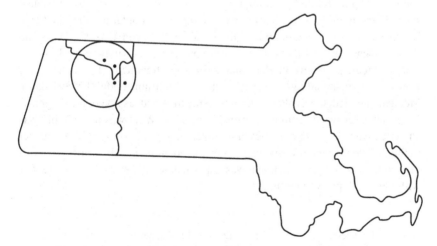

Fig. 8.2. Map of the study area in Massachusetts with the center villages of the four towns of primary interest. Working clockwise from the upper left quadrant: Shelburne Falls, Greenfield, Montague, and Deerfield. Rendered by authors.

Scarlet fever
The scarlet fever, which has very seldom prevailed in this town, is now
prevailing quite extensively among children. We hear of its prevalence also
in several of the neighboring towns.
(*Greenfield Courier*, December 27, 1858)

On the eve of this first epidemic, in December 1858, a growing uneasiness
overtook the Valley. By means of the local newspapers, and no doubt word-of-
mouth, knowledge of the epidemic's presence elsewhere was widespread. By
then, numerous individuals were being diagnosed with scarlet fever locally. In
January 1859, things became much worse, with many more infected and several
children dying in the four towns. A full-blown epidemic was at hand, and, as
we see below, in Deerfield the main street was a site of considerable angst for
the families residing there.

There were 15 deaths in the town of Greenfield in the month of January,
1859–3 of consumption, 9 of scarlet fever, 1 of whooping cough, 1 of
inflammation of the brain, and 1 accidental.
(*Greenfield Courier*, February 7, 1859)

Medical history

Physicians were ubiquitous in western Massachusetts by the mid-nineteenth
century. Each of the study communities had at least one doctor, Greenfield had
several, and the relatively small town of Deerfield (population 3073 in 1860)
had as many as five doctors practicing between 1858 and 1870 (L. Dame, no
date). At least three of these doctors were present at meetings of the Franklin
County Medical Society in 1858 and 1859 when treatments for scarlet fever
were discussed, including the use of 'tincture of veratrum vivide' (FMS no date).
Medical knowledge was also increasing at the time and, as noted, the diagnosis
of scarlet fever was becoming reasonably reliable. What was not reliable was
any effective level of therapeutics or treatment that physicians could administer
to the afflicted. Even after the bacterium was identified in the third quarter of
the century, an effective treatment was not available until many years thereafter
with the development of antibiotics.

For scarlet fever:
* keep the bowels free by a free use of fruits, berries and cracked wheat.
* keep out the rash by the prevention of chilliness and looseness of bowels.
* keep down thirst and fever by acid drinks, lemonade, buttermilk, etc.
* keep the room cool and well ventilated.

- if there is a tendency to debility, add some meat, poultry, and soups, with bread crust, to the diet.
- In great heat of the skin, sponge it frequently with tepid water.

(Hall's *Health at Home* 1872)[3]

At the time of this epidemic, however, prescriptions were abundant. By consulting doctor's journals and reading the home manuals written by the physicians of the period we noted many concoctions available for scarlet fever:

> Carbonate of ammonia, sixty grains; paregoric, two oz.; wine of ipecac, forty drops; water, six oz. Give one tablespoon in some lemonade, or vinegar and honey every three hours.

(Hall 1872: 756)

Because scarlet fever was by no means always fatal, and because the opiates, alcohol, and other ingredients might make patients feel some relief from symptoms, there were occasions when physicians were highly valued for their efforts, despite their limited capacities to treat and certainly to cure.

> Northfield – The scarlet fever is prevailing to considerable extent, but our physicians thus far have managed it very successfully. We think it no more than a just compliment to them to say they are as good as the county affords, attending early and late and at all hours, faithfully, to their not infrequently arduous duties. Success attend them.

(*Greenfield Courier*, February 12, 1859)

Epidemiology

The 1858–59 epidemic of scarlet fever impacted Massachusetts severely. This epidemic caused 2089 deaths, 95% of which were of children aged 15 years or younger. In our four-town analysis the percentage aged 15 or younger was 96.9%. There were 227 deaths from categories that likely were reducible to scarlet fever (e.g. scarlatina, etc.) in the four towns between 1850 and 1910. Of these, 220 cases occurred among children under the age of 16; 206 of them to children less than 10 years of age. From the total of 206 childhood deaths in the four towns between 1850 and 1910, we focus on 89 deaths to children for the epidemics of 1858–59 and 1867–68, and then on a small number of those occurring specifically in the town of Deerfield. In the four study towns, it should also be noted that 42% of all childhood deaths during the years 1858–59 were due to scarlet fever.

The mortality trend for Massachusetts during 1858–59 by month is depicted in Fig. 8.3. Despite a relatively small number of cases, our results

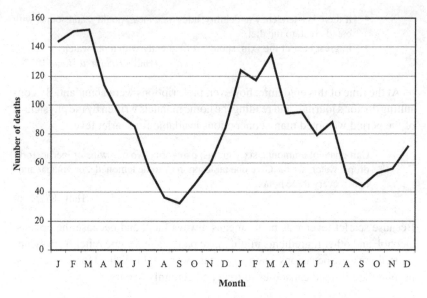

Fig. 8.3. Number of deaths due to scarlet fever, by month: Massachusetts, 1858–59.

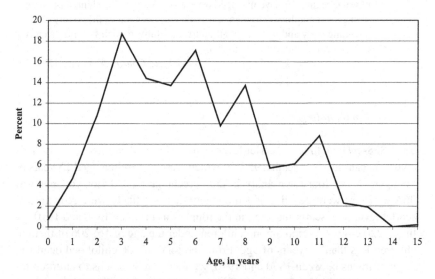

Fig. 8.4. Percentage of deaths among children <16 years due to scarlet fever, by age: selected Connecticut River Valley towns, 1850–1910.

tend to fit the patterns outlined both in the Commonwealth as a whole and in the epidemiological literature. The age distribution of fatal cases, illustrated in Fig. 8.4, showed remarkably few infants, with the highest proportion of deaths occurring between the ages of 2 and 8 years. A low rate among infants

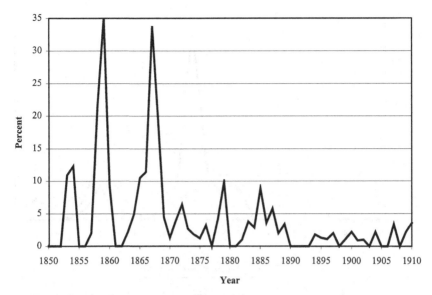

Fig. 8.5. Percentage of deaths among children <16 years due to scarlet fever, by year: selected Connecticut River Valley towns, 1850–1910.

is consistent with the presence of maternal antibodies in children in their first 12 months. An analysis of the sex distribution of cases reveals a slight but significant difference by sex, with somewhat more boys dying than girls ($p \leq 0.01$, $\chi^2 = 6.37$).[4]

The two principal epidemics of the region show up in sharp relief for the towns (Fig. 8.5), and the reduction in epidemics in these towns by 1890 is consistent with the reduction of scarlet fever deaths throughout the entire region at this time. An analysis of the linked data from the Census and the mortality data for the following two years in the nearby cities of Northampton and Holyoke mirrors this trend. During the period 1850–1910, the proportion of deaths in the entire population due to scarlet fever ranged from 0 to 6.4%, with the peak in both towns during the 1870–72 period and diminishing to less than 0.4% by 1910. The age distribution of deaths by scarlet fever in Northampton during 1870–72 (the period for which there were enough deaths to yield a large enough sample size) was very similar to that of the four study towns. During these years, children aged 2 through 8 had the highest proportion of deaths due to scarlet fever. In 1870–72 in Northampton, 17.2% of deaths among children under 16 years was due to scarlet fever. The 1867–68 epidemic in the four study towns does not show any particularly strong seasonal patterns, but the 1858–59 epidemic shows a sharp peak in January 1859 and a strong secondary peak in July, as illustrated in Fig. 8.6.

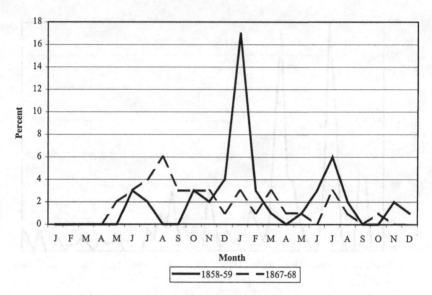

Fig. 8.6. Number of deaths among children <16 years due to scarlet fever, by month: selected Connecticut River Valley towns, 1858–59 and 1867–68.

Scarlet fever in Deerfield, Massachusetts

In Deerfield, Greenfield, Montague, and Shelburne we see a microcosm of the larger State and County events. The rank order of the proportion of deaths among children less than 16 years due to scarlet fever in these four towns during the study period was Greenfield (5%), Deerfield (6%), Shelburne (7%) and Montague (2%). It is interesting to note that the town of Montague lies across the Connecticut River from the other three, and this may have provided some natural barrier to the spread of the infection, given the reduced frequencies apparent for Montague.

Our current research is focused on the systematic tracking of individual households in which scarlet fever deaths occurred, and an example from Deerfield illustrates this approach. In the Manuscript Census of Massachusetts for the town of Deerfield in 1860, we found 12 households altogether in which children were lost to scarlet fever in 1859. The average age of the children was 3 years 11 months and the age distribution fits the profile for risk of death from scarlet fever very well, which is often given as being between 2 and 8 years. Somewhat serendipitously, these are evenly divided between farmers and laborers (mostly mill operatives in the local cutlery factory). There are six households of each category. In this sample the farmers' average wealth in real and personal property is US$4354.33. For the laborers it is US$635.50. In other words, the farmers were on average 6.8 times wealthier than the laborers – and among the wealthiest citizens of the town – yet both lost equal numbers of children.

Of the seven deaths occurring to children in January 1859, two occurred to partially linked families who were most likely recent immigrants, and who resided in Cheapside, the most working-class district of the town. Four occurred to farmers, and one we have as yet been unable to link. Of the deaths that occurred in farming families, we illustrate the locations of three of the residences in Fig. 8.7. A cluster of deaths occurred at the north end of the old village center in January, claiming three boys who lived in close proximity to each other, two of whom were very close friends. On January 10, Elihu Ashley, age 8, died, on January 18, Jesse Stebbins, age 2 years and 4 months succumbed to the disease, and then, on January 20, Frank Sheldon, age 6, joined his friends in death.[5]

Early photos or renderings of the homes of the street (see McGowan and Miller 1996), and other records reveal that these were not only the sons of successful gentleman farmers, but also among the most prominent families in the community. By way of contrast, Elihu Ashley and Frank Sheldon were buried in Laurel Hill Cemetery sitting on a hill above the prosperous section of Deerfield, whereas Gustavus Roos and Maria Logan were buried in the 'German section' of the Green River cemetery in the neighborhood of Cheapside. All were victims who died in Deerfield during the month of January.

As we slowly add to the list of individuals linked in the 1867–68 epidemic, and in the town of Greenfield, we are seeing similar trends: deaths occurring in households from middle-class neighborhoods, or productive farming sectors, as well as those occurring in neighborhoods known to be much less well-off. In the case of Deerfield, the former are not households that would lack for food, warmth, or other economic amenities. Neither of these neighborhoods are characterized by large tenement dwellings, but a significantly larger number of poorer citizens likely resided in the Cheapside district of Deerfield, and in multi-family dwellings.[6]

Discussion

Based on typical epidemiological patterns for scarlet fever from accounts of the past, it would be common for many people in any given community to be infected and symptomatic. Within families several members might be in various stages of the disease, with the children aged 2–8 most vulnerable to severe symptoms, and perhaps death. Even from our very small samples these patterns are apparent.

The data component of this study has been derived from a predominantly rural area of western Massachusetts in the nineteenth century. Unlike previous

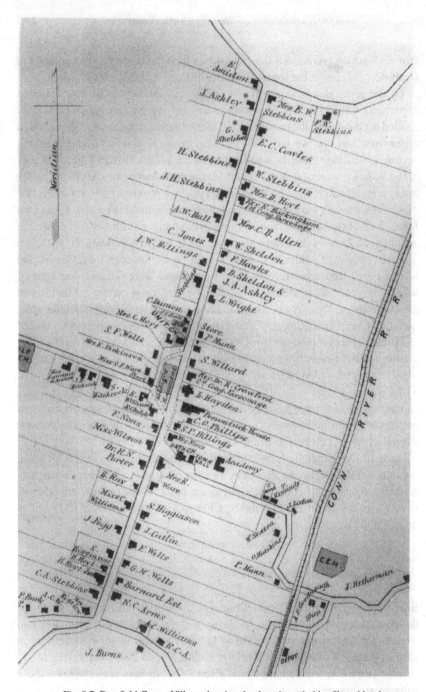

Fig. 8.7. Deerfield Centre Village showing the three households affected by the epidemic (arrowed). From *Atlas of Franklin County, Massachusetts*, F. W. Beers and Company (New York, 1871). Reproduced with kind permission of The Henry N. Flynt Library of Historic Deerfield.

analyses of scarlet fever epidemics (see, for example, Hardy 1993; Duncan *et al.* 1996), this population was less subject to urban patterns of disruption in food supplies as a function of economic recessions. While the Massachusetts population was residentially and occupationally diverse at the time of study, we are confident that nutritional stress would be much less severe than in poorer, urban populations in North America and Europe at the time. The limited but detailed evidence available in our preliminary analysis suggests to us that neither nutritional nor socioeconomic factors can be considered sufficient risk factors for explaining the distribution of scarlet fever deaths in our sample during the epidemics of 1858–59 and 1867–68 in the Connecticut River Valley of Massachusetts.

Research on England and Wales over the last decade or two has re-evaluated McKeown's classic study (1976) of mortality decline in Britain, in which he argued that this decline was due primarily to improved standard of living, which translated into improved nutrition for those at greatest risk for infectious disease. More recent research (see especially Szreter 1988; Hardy 1993) has carefully evaluated the historical record and McKeown's arguments. These studies find a much more compelling explanation in the rise of social interventions inaugurated primarily by public health reformers in the mid-nineteenth century. And, in the case of scarlet fever, they see a case of a disease that may be the best candidate for attenuation of virulence among all the major infectious diseases investigated by McKeown and others.

The strength of the arguments now seem to favor the health reform/social intervention hypothesis over the standard of living hypothesis for the greater proportion of infectious disease mortality in England and Wales during this period. However, an important element of nutritional status should not entirely be dismissed in the British case, nor in North America, even though its correlation with standard of living may be more complex than that encompassed in the traditional arguments and debates (see, for example, King and Ulijaszek 1999: 164–71). We are referring to the close and perhaps synergistic interrelationship between nutrition and infection that can occur in a host of disease contexts and which can be influential not so much in who becomes infected, but in who survives. A mother's ability to pass on sufficient antibodies to her child surely depends on her own nutritional status and on her ability to effectively nurse. Likewise, an infant or child who is compromised nutritionally – when 'challenged' to mount an effective immune response – is perhaps less able to do so than his or her well-nourished counterpart. Important in this regard are key micronutrients that may or may not be available in the impoverished diet.

A growing body of clinical research is demonstrating that infections in infancy and early childhood, particularly dysentery and enteric infections, affect the child's ability to resorb key micronutrients and can then leave the child more susceptible to severe outcomes from secondary or later infections (see, for example, Guerrant *et al.* 2000). Although there do not seem to be studies specifically referring to this sequela in relation to scarlet fever, we consider it an important problem for many of the diseases we are currently investigating in the historical context, including scarlet fever. Moreover, this scenario is a very likely one for urbanizing and industrializing communities in nineteenth century Britain and North America, where dysentery among infants and children was ubiquitous due to unsanitary living conditions, artificial feeding of infants and substandard working conditions. We do not raise these issues of nutrition here because we have the data to test a specific hypothesis. Rather, we believe that nutrition is important for the reasons described above and should not be dismissed entirely because of the somewhat naive way in which it was proposed by McKeown and others. McKeown's (1976) model was developed to explain a long secular trend in declining mortality that was believed to have occurred in Britain (and the US), but, in fact, during rapid industrialization in the early nineteenth century UK and in the later nineteenth century US, mortality for children was not showing a systematic decline, but in fact was on the increase at various points in time. The water- and food-borne diseases to which many children and some adults succumbed certainly can be expected to have been accompanied by an important set of nutritional risk factors.

Conclusions

While lacking sufficient data at this point to make the strongest possible case, our preliminary findings lead us to the following provisional conclusions.

1. Molecular biology. From molecular biology we find clear evidence for the capabilities of evolving virulence and attenuation in *Streptoccocus pyogenes*, particularly with regard to what is being learned about the *speA* gene and associated M proteins. The fact that a virulent strain was cultured from tissues preserved from the early part of this century pushes our knowledge back closer to the time of the pandemic in the nineteenth century.

2. Historicity. The time, place, and nature of the epidemic in western Massachusetts is well established by vital statistics data derived from the target communities and from the historical record as revealed in newspapers and regional writing of the day. We can date the incidences quite specifically and graph their trajectories.

3. Medical history. Documents from the time suggest that the medical community could offer little in the way of treatment of those infected with scarlet fever during the time period under consideration. Indeed, many medical historians (e.g. McKeown and Record 1967; McKeown 1976; Bynum 1994) suggest

that the decline in scarlet fever occurring after 1880 was primarily related to improved preventive public health measures such as quarantining, improved nutrition, and antiseptic practices. To this, however, we can add one important caveat: if the children of well-off families are consistently under the care of physicians because the family can afford it, then those children may be vulnerable to the complications of treatment as illustrated in the historical example quoted below.

> Homeopathy gives belladonna as soon as any dryness or burning is noticed in the mouth and throat, and there is a desire to drink but no ability. Give mercurius in six hours after the second dose of belladonna; and six hours later, arsenicum, if there is great prostration and the ulcers emit an offensive odor. If arsenicum does not restore reaction, then give nux vomica. If inflammation, give aconite, followed by belladonna, if the pulse falls and fever abates. If the skin burns and there is drowsiness and stupor, give opium. If convulsions are present and are not relieved by opium, give zincum. If the eruption is intense, give sulphur.
>
> (Hall 1872: 451)

4. Epidemiology. The epidemiological literature does suggest that there are associations with adequate nutrition and other advantages that can be purchased by a higher socioeconomic status. Duncan *et al.* (1996),[7] looking at time series analyses of scarlet fever deaths in England and Wales, show a lag effect of deaths that they believe is associated with poor maternal nutrition and subsequent loss of their young children. Since resistance to the *Streptococcus pyogenes* bacteria would primarily be due to the immune response, with the presence of the proper antibodies, nutrition would play a reduced role. It could only serve either in weakening the mother's own ability to transmit those antibodies through placental transfer neonatally and nursing postpartum, or by preventing a malnourished child from mounting a sufficient immune response or becoming more susceptible to other opportunistic infections. In contrast, in our microcase study, as many as half of the victims we have surveyed thus far come from families that would not appear to be at risk of under- or malnutrition and whose families could have provided the best level of nutrition and care available at the time. Since streptococcal bacteria can be both respirated and ingested, contamination may be even more significant a problem in this population than issues of malnutrition.

Although we cannot claim that the microepidemiological approach illustrated above can resolve the answer, in light of the evidence presented above, the virulence argument remains to us the stronger residual explanation. It is also the conclusion of Szreter (1988) and Hardy (1993), after extensive review of the historical context in England. Given what is being discovered about virulent and antibiotic-resistant streptococci in contemporary populations, we have no reason to doubt that virulent strains evolved in the past. Medical historians and epidemiologists have certainly acknowledged this without providing much evidence to demonstrate its plausibility. Spontaneous and cyclical changes in disease patterns, often resulting in epidemics, have been noted historically not only in the case of scarlet fever, but also with measles, syphilis, and tuberculosis (Dubos 1980). Continuing research on this topic which bridges the work of epidemiologists, historians, and anthropologists, can only serve to expand the understanding of scarlet fever virulence and of epidemic disease in general.[8]

Acknowledgements

Alan Swedlund originally undertook the analysis of infectious disease mortality in the Connecticut Valley with Drs Richard Meindl and Alan McArdle, whose dissertations were based on this research and to whom he is grateful and indebted. We thank Susan Hautaniemi, Deborah Rotman, Douglas Anderton, and Ann Herring for assistance or comment on various aspects of this chapter. We acknowledge the European Studies Program of the Department of Anthropology, University of Massachusetts, and Dr Anthony Boyce, St John's College, Oxford University, for support of a research leave for Dr Swedlund in 2001 that enabled completion of the writing of this chapter. Final analyses of the Franklin County data, and the data on Northampton and Holyoke, Massachusetts, were provided by the Connecticut Valley Demography Project, which is currently supported by NSF Grant # 9910755.

Notes

1 Our understanding of events in the nineteenth century is frequently dependent on reports from the popular media and particularly local newspapers. This contemporary analogy reminds us that our understanding of disease events is strongly shaped by that same media today.

2 At least two strains of *Streptoccocus pyogenes* are recognized, M and T. The M and T antigens are cell wall surface proteins eliciting type-specific protective antibodies (Katz and Morens 1992: 298).

3 Hall's *Health at Home* went through many editions, was readily available to literate citizens of western Massachusetts, and was typical of such advice manuals of the day.

4 Although in the town of Shelburne the reverse appears to be true, with more boys dying than girls. It should be added that the results presented here are not affected

by differing age structures in the four towns, Franklin County, or the Commonwealth of Massachusetts. Censuses reveal very consistent age structures for the age groups at risk.

5 Biological anthropologists continue frequently to avoid putting names and faces on their subjects. When this is done to protect the anonymity of subjects who desire or require it, this is commendable. When it is done with the intent of giving the appearance of some level of greater detachment or objectivity, it is not. The project from which this chapter stems is, in part, about giving 'statistics' a face and an identity.

6 Cheapside was a large district at the north end of Deerfield along the confluence of the Connecticut, Deerfield, and Green Rivers. Cheapside was annexed to Greenfield in 1896, and Green River Cemetery is in the Greenfield portion. In addition to some farming areas within Cheapside, there was also railroad access, manufacturing, and commercial activity.

7 There are several other problems with this analysis; for good reviews see Luckin (2000) and Mooney (2000).

8 Although clearly beyond the purview of this chapter, we caution against a third hypothesis that is occasionally preferred by adaptationists, and that is that the human host populations have coevolved genetic resistance to the streptococcal pathogen, as well as others. As Levin and others have shown, the evolutionary odds are generally in favor of the rapidly reproducing pathogen at the expense of the slowly reproducing human host. General immune competence is strongly affected by environmental factors and genetic factors are thought to be present as well. However, since 'specific immune defenses are adaptive at the somatic level and, as such, reduce the intensity of (and need for) selection leading to germ-line evolution,' infectious disease is very limited as an effective agent of natural selection in human populations (Svanborg-Eden and Levin 1990: 43).

References

Bynum, W.F. (1994). *Science and the Practice of Medicine in the Nineteenth Century.* Cambridge: Cambridge University Press.

Centers for Disease Control and Prevention (1999). Summary of Notifiable Diseases, United States, 1998. *Morbidity and Mortality Weekly Report* **47**(53), 1–93.

Cimolai, N., Trombley, C., Adderley, R.J. and Treddwell, S.J. (1992). Invasive streptococcus pyogenes infections in children. *Canadian Journal of Public Health* **83**(3), 230–3.

Cleary, P.P., Kaplan, E.L., Handley, J.P., Wlazlo, A., Kim, M.H., Hauser, A.R. and Schievert, P.M. (1992). Clonal basis for resurgence of serious Streptococcus pyogenes disease in the 1980s. *Lancet* **339**, 518–21.

Colman, G., Tanna, A., Efstratiou, A. and Gaworzewska, E.T. (1993). The serotypes of Streptococcus pyogenes present in Britain during 1980–1990 and their association with disease. *Journal of Medical Microbiology* **39**(3), 165–78.

Dame, Lawrence, M.D. (no date). Biographical papers on the doctors of Franklin County Massachusetts. Pocumtuck Valley Memorial Association Library, Deerfield, MA.

Darmstadt, G.L. (1998). Scarlet fever and its relatives. *Contemporary Pediatrics* **15**(2), 44–63.

Dowling, H.F. (1977). *Fighting Infection: Conquests of the Twentieth Century.* Cambridge, MA: Harvard University Press.

Dubos, R. (1980). *Man Adapting.* New Haven, CT: Yale University Press.

Duncan, C.J., Duncan, S.R. and Scott, S. (1996). The dynamics of scarlet fever epidemics in England and Wales in the 19th century. *Epidemiology of Infectious Disease* **117**, 493–9.

Efstratiou, A. (2000). Group A streptococci in the 1990's. *Journal of Antimicrobial Chemotherapy* **45**(Suppl.), 3–12.

F.M.S. (no date). Records of the Franklin District Medical Society, 1851–1923. Transcripts of Dr. Lawrence Dame. Pocumtuck Valley Memorial Association Library, Deerfield, MA.

Guerrant, R.L., Lima, A.A. and Davidson, F. (2000). Micronutrients and infection: interactions and implications with enteric and other infections and future priorities. *Journal of Infectious Diseases* **182** (Suppl. 1), S134–8.

Gunzenhauser, J.D., Longfield, J.N., Brundage, J.F., Kaplan, E.L., Miller, R.N. and Brandt, C.A. (1995). Epidemic streptococcal disease among Army trainees, July 1989 through June 1991. *Journal of Infectious Disease* **172**(1), 124–31.

Hall, W.W. (1872). *Health at Home, or Hall's Family Doctor.* Hartford: S.M. Betts Company.

Hardy, A. (1993). *The Epidemic Streets: Infectious Disease and the Rise of Preventive Medicine, 1856–1900.* Oxford: Clarendon.

Katz, A.R. and Morens, D.M. (1992). Severe streptococcal infections in historical perspective. *Clinical Infectious Disease* **14**(1), 298–307.

King, S.E. and Ulijaszek, S.J. (1999). Invisible insults during growth and development: contemporary theories and past populations. In *Human Growth in the Past: Studies from Bones and Teeth*, ed. R.D. Hoppa and C.M. Fitzgerald, pp. 161–82. Cambridge: Cambridge University Press.

Luckin, B. (2000). Review of S. Scott and C. Duncan (eds) Human demography and disease. *Medical History* **44**(3), 426–48.

McGowan, S. and Miller, M. (1996). *Family & Landscape: Deerfield Homelots from 1671.* Deerfield, MA: Pocumtuck Valley Memorial Association.

McKeown, T. (1976). *The Modern Rise of Population.* New York: Academic Press.

McKeown, T. and Record, R.G. (1967). Reasons for the decline of mortality in England and Wales during the 19th century. *Population Studies* **16**, 94–122.

Mooney, G. (2000). Review of S. Scott and C. Duncan (eds) Human demography and disease. Cambridge, Cambridge University Press. *Progress in Human Geography* **24**(3), 515–16.

Musser, J.M., Nelson, K., Selander, R.K., Gerlach, D., Huang, J.C., Kapur, V. and Kajilal, S. (1993). Temporal variation in bacterial disease frequency: molecular population genetic analysis of scarlet fever epidemics in Ottawa and Eastern Germany. *Journal of Infectious Diseases* **167**, 759–62.

Spencer, R.C. (1995). Invasive streptococci. *European Journal of Clinical Microbiological Infectious Disease* **14** (Suppl.), S26–S32.

Stevens, D.L. (1992). Invasive group A streptococcus infections. *Clinical Infectious Disease* **14**(1), 2–11.

Stevens, E.L., Tanner, M.H., Winship, J., Swarts, R., Ries, K., Schlievert, P. and Kaplan, E. (1989). Severe Group A streptococcal infections associated with a toxic-shock-like syndrome and scarlet fever toxin A. *New England Journal of Medicine* **321**(1), 1–7.

Stollerman, G.H. (1988). Changing group A streptococci: the reappearance of strepto-coccal 'toxic shock.' *Archives of Internal Medicine* **148**, 1268–70.

Svanborg-Eden, C. and Levin, B.R. (1990). Infectious disease and natural selection in human populations: A critical examination. In *Disease in Populations in Transition*, ed. A.C. Swedlund and G.J. Armelagos, pp. 41–6. Westport, CT: Bergin and Garvey.

Swedlund, A.C. (1990). Infant mortality in Massachusetts and the United States in the nineteenth century. In *Disease in Populations in Transition*, ed. A. C. Swedlund and G. J. Armelagos, pp. 161–82. Westport, CT: Bergin and Garvey.

Swedlund, A.C., Temkin, H. and Meindl, R.S. (1976). Population studies in the Con-necticut Valley: Prospectus. *Journal of Human Evolution* **5**, 75–93.

Szreter, S. (1988). The importance of social intervention in Britain's mortality decline c. 1850–1914: a re-interpretation of the role of public health. *Social History of Medicine* **1**, 1–37.

9 The ecology of a health crisis: Gibraltar and the 1865 cholera epidemic

LAWRENCE A. SAWCHUK AND STACIE D.A. BURKE

Introduction

It was perhaps no exaggeration that nineteenth century Gibraltarians achieved the dubious distinction of being 'a population, filthy in themselves, and over-crowded, perhaps, beyond any community in the world' (Hennen 1830: 71). As was the case in many other urban centers of westernized countries, the inhabitants of Gibraltar faced insults from their poor environmental situation and, all too frequently, their daily exposure to infectious diseases. Not surprisingly, poor health and low survivorship were normal by-products of life for all but the very privileged in these urban deathtraps. For Gibraltar, the 1860s captures this dynamic and has been selected here to examine the forces and factors that affected health under a variety of environmentally influenced stressors.

Numerous factors operated to create unhealthy conditions in Gibraltar. Population growth was a particular problem as it had occurred at such an unprecedented rate within an infrastructure that was woefully inadequate to meet its demands. Housing was in limited supply and was further aggravated by the fact that landowners were more concerned with making large and quick profits than with providing proper accommodation for the poor working classes. Aside from the burdens of a growing local population, Gibraltar was also host to a number of itinerant laborers, immigrants, and visitors. As a strategically located free port, Gibraltar proved to be an important node in a trade network linking numerous urban centers throughout Europe, the Mediterranean, Africa and the New World. Commerce and an efficient transportation system facilitated the creation of a vast reservoir of potential hosts for the exchange of infectious diseases. The nineteenth century saw human populations becoming more mobile, covering long distances over relatively short periods of time: the global pattern of disease transmission was no longer in its infancy (see Curtin 1989; Dobson 1989; McNeil 1977). As a garrison town, Gibraltar was home to several thousand military families on tours of duty lasting for up to two years. The large-scale and sometimes rapid group relocations of military personnel

178

provided an ideal medium for the spread of infectious disease. Gibraltar's position as an important market town in the southern Iberian peninsula attracted the daily movement of thousands of Spanish foreigners to the Rock to sell their wares and provisions. Gibraltar was in a constant state of demographic flux and it was precisely this co-mingling of individuals from 'near and far' with an already overcrowded and densely packed population which presented the ideal conditions for the introduction and spread of both 'old and new' infectious diseases. While it is debatable whether nineteenth century hospital treatment at this time conferred any real therapeutic value (see, for example, Cherry 1980), Gibraltar's Colonial Hospital was at this time poorly administered with patients segregated not according to illness but, rather, religious denomination (Benady 1994). Owing to its unusual configuration of port city, naval base, garrison town and commercial center, residents of this British colony would ultimately sacrifice health for their participation in the global network.

This study also examines the impact of a major infectious cholera epidemic on a population that typically experienced elevated levels of normal background mortality. The cholera epidemic of 1865 punctuated Gibraltar's dark times with an episode of what demographers refer to as 'crisis mortality.' Defined as 'the sudden and dramatic increase in the death rate arising from a common, usual causal factor operating for a limited period of time' (Bouckaert 1989: 218), crisis mortality can directly and indirectly modify the demographic parameters of large populations (e.g. reproduction, marriage, and migration), as well as affecting both their genetic and cultural continuity (De Vries *et al.* 1979; Sawchuk 1996). This chapter focuses on the broad ecological factors that contributed to the heightened expression of mortality in the time of cholera among residents of the Rock.

Gibraltar as a research site for human biology in the archives

Often mistaken as an island settlement, Gibraltar is actually located on a peninsula of oblong form, what Aldrich and Connell (1998) have referred to in a political sense as a continental enclave (Fig. 9.1). Connected with the southern tip of Andalusian Spain by a flat sandy strip of land, Gibraltar is some 3.2 miles in length and a total of about 7 miles in circumference (Rose and Rosenbaum 1991). The entire territory covers approximately 1266 acres, or a mere 3.6 square miles. Historically, Gibraltar was distinguished for its strategic location at the western entrance of the Mediterranean, though the Rock had little to offer in the way of natural resources. Since the greater part of Gibraltar was ill-suited for either agriculture or pasturing, the local population was dependent on other countries, such as Spain and Morocco, for food and provisions. About

Fig. 9.1. Location of the territory and town of Gibraltar (after Gentleman's Magazine 1762).

nine-tenths of the territory available for housing was occupied by naval or military forces. Consequently, civilians constantly struggled to find affordable and decent housing. This physical setting and periodic sociopolitical isolation from Spain imposed constraints upon the spatial and demographic development of the Gibraltarian population.

Gibraltar's singularity and importance as a site for supporting historic community health studies stems from a number of unusual attributes that are seldom found in typical urban areas. Many of these features are directly linked to its status as a military garrison and important colonial outpost. First, Gibraltar was at best only capable of supporting light industry at a very modest level;

it can therefore be argued that the health of this urban community was primarily susceptible to high-density living and poor sanitary conditions (see, for example, Friedlander *et al.* 1985; Woods 1982; Wrigley 1969) and not to industrialization *per se.* Second, the community has always maintained clearly fixed geographical boundaries, which allow for comparisons of unchanging spatial units over time. Third, migration, arguably the most vexing problem in studies of urban centers, is less of a problem in Gibraltar since all immigrants were carefully monitored under a strict permit system and emigration was generally not preferred once locals were established on the Rock. Fourth, a high-quality registration system established in the nineteenth century provides the opportunity and ability to readily identify and track individuals, families, and groups over time. Fifth, the population of Gibraltar is sufficiently large and diverse to sustain detailed year-by-year analyses, yet small enough to be manageable without incurring the necessities of sampling strategies. Finally, extensive archival sources of colonial correspondence, newspapers, and military documents are available from both the Gibraltar Government Archives and the Public Record Office in London, England. Being able to contextualize nineteenth century Gibraltar through these sources means a potentially greater understanding of the ecological context within which health and illness were negotiated.

While the characteristics of a colonial outpost, port city, important commercial center, garrison town, and strategic location can be found elsewhere (for example, Malta, Hong Kong), the collective expression of these features in a location of limited and inhospitable terrain is singular in the case of Gibraltar. Although it is unlikely that we can make broad generalizations of the impact of disease on human populations from any single community, this study is useful in illustrating just how important local ecology and history are in the determination of the expression of disease in human communities. Further, there may be some merit in the Gibraltar study providing unusual grist for testing the reigning paradigms of epidemiological theory.

The Gibraltar archives

Our quantitative assessments of health in nineteenth century Gibraltar develop out of a number of sources, including the Gibraltar Police Death Registry, the Gibraltar Government Census of 1868, and the Colonial Blue Books. Collectively, these sources provide a reasonable reflection of vital events and the raw data necessary to establish measures of mortality. Richly detailed and labor-intensive record-keeping, one of Gibraltar's most remarkable attributes, is largely a by-product of colonialism and the strong mandate to monitor and safeguard the Crown's territories. In Gibraltar, this was particularly important given its significant military value and, as trade increased in scope, commercial importance. Gibraltar's extremely limited geographical space led to a particular concern over tracking the local populations' evolving demographic character.

Since the 1865 cholera epidemic predates the official Registration of Deaths Ordinance of 1868 and the subsequent appointment of W. Duffield as Registrar of Deaths effective the 1st of January, 1869, it was necessary to turn to alternate sources of death information. The Gibraltar Police maintained a register of deaths from September 30, 1859 to January 30, 1869. Before this time crude death counts were merely recorded in the Gibraltar Blue Books. Consequently, no nominal information can be extracted from government sources prior to 1859. The police records documented deaths on a daily basis according to name, age, sex, residence, birthplace, cause of death and attending physician. While death registration was not compulsory at this time, the police registry does afford the best opportunity to understand the cholera epidemic of 1865 both unto itself and within the context of Gibraltar's general background mortality of the 1860s.

In order to assess the mortality experience of the population during the 1860s, it is critical first to determine the demographics of the population which was potentially at risk of dying. An extensive set of nominal census rolls for the civilian population (e.g. 1777, 1791, 1834, 1868, 1871, 1881 and 1891) provides these often-elusive estimations. One of us (L.A.S.) has computerized the 1868 census of Gibraltar, thereby providing the basis for the tabulations necessary to ground the 1860–68 study within the context of the 'population at risk.'

As with any quantitative data from the nineteenth century, registers of vital events and census rolls are not immune to problems of under-reporting and other general reporting errors or bias. As a case in point, even at the time, there was no easy consensus regarding the total number of cases and deaths during the 1865 cholera epidemic. Dr Sutherland, sent by the colonial authorities to investigate the outbreak of cholera at Gibraltar, was keenly aware of such problems, comparing the Medical Officer of Health's figures with those of the police registry and others published in the local newspaper (Table 9.1). These differences led Sutherland to conclude that:

> the statistics of deaths from cholera among the civil population are only
> approximations to the truth. Three separate accounts of mortality, derived
> from three different sources . . . vary so materially from each other as to
> convince me that, small as the population is, the actual number of epidemic
> deaths in Gibraltar during the late cholera is unknown.
>
> (Sutherland 1867: 2)

There is no doubt that certain problems can stem from a reliance on a single source for deaths, particularly since the police registry appears to have

Table 9.1. *Total number of cholera cases and deaths for Gibraltar's civilian population in 1865*

Source	Cases	Deaths
Police Registry	Not recorded	421
Principal Medical Officer's Return	902	477
Gibraltar Chronicle	821	420

Source: modified from Sutherland (1867).

under-enumerated the total number of cholera deaths in 1865. It is the intention of this study, however, not to establish precise measures of mortality, but rather to gauge the general trends in mortality during this period of non-compulsory registration in Gibraltar.

Situating the 1865 epidemic in measures of population health

To more fully understand the impact of the 1865 cholera epidemic on Gibraltar's local population, it is necessary to first situate the crisis year within the context of prevailing conditions of the times. Three measures of population health are investigated here: early fetal mortality, infant mortality, and overall life expectancy.

One common measure of fetal mortality is the stillbirth rate, typically defined as the number of fetal deaths at or beyond 28 weeks of completed gestation per 1000 livebirths and stillbirths. The underlying causes of stillbirths are complex and often correlated with an array of biological (sex, age and parity of mother, birth spacing, father's age, multiple births, Rh-incompatibility), social (socioeconomic status, nutrition), and environmental factors (see Gruenwald 1969; Sawchuk *et al.* 1997; Sutherland 1949; Zusman and Ornoy 1990). Stillbirths are an indication that, even in the womb, fetuses can be susceptible to endogenous factors stemming from women's experiences and exposures during pregnancy.

Following parturition, infants first come into direct contact with their environments. The infant mortality rate (IMR or the number of deaths under one year of age per 1000 livebirths) has traditionally been used as a proxy measure for the prevailing social and sanitary state of affairs in a community. There is further evidence to suggest that infant mortality, rather than childhood mortality, is the most sensitive indicator of the effects of malnutrition (Scott *et al.* 1995). Variation in infant mortality rates can also be influenced by culturally rooted practices such as breastfeeding, age and season of weaning, and family planning.

For adults, the pace and relative impact of mortality can be traced through the lifecourse and is reflected in measures of life expectancy. Assessments of overall health are derived from abridged life table analyses (for a more complete discussion of fundamental aspects and assumptions of the life table, see, for example, Shryock *et al*. 1973). The life table is an empirical measure of mortality that takes into account deaths over a number of discrete age categories, for example under 1, 1–4, 5–9, 10–14, and so on. This study adopts the current life table approach. The current life table, sometimes referred to as a period life table, is based on a short period of time, typically a year or an intercensal period, in which mortality has remained stable. This type of life table structures mortality patterns for a cross section of the population under study, or what demographers refer to as a synthetic cohort. Mortality estimates for the current study are derived following the methodology of Chiang (1984). The 'population at risk of dying' was calculated from the 1868 nominal census, and the number of births for each year under study was drawn from the Blue Books for the Colony of Gibraltar.

To gain a better understanding into the manner in which a selection of diseases influenced the health of Gibraltarians in the 1860s, the multiple decrement life table was employed (see Preston 1976; Preston *et al*. 1972). Given the time period under investigation, groupings of reported causes of death focused heavily on infectious diseases, specifically weanling diarrhea, respiratory tuberculosis, 'other respiratory diseases,' scarlet fever, smallpox, measles, and, of course, cholera. With the exception of weanling diarrhea, the disease classification used here follows that established by Preston *et al*. (1972). Problems that emerge in this type of analysis include under-reporting or misclassification (e.g. of those diseases which implied some social disgrace such as syphilis), and vague (e.g. old age), or symptomatic (e.g. colic) causes of death. Nonetheless, when treated cautiously, analysis of cause-specific data for the nineteenth century can provide fresh insights into prevailing health conditions (see Alter and Carmichael 1996; Williams 1996).

The stillbirth rate

The magnitude of late fetal deaths in Gibraltar was very high during the 1860s. The police register provided monthly tabulations on the number of stillbirths and the Gibraltar Blue Books provided the necessary denominator. Between 1860 and 1868 there was an average 70.1 stillbirths per 1000 pregnancies in the population (Table 9.2).

Given the seasonal nature of the 1865 cholera epidemic, our attention focused on the months of August through November. Stress was particularly

Table 9.2. *Mortality estimates according to stillbirth rate, infant mortality rate, and life expectancy, Gibraltar 1860–1868*

Year	SBR	IMR	Life expectancy at birth	Life expectancy at age 5	Epidemic
1860	82.2	174.9	30.2	34.2	Cholera / smallpox
1861	72.2	176.1	36.1	41.9	—
1862	73.3	188.3	30.2	35.4	Scarlet fever
1863	57.4	179.7	34.4	40.1	—
1864	61.2	200.8	27.1	32.1	Measles
1865	82.6	187.0	18.9	21.7	Cholera / smallpox
1866	74.6	135.3	41.1	46.0	—
1867	59.3	200.5	32.2	38.2	Scarlet fever
1868	67.7	186.7	34.1	40.1	Measles
Average 1860–68	70.1	179.3	30.6	35.6	—

acute during these months as cholera was prevalent and a sanitary cordon was thrown around the community, effectively cutting all social and economic ties with neighbouring states. The 'normative period' in this analysis includes the years 1861–64 and 1866–68 when cholera was not present in Gibraltar. Owing to the epidemic, 1865 is isolated as the 'crisis period' with all of its associated stresses. The stillbirth rate during the 'crisis period' (August–November of 1865) was 100.5 per 1000 births compared with 64.9 per 1000 births during the 'normative period' (August–November of all years excluding 1865). The mean difference of 35.8 stillbirths per 1000 births is statistically significant ($Z = 2.271$, 30 d.f.). Rabbani and Greenough (1992) have found that women in their third trimester who develop cholera relative to those free from cholera have a 50% higher fetal mortality risk, though the reason for this increase in fetal mortality is largely unknown.

The infant mortality rate

Despite the high stillbirth rate, the majority of pregnancies culminated in a liveborn infant. The survival of these infants was challenged on a daily basis by the extremely hostile environment characterizing nineteenth century Gibraltar. Infant mortality rates are calculated as the ratio of the number of deaths under one year of age divided by the total number of births for that year and standardized by 1000. Over the study period IMRs were high, with an average of 179.3 deaths per 1000 livebirths (Table 9.2). Consistent with a population experiencing high mortality under a regime of infectious disease typical of the nineteenth century, there was yearly variation in infant mortality and, at times, the differences were considerable (see also Omran 1971, 1983). Throughout the 1860s, infant mortality rates varied from a low of 135.3 in 1866 to a high of 200.8 in 1864. The cholera epidemic of 1865 was not correlated with any noticeable increase in infant mortality.

Although these figures are derived from non-compulsory death registration, the observed IMRs appear to be reasonable and consistent. Gibraltar's estimations correspond closely with those reported for the neighbouring Spanish province of Andalucia for the 1860s (see Dopico 1987). The values observed for the 1860s are also in line with the average rate of 174.8 infant deaths per 1000 births calculated for the period 1870–79, a period of compulsory birth and death registration. Further analysis of the structure of infant mortality by age shows that there was no significant difference in the shape of the mortality curve for these two decades, attesting to the reliability of registration in the non-compulsory years. While important in itself, accurate reporting of infant births and deaths is also particularly critical to determining the quality of life table analyses.

Life expectancy

Overall survivorship, as measured by life expectancy, basically mirrored the pattern established by infant mortality. Life expectancy at birth ranged from a low of 18.9 years during the cholera epidemic of 1865 to a high of 41.1 years in the year directly following the epidemic (Table 9.2). This astute observation was not lost on Horatio Stokes, Gibraltar's Medical Officer of Health in the 1860s, as he explains the upswing in life expectancy:

> The cholera epidemic of 1865 having swept off a large number of the weak, the intemperate and those disposed to disease, there was therefore a more favourable condition of the population in the following year.
>
> (Stokes 1867: 47)

Overall, Gibraltar's estimates were compatible with international life expectancies of the time (Table 9.3). Outside of the 1865 epidemic year, Gibraltar's typically higher survivorship figures relative to those of Spain can, in part, be attributed to a higher standard of living enjoyed by most Gibraltarians, at least in terms of better wages and year-round employment opportunities.

An analysis of cause-specific mortality, summarized in Table 9.4, indicates that the overall pattern of mortality in Gibraltar was most strongly and consistently shaped by three endemic disease clusters (average years removed from life expectancy): weanling diarrhea (5.6 years), respiratory tuberculosis (2.6 years), and other respiratory infections (1.1 years). Of these and the most common

Table 9.3. *Life expectancy in Gibraltar and selected countries*

Location	Year	Males	Females
Gibraltar	1860–68		30.6 (combined)
England and Wales[a]	1871–80	41.4	44.6
Norway[a]	1856–65	47.4	50.0
United States[b]	1860	43.2	44.1
Spanish Provinces[c]			
Cadiz			29.9 (combined)
Malaga	1863–70		33.4 (combined)
Sevilla			31.8 (combined)

[a] *Source:* Lancaster (1990).
[b] *Source:* Haines (1994).
[c] *Source:* Dopico (1987).

Table 9.4. *Potential years gained if specific disease or disease cluster were eliminated*

Year	$e(x)$[a]	Weanling diarrhea	TB	Other respiratory infections	Smallpox	Measles	Scarlet fever	Cholera
1860	30.2	5.2	2.4	1.0	2.1	—	—	1.6
1861	36.1	7.4	3.9	1.0				
1862	30.2	5.2	2.8	1.0			4.5	
1863	34.4	5.6	3.8	1.7				
1864	27.1	6.7	1.9	1.0		4.9		
1865	18.9	4.2	1.0	0.5	2.5			11.9
1866	41.1	6.5	4.1	1.0				
1867	32.2	5.4	3.9	1.3			1.1	
1868	34.1	6.5	3.3	1.8		2.5		
Mean	30.6	5.6	2.6	1.1	0.8	0.9	0.9	2.3

[a] Life expectancy at birth.

epidemic diseases to periodically visit the Rock, however, cholera exerted the single greatest insult to life expectancy in 1865, responsible for deducting some 12 years from the average life expectancy in that year.

Pathogens of everyday life

Upon weaning and during the first fragile years of life, the most consistent and serious threat to life among young Gibraltarians during the 1860s was weanling

diarrhea. Weanling diarrhea represents a cluster of undifferentiated acute diarrheal diseases associated with early childhood (Sawchuk *et al.* 1985). Historically, in Gibraltar and elsewhere, deaths due to 'teething' or 'difficult dentition'
were commonplace and recognized as legitimate and certifiable causes of death
(see for example Neaderland 1952; Radbill 1965; Smith 1889). As a turn of the
century physician in Gibraltar remarked:

> the terms 'Diarrhoea' and 'Gastro-enteritis' are very loosely used and it is
> unsatisfactory to have to return all cases such as these under this heading as
> it is probably that in the majority of instances they are due to dentition or
> irregular feeding.
>
> (Elkington 1901: 8)

The diarrheal diseases resulted from exposure to microbes, either through the
consumption of contaminated food, milk, or water, or through handling and
bathing. Nutritionally compromised following diarrheal disease, even infants
who survived the infection were prone to other opportunistic infectious diseases
(see Lunn 1991). While weanling diarrhea was endemic, its relative impact
could vary on a yearly basis depending on prevailing climatic conditions and
the periodic outbreak of synergistic infectious diseases such as measles.

   Early hospital returns indicate that, in the absence of epidemics, respiratory
tuberculosis was a leading cause of illness in Gibraltar and, ultimately, the
primary cause of death during adulthood. Endemic to the Rock, tuberculosis
was greeted with a fatalistic attitude and the hospital became the receiving
ground for those nearing death:

> We have to still regret the large proportion of Consumption of the Lungs,
> which are brought to the Hospital in the latter stages of this malady. You are
> aware, that the Spaniards and other Foreigners believe this disorder to be
> infectious, and generally refuse to receive into their dwellings persons so
> afflicted; and the friends and relatives urge their admission into the Hospital
> with a view of relieving themselves from the burden and expense attendant
> on the last stages of an incurable malady.
>
> (*Gibraltar Chronicle* 1822)

The relative impact of respiratory tuberculosis, as judged by the cause-deleted
method of life expectancy analysis, falls in line with that reported for England and Wales (Table 9.5). Age and sex were important mitigating factors
in the expression of tuberculosis (TB) mortality in Gibraltar. Among women,
there was a classic pattern of TB mortality with the probability of death rising sharply through the teenage years and reaching a peak at 20–24 years of
age. A slightly higher rate was observed for males with a notable difference
of peak mortality occurring later in life between 40 and 44 years of age. This
pattern of TB mortality is suggestive of a community with a high annual rate

Table 9.5. *Potential years gained if*
respiratory tuberculosis is eliminated:
1860–64, 1866–68

Location	Males	Females
Gibraltar (1860–68)	2.3	2.4
England and Wales[a]		
1861	3.4	4.0
1871	2.1	2.3

[a] *Source:* Preston *et al.* (1972).

of infection, where virtually everyone has been infected with the bacilli by
twenty years of age. The rise in tuberculosis mortality in the older age groups
has been interpreted by some as indicating the ebb of an epidemic with a long
periodicity, spread through increasing exposure among segments of the population and ultimately diminished by genetic selection (Grigg 1958). Other
researchers have suggested that the decline in tuberculosis mortality was related to better nutrition, improvements in occupational health, better housing,
and public health interventions (specifically the introduction of the sanato-
rium and pasteurization) (see, for example, Fairchild and Oppenheimer 1998;
Hardy 1993; McFarlane 1989; McKeown 1976, 1979; Szreter 1988; Woods and
Woodward 1984). Given our study period (1860–69) and Gibraltar's social and
medical history, it is unlikely that the non-genetic factors cited above constituted a major factor in the expression of the age-specific pattern of tuberculosis
mortality.

The 'other' infectious respiratory disease cluster, including pneumonia,
bronchitis, and inflammation of the lungs, displayed low levels of year-to-
year variation, a finding to be expected in a temperate Mediterranean climate.
Though this disease cluster proved to be a year-round health hazard, close to
half of the deaths in this category occurred in the months of January through
April. With no central heating and with damp living quarters, such an association is not unexpected. Although one-quarter of these deaths were concentrated
in the first five years of life, this group of infectious respiratory diseases posed
an underlying threat to all age groups in the community.

The trials and tribulations of daily life

Many features of the sanitary and housing infrastructure of the town and garrison
supported the existence of the pathogens of everyday life. The impact of

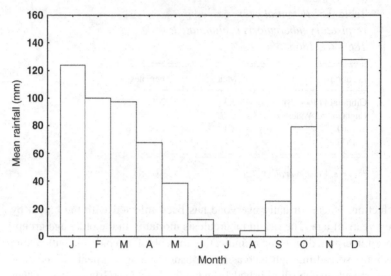

Fig. 9.2. Average monthly rainfall in Gibraltar for the period 1850 to 1995.

weanling diarrhea was largely exacerbated by a long-standing water supply problem typical of Gibraltar's summer and fall months. Historically, rainfall was a topic of great concern in the Spanish Mediterranean region because of its scarcity, typically threatening long periods of drought (Romero *et al.* 1998). Aside from the unanticipated shortages which occurred from year to year, Gibraltarians did expect the normal annual summer drought (see Fig. 9.2). While Gibraltarians developed adaptive strategies of water rationing and communal sharing to cope with the yearly summer drought, each year was met with great anxiety as to whether there would be a sufficient supply of this first necessity of life (Sawchuk 1996). Intensifying the stress arising from this seasonal shortage of water was the inevitable rise in temperature and humidity during the summer months (see Fig. 9.3).

   With little government financial support and for the most part left to their own ingenuity, the civilian community gradually developed its own decentralized water supply system. The oldest and simplest means of obtaining water was by catching or 'harvesting' rainfall from roof tops and terraces and storing the water in large underground tanks or cisterns. Contamination of this water supply was a constant problem, particularly because of the inability to keep the harvesting sites clean. Organic pollution of water was often risked by the excreta of birds and cats, typical rooftop dwellers, and by the accumulation of dust and dirt. Further problems could arise through the leakage of surrounding polluted water into poorly constructed cisterns or even contamination during the

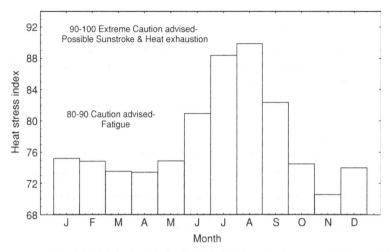

Fig. 9.3. Monthly heat stress index values in Gibraltar for the period 1850 to 1995 (concept from Matzarakis and Mayer 1997).

process of drawing the stored water. As Gibraltar's Medical Officer of Health observed,

> But whatever may be the impurities arising from the collecting area, they are nothing, when compared with those that are liable to occur from the position of the tank itself. Placed, as I have constantly seen it, with unprotected covering, under dirty bedrooms, greasy kitchens, filthy storerooms, patios, shops and public passages and often surrounded on all sides by defective drains, placed in such a position, it must be considered one of the gravest sanitary defects here, and one of the chief channels, by which enteric fever, cholera and many less apparent and milder diseases are propagated.
>
> (MacPherson 1890: 21)

The potential for the contamination of Gibraltar's potable water supply during the nineteenth century remained a constant threat to health.

While Gibraltar's residents could obtain water from local wells, its brackish and highly contaminated nature generally discouraged its consumption. Dubbed 'sanitary water,' well water was typically relegated for cleaning and other non-consumptive purposes. The purity of well water was constantly threatened by the disposal of sewage and refuse into earthen pits dug in close vicinity to where wells were situated. The simple construction of patio wells yielded ineffective barriers to ground water which filtered through polluted soil and then added to the wells' reserves. Gibraltar's ground water contamination was enhanced by the porosity of the Rock's limestone foundations, the pattern of high-density residential living, and the practice of keeping animals and poultry in close proximity to living quarters. Contamination of well water with salts of lime, magnesium and organic matter became progressively more problematic as hot dry summer days wore on (Cox 1877). Medical authorities were only too aware of the potential sources of well and ground water contamination on the North Front of the Rock:

a. The cemetery of a population of about 24,000 added to . . . at the rate of 600 bodies a year.
b. The kennels of the garrison fox hounds.
c. The wash houses for washing hospital bedding and clothing.

All the above were located within a distance of 500 yards from the wells.

d. The cattle sheds and slaughter-houses for the meat supply of the town and garrison are within 1,100 yards.
e. Highly manured market garden; distance 200 yards.
f. Wooden barracks for 400 men; distance 550 yards.
g. The rapidly growing and wholly undrained town of Linea in Spanish territory; distance 1,200 yards.

(Cox 1877)

Similar problems existed in the water supply stemming from the nearby Spanish town of La Linea. A report on samples drawn from four wells located on the Spanish Lines, which provided water for sale in Gibraltar, revealed the following alarming results:

No. 1. A well near San Pedro Alcantara shows considerable pollution. It contains an amount of organic matter – Chlorine – highly dangerous to public health.
No. 2. This came from a well near the cemetery. It is still more polluted than No. 1. In addition to excess of organic matter it contains very numerous living animals, also pieces of hair, woolen fibers, particles of epithelium from the human skin indicating pollution from household refuse.
No. 3. From Tarifeno Well. This is very fair water. It shows no sign of pollution. It contains fine sand. Small pieces of decayed vegetable matter apparently the result of carelessness in the collection on transit. With ordinary care with filtration it might be used with safety.
No. 4. From a well near the "Bull Ring". This is the most polluted of all the specimens. It contains ten (10) times more organic matter than ought to be in good water. It contains a large quantity of matter in suspension, hair, bits of wool dried, bits of mucus – indicating great pollution from excrete of men or animals.

(Collins 1885)

Unsterilized barrels used for the distribution of water from house to house only aggravated the problem of contamination since there was no sanitary control over the condition of the containers or the handlers of these vessels.

These contaminant-prone water supply systems posed a serious health risk to Gibraltar's infants. Despite risks of contamination, however, water consumption was unavoidable. In 1866 Gibraltar's Medical Officer of Health wrote that 'most of the water used by the inhabitants for drinking, whether derived from wells or house cisterns, is so highly charged with impurities as to be unwholesome' (Stokes 1867: 48). The hazards of impure water were not confined to weaning or artificially fed infants, as all babies would come into contact with water over the course of feeding, handling, washing, and bathing.

Gibraltar's milk supply was also more or less devoid of sanitary control and therefore constituted a serious health risk to young children, particularly infants in the process of being weaned. Most of the of milk sold in Gibraltar came from southern Spain, where both the milking operations and the storage and transportation of milk fell outside of the control of local authorities. Bacteriological contamination and the adulteration of milk (with, for example, impure water) was commonplace in many European countries (Atkins 1992; Beaver 1973; Dwork 1987; Moring 1998). A deficiency in legislating or enforcing any food-related bylaws further increased the exposure of young and old alike to the pathogens of everyday life. It was here that weaning foods could pose certain risks to susceptible infants. In his correspondences, the Medical Officer of Health notes

> With reference to articles of food, I regret I cannot report so satisfactorily. Owing to a reluctance on the part of the public to give information, which may affect the interests of retailers, much unwholesome food is imported and consumed. Complaints are made known when it is too late to remedy the evil. The filthy and disgraceful state of many of the licensed bakeries has been brought to your notice ... The public sale of condemned pork, damaged fruits, putrid and rotten fish, unwholesome sausage, mildewed cheese, bad flour, &c., &c., has been checked and prohibited, ... [but] there is reason to believe that many of the condemned articles were afterwards brought into the City for consumption in a clandestine manner.
>
> (Stokes 1867: 48)

The problem of unwholesome food was particularly acute during the hot summer months when the population consumed increasing amounts of fruit and vegetables owing to their low cost and great abundance. The common practice of using human and animal waste as fertilizer often meant that green vegetables and fruits were grown in close association with untreated fecal matter.

While the complex relationship between inadequate nourishment and disease remains imperfectly understood, it is safe to say that 'inadequate nutrition increases susceptibility to some infective agents and increases the severity of some infections once contracted, especially in children' (De Bevoise 1995: 57). Relative to the better-off Gibraltarians, the poor were at a disadvantage nutritionally. Lower purchasing power, less free time and the inability to travel greater distances meant that the poor were restricted to purchasing their food items from local shops or peddlers. The high price of imported charcoal frequently meant that, for the poor, food went improperly cooked. This, coupled with poor storage and the lack of refrigeration, particularly in the hot summer months, added an additional risk of consuming contaminated food.

Given the diverse background of its peoples, nineteenth century dietary customs in Gibraltar undoubtedly varied according to place of origin, length of residence, religious affiliation and wealth. One account from Dr Kelaart (1846) describes the diet of the poor Gibraltarian as consisting 'chiefly on fish and vegetables which are to be had, in great abundance, both cheap and good.' The vegetarian diet of the working poor in Gibraltar was confirmed in later quantitative studies (Government of Gibraltar 1906, 1939) that also pointed out that the diet, while adequate in caloric value, was deficient in protein. The single most important food item was bread, making up 31% of total caloric intake. In contrast, meat and fish collectively made up only 7.8% of the total intake. Meat was only consumed in very small quantities twice weekly. It was regarded as a luxury and only the cheapest cuts were eaten owing to the cost. Fish, typically very inferior varieties such as sardines and shellfish, were fried in oil with onions and garlic. Cooking oils, usually of an inferior class, contributed 685 calories, or 21.4% of the total caloric intake. Eggs, milk and cheese were viewed as luxury items and seldom eaten.

If procuring wholesome and sufficient food was a frequent problem for the poor working classes, finding suitable affordable accommodation on the Rock was a constant dilemma. Person-to-person transmission of infectious disease was greatly facilitated at the household level, where families were packed into cramped, dirty, dark, and ill-ventilated one-room apartments. In Gibraltar, communal living in patio-style buildings was the norm; 'patios' are low-rise apartment buildings built around an open communal courtyard. Approximately one-third of the population occupied buildings with more than ten households, totaling an average of 60 residents per unit. Few families owned their homes and few resided in single family dwellings. Housing for the poor was of a universally bad description and wholly inadequate in terms of the most basic necessities.

> The ground for building being very dear, and house-rent excessively high, cellars and stables have been converted, in many places into dwelling-houses, to receive numerous families, without any regard to their health or accommodation; and in other places, a great number of sheds have been constructed, in such a manner to preclude access to ventilation, affording, besides, materials for putrefaction, by the decayed state in which they are frequently left. To overcome the excessive price of house-rent, the poor labouring classes of people have been compelled to crowd themselves

> in the same apartments, where it is not unusual to see three rows of beds one
> above another, while some are living on the floor, which is generally badly
> paved, and always damp.
>
> (Hennen 1830: 71)

For the very poor, the bare dirt or brick floor served as the sleeping ground with
little or no furniture or other comforts.

Deficient waste and sewage disposal was of particular concern to colonial and
military administrators in nineteenth century Gibraltar since filth and noxious
odors were not only offensive, but they were also perceived by some as a
common cause of illness, particularly among those upholding the miasmatic
model of disease transmission (see Sawchuk and Burke 1998). Conditions in
the communal courtyard were often no better than what was available inside
the apartment.

> Some of the areas are crowded with water-butts, old mats, oil-jars, and
> lumber of all descriptions, affording a nest for filth, and a fruitful source of
> putrescent exhalations, independent of their seriously diminishing the cubic
> mass of air, the circulation of which is still further obstructed, by lines and
> poles crossing the areas for the purpose of drying linen.
>
> (Hennen 1830: 71)

In the absence of a proper scavenging system and a lack of perceived responsibility over public patio space, the condition of the courtyard offered little relief
from the overcrowded unsanitary living conditions within the apartment itself.
Privies or necessaries were either absent or, at best, inadequate for the needs of
the tenants. The custom of keeping working animals (horses, mules and goats)
and poultry in close proximity to living quarters added further insult to the
sanitary condition of Gibraltar's housing.

Gibraltar's sewage system, laid down in 1815, was of a primitive construction, its main accomplishment being to simply direct untreated human waste
into the waters of Gibraltar Bay at Line Wall. In his assessment of the sewage
system in 1866, Sutherland remarked:

> The sewers are of very large dimensions, generally they have flat bottoms,
> their construction is of the roughest character, and sometimes no mortar is
> used for the purpose; there has been no attempt at uniformity at all; some
> lines are laid almost horizontally, while in others the fall is excessive;
> sometimes the cover of the sewer is open and allows foul gases to escape; no
> means of flushing were provided; and some of the sewers were found loaded
> with deposit; except where the house drains were choked up and retained the
> foul matter in them, so as to keep it out of the sewers.
>
> (Sutherland 1867: 19)

Even in their own homes, Gibraltarians could not escape these noxious odors:

> In most instances, the patios are entered by a small door, shut at night, and as
> the patios are rarely four or five yards square, and the other ventilation of the
> rooms is from the patios, the entire drainage system of the town appears
> almost to have been especially contrived to supply the houses with
> sewer air.
>
> (Sutherland 1867: 19)

In sum, house drains were characterized as 'nothing but cesspits opening into
the patios by gratings, privies or sinks' (Sutherland 1867: 19).

Even after the turn of the century, Dr Horrocks, Gibraltar's Medical Officer
of Health, was critical of the quality of the local home environment, particularly
for its impact on tuberculosis illness:

> It is now generally admitted that overcrowding in close damp rooms is a
> most fertile source of pulmonary consumption. In the upper [and poorer]
> parts of town, and in many houses where there are tenants in common, very
> frequently a whole family of ten persons may be found sleeping in one close
> damp room, with windows and door closed and an oil or petroleum lamp
> burning on the table. There can be no doubt that children growing up under
> such unsanitary conditions must have weak lungs.
>
> (Horrocks 1904)

Despite their awareness of the intimate relationship between illness and poor
living conditions, medical authorities were in no position to impose reforms in
a colony where housing was already in incredibly short supply.

While overcrowding undoubtedly contributed to the spread of TB and the
other air-borne diseases, the local custom of sick visiting also contributed to
the ultimate spread of everyday pathogens. A foreigner describes this custom
of calling on the sick in turn-of-the-century Gibraltar:

> On entering one of these Patios [apartment buildings] you will probably find
> a medley of people squatting in the courtyard round an itinerant vendor of
> goods, all chattering away at the tops of their voices... Ascend the narrow
> wooden steps and you may pass a room in which lay a sick person who will
> be attended and half-suffocated by a large number of men, women, and
> children who are crouching round the bed with sympathetic motives... The
> room, moreover, is clothed in filth from floor to ceiling, and the atmosphere
> vitiated; in fact, the general condition (as I have often witnessed) and state of
> overcrowding in these Patios inhabited by the poorer classes, is certainly
> neither wholesome nor a credit to civilization and sanitary reform.
>
> (Thomsett 1902: 156–8)

Ephemeral visitors: pathogens on the move

There were several pathogens which were only capable of establishing periodic epidemics in Gibraltar. Their more transitory appearances were the by-product of two main features: Gibraltar's small population size (and, therefore, with acquired host-resistance, only a small number of susceptibles at any given time) and the generally self-limiting properties of the pathogens themselves (e.g. susceptibility to climate change). In addition to the epidemics of cholera in 1860 (less severe) and 1865 (more severe), the 1860s also saw numerous epidemics of smallpox, measles, and scarlet fever.

Smallpox

Despite a long-standing awareness of the protective benefits of vaccination among Gibraltar's medical community, deaths due to smallpox remained commonplace during the 1860s. Each of the two smallpox epidemics coincided with outbreaks of cholera, for the first time in 1860 and then in 1865. The 1865 smallpox outbreak was the most serious, killing some 56 individuals. Statistics compiled through a house-to-house survey by the Police Magistrate indicated that in at least two of the districts in the upper and poorer part of the town, less than half (41%) of children under 16 years of age were vaccinated against smallpox (Flood 1866). In addition to the fact that vaccination was not compulsory at this time, there was also the issue of concealment. In Gibraltar, in order to secure a vaccination from a physician, it was necessary to first take out a Vaccination Order at the Police Office. Since a number of parents and guardians were living in Gibraltar illegally, their children were not vaccinated in order to maintain the family's anonymity. Removal from the garrison was always a risk for the foreign-born poor who depended on their employment in Gibraltar to support their families. As a result, these families may have preferred to 'take a chance' with the pox, rather than face possible expulsion. According to Scott and Duncan (1998), among unvaccinated individuals, case fatality rates due to smallpox have been estimated at between 15% and 25%, though rising to 40–50% among those who are very young or very old.

Measles

Unlike the other infectious diseases considered here, measles is unique as an 'acute community infection' and as such particularly sensitive to population size. Black (1997) has shown that measles cannot remain endemic in any population with fewer than 5000 to 10 000 births per year. As a result, Gibraltar's small population was incapable of supporting measles in a year-round fashion and the disease would typically strike every four to five years, often in the winter season (MacPherson 1895). Case fatality would vary according to the virulence of the virus, the dose at exposure, synergism between infections, sex, malnutrition, age at infection, and genetic susceptibility (see for example, Aaby 1992; Black 1997; Black *et al.* 1977; Morley 1973; Reves 1985). Epidemiological evidence from developing countries suggests that the transmission of measles is also greatly facilitated by crowding of susceptible individuals living together (Aaby 1992). High concentrations of the population living together in patio buildings facilitated the spread of the measles virus, though the small population denied the disease a permanent residence or endemicity within the community of Gibraltar. The spread of measles was also facilitated through the local custom of attempting to intentionally infect children,

> There is little doubt that the wide spread distribution of the disease here is to some extent due to ignorant and almost criminal custom of purposely exposing children to the infection, in the hope that they may get it and have done with it. Perhaps, if parents realized what a fatal disease it has proven to be in Gibraltar in the past, if they realized that it is also a frequent cause of permanent disability and chronic suffering, they would be more ready to assist in preventing its spread in the future.
>
> (MacPherson 1892: 11)

Scarlet fever

The 1860s were also notable for two epidemics of scarlet fever – a major occurrence in 1862 and a slight bout in 1867–68. The 1862 epidemic ravaged Gibraltar's youth throughout the winter, spring, and summer months. Owing to its concentration during childhood, deaths due to scarlet fever made a substantial impact on overall survivorship in that year. The life table analysis reveals that approximately 4.5 years would have been gained through the elimination of this cause of death (Table 9.4). From the perspective of the 1860s, scarlet fever ranked high in impact, alongside measles and smallpox, as a major childhood killer. This disease, however, began to wane in importance in Gibraltar as early as the 1870s, as it did in other westernized countries in the latter half of the nineteenth century (Lancaster 1990; Scott and Duncan 1998; Wohl 1983).

The dreaded cholera

Throughout the nineteenth century, cholera was one of most feared scourges in the western world, spreading both misery and death in its wake (see, for

Table 9.6. *Nineteenth century cholera pandemics in Gibraltar*

Cholera pandemic	Pandemic years	Gibraltar outbreak	Cholera deaths in Gibraltar			
			Civilians	Military	Convicts	Total
1st	1817–1824	No	—	—	—	—
2nd	1829–1837	1834	252	162	N/P[a]	414
3rd	1840–1860	1855	—	—	—	a few
		1860	36	41	13	90
4th	1863–1875	1865	421	98	57	576
5th	1881–1896	1885	24	2	N/P[a]	26

[a] There were no convicts present in Gibraltar in this time period.

example, Barua 1992; Evans 1987; Longmate 1966; McGrew 1965; Patterson 1994; Pollitzer 1959; Rabbani and Greenough 1992; Raufman 1998; Snowden 1995). Epidemics of cholera reached Gibraltar five times during the nineteenth century, and with enough ferocity to fulfill the criteria of crisis mortality in both 1834 and 1865 (see Table 9.6) (Sawchuk 2001).

Capable of killing a person within hours through rapid dehydration caused by violent vomiting and diarrhea, cholera has attained the distinction of being one of the most fatal infectious diseases in history (Watts 1997). Caused by *Vibrio cholerae*, cholera was not endemic to Gibraltar but was instead a disease of importation. With masses of people entering Gibraltar on a daily basis, either by land or by water, it would have been impossible to filter out all of those who carried the potential threat of epidemic disease. Cholera carriers, or those with subclinical infection, could escape even the most trained medical eye. The carrier state for classic Asiatic cholera is believed to last from three to six days, although convalescent carriers may shed vibrios for up to two or three weeks. Cholera transmission is further facilitated by the fact that vibrios can survive in water for extended periods of time. Smallman-Raynor and Cliff (1998) estimate that vibrios are capable of persisting for upwards of three weeks in well water, given the right local conditions.

Recurrent infections of cholera in an individual are rare since natural infection typically confers effective and enduring immunity to the disease. Although the course of the disease is now generally understood, the problem still remains to explain why some individuals suffer only mild discomfort while others succumb to cholera. Known factors that increase host susceptibility include reduced stomach acidity, the immunological burdens of other coexistent gastrointestinal diseases, and undernourishment (Richardson 1994). While research does not support the common belief that chronic malnourishment can increase the risk of contracting cholera, it can increase the risk of complications after infection, especially among children (see, for example, Glass and Black 1992; Rabbani and Greenough 1992). Historically, cholera was associated with destitution and squalor, overcrowding, the recycling of infected clothing, defective sewage treatment, unwashed hands, contaminated produce, and impure water (Snowden 1995).

Cholera entered Gibraltar in the summer of 1865 and within weeks claimed nearly six hundred lives. Despite the insidious threat of this disease, there was no widespread panic, no anarchy, and no riots in the streets, for cholera was no stranger to Gibraltarians, who had endured earlier epidemics in 1834, 1849, 1854, and 1860. The first Gibraltarian case of this fourth pandemic of cholera was identified on August 19, 1865. In this particular epidemic, it is likely that cholera entered Gibraltar with the 22nd Regiment restationing from Malta. Previous cases of sporadic cholera had been reported in the Gibraltar Chronicle earlier in the year and were met with considerable dismay by the local merchants. The following memorial was sent to the Colonial Secretary:

> That the Inhabitants of this City generally and the Mercantile portion of the community in particular view with grave apprehension and alarm the very serious injury which has been inflicted upon trade by the notices recently published in the *Gibraltar Chronicle* by the Board of Health and also by reason of the Captain of the Port having on several occasions as herein after forth, issued foul bills of Health to vessels leaving Gibraltar.
> (Exchange Committee and Chamber of Commerce 1865)

Although local medical authorities pronounced that Gibraltar was free of cholera on November 1, connections remained severed with Spain until November 23, 1865. The inhabitants of the Rock suffered 13 long weeks of social and economic isolation under the threat of cholera.

Cholera and the decentralized effect

The distribution of cholera deaths over the space of the 1865 epidemic differs from the pattern seen in other European centers where there was typically an abrupt rise in the mortality rate, a peak, and then a rapid decay in the death rate. This type of patterning in a cholera epidemic is often explained by highly centralized water supply systems. The pattern was somewhat different in Gibraltar, where the distribution of deaths rose and decayed more gradually (see Fig. 9.4). For once, Gibraltar's 'primitive' decentralized water supply system was advantageous since, under these circumstances, it was understood that 'the pollution of one or two, say with the specific germ of enteric fever, is not likely to set up an extensively distributed epidemic, such as has been recorded frequently in cities, whose inhabitants are supplied with drinking water from a common source' (MacPherson 1890: 21).

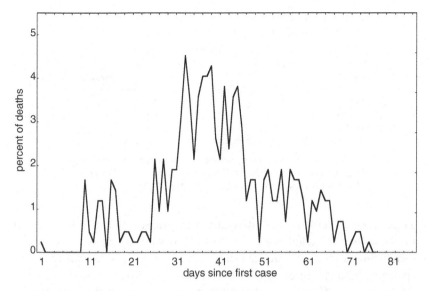

Fig. 9.4. Daily distribution of cholera deaths during the 1865 epidemic in Gibraltar.

Cholera mortality by age

In the absence of a centralized water supply, the mode of transmission in Gibraltar that eventually led to the epidemic was influenced more by the high degree of interpersonal contact which was common in this small and crowded community. Yet not all individuals in Gibraltar were at equal risk of contracting the disease. For example, the cholera death rate by age showed a U shape distribution with individuals over the age of 49 suffering the highest death rate due to cholera at 45.0 per 1000 at risk. At the other end of the age spectrum, those under five years of age suffered cholera death rates of 34.8 per 1000 at risk. Older children and young adolescents exhibited the lowest mortality burdens with approximately 13 deaths per 1000 at risk. Studies done by Morris (1976) on the British experience with the 1832 cholera epidemic displayed a similar pattern, regardless of class.

Cholera mortality at the community level

The toll of lives taken during the 1865 cholera pandemic were unequally distributed across Gibraltar's various communities (see Table 9.7). The men of Gibraltar's Convict Station experienced the highest death rate in the community. Overall, the adjusted mortality rate for cholera alone stood at 18.6 per 1000,

Table 9.7. *Measures of crisis mortality due to cholera among the residents of Gibraltar*

Group	Cholera death rate per 1000 living[a]	Hollingsworth index[b]
Military	15.6	53.4
Civilians	23.3	69.1
Convicts	63.5	310.9

[a] Number of deaths, estimated population size, duration in weeks: military (98, 5000, 14); civilians (421, 18 000, 12); and convicts (57, 750, 7).
[b] Modified Hollingsworth index (Rodriquez Ocaña 1992).

a value comparable to the total adjusted mortality rate of 18.0 per 1000 observed for the normative period between 1861 to 1863 when no major epidemic occurred. Simple mortality rates cannot adequately capture the full impact of this epidemic, however, since they are unable to factor in the effects of the duration of the crisis. One empirical measure that can accomplish this objective is a modified Hollingsworth index (H), where $H = [(\text{deaths}/(\text{population}-\text{deaths})) \times (\text{duration of epidemic measured in weeks})^2] \times 1000$ (Rodriquez Ocaña 1992). The extraordinarily high H index (310.9) among the convicts is due not only to a high number of deaths, but also to a relatively rapid transmission of the disease facilitated through the overcrowded and highly communal prison conditions.

In stark contrast, the military community fared considerably better than either the convicts or the civilians. This finding may be attributed to a number of factors. First, the military had a more favorable age distribution with far fewer older individuals who were, at least in the civilian population, the group that exhibited the highest mortality rate. Second, the military population was probably better off nutritionally than the working-class Gibraltarian because of reliable ration-based food supplies. Finally, while it is unlikely that even the military segment of Gibraltar had access to a significantly better **quality** of water, there is evidence to suggest the **quantity** of water available was higher. Owing to the known scarcity of potable water, the military population stationed at Gibraltar was placed at all times on a water allowance just as if the garrison was in a state of siege. Water was rationed according to rank, age, and sex (the females being the wives and daughters of military men).

Cholera mortality at the district level

As the number of cholera deaths began to accumulate it soon became apparent that there was also considerable variation in rates of attack and recovery from one patio to the next. While the capricious nature of the spread of cholera remained perplexing to authorities and the public alike, it was generally understood by the 1860s that poverty put some Gibraltarians at greater risk than others. There is considerable evidence today to suggest that poverty acts negatively on health (see, for example, Antonovsky 1967; Robinson and Pinch 1987), even with respect to infectious diseases. In Gibraltar, the pattern of cholera was clear: residents in the upper portions of the Rock suffered dramatically higher death rates than those located in the lower part of the town. Whereas some districts in the upper Rock experienced mortality rates as high as 59.1 cholera deaths per 1000 at risk, other districts in the central town area escaped with a conspicuous absence of deaths.

From the perspective of local authorities, the association between health, poverty and cholera was perfectly obvious:

> The sudden cessation of labour and loss of wages to the poor caused by the imposition of the [sanitary] cordon, was no doubt a predisposing cause of the disease, but it is impossible to say whether there may not also have been a more lax observance of sanitary requirements in the dwellings and habits of the poor, which may have tended to make the present epidemic more fatal than its predecessors.
>
> (*Gibraltar Chronicle* 1865)

While such observations may have held a natural appeal among the members of the upper classes, cholera mortality differentials were not simply the result of a lack of wealth or lax hygiene but, rather, a constellation of factors that were experienced together in a 'poverty complex.' An inadequate and polluted water supply, poor housing, deficient sewage and scavenging systems, an improper and insufficient diet, and poor education were all facets of this complex. Collectively, this set of interrelated features increased susceptibility to infectious disease, including cholera.

The methods of ecological regression (see, for example, Buffler *et al.* 1988) were adopted to quantitatively assess the impact of some of these 'poverty complex' factors on the observed pattern of cholera mortality. The focus of the analysis is the comparative aggregate experience of Gibraltar's 27 administrative districts and the unit of investigation is the collective mortality experience of all members in a given district, not the mortality experience of individuals. This ecological approach aids in structuring the district-level context in which the cholera epidemic was played out. In the statistical analysis, the objective was to predict variation in district-level cholera mortality rates (the dependent variable) through a number of independent predictor variables. The independent variables focused on a selection of epidemiologically important district-level indices, as follows.

1. percentage of houses with a well
2. percentage of houses with a cistern
3. square yards per person (density measure)
4. percentage of patios with 10 or more families (crowding measure)
5. diarrhea mortality

Before proceeding to the results, it is important to acknowledge that any study modeling through ecological regression must consider the phenomenon of the 'ecological fallacy' (for a comprehensive discussion, see Huston 1991; Schwartz 1994; Susser 1994; Tranmer and Steel 1998). According to Tranmer and Steel (1998: 817), 'the ecological fallacy arises when the results of an analysis based on area-level aggregate statistics are incorrectly assumed to apply at the individual level . . . Area-level and individual-level statistics are often quite different because the area-level statistics are subject to aggregation effects.' The variables used in the present study conform to what Susser (1994: 826) has referred to as 'contextual variables,' that is the 'median, mean, or proportion of an attribute . . . derived from a measured attribute of individuals within each group [which] characterizes the group and not the individual.' Echoing Schwartz's (1994) views, the ecological analysis presented here provides a means to examine variables which are not available at the individual level and thereby serve to illuminate the broader structural, contextual, and sociological context within which, in this case, cholera mortality was manifested upon the residents of Gibraltar.

Initial examination of district-level cholera mortality rates indicated that values were positively skewed and that a transformation of the rates would be necessary to satisfy the assumption of a normally distributed dependent variable that is inherent in regression analysis. Accordingly, the natural logarithm of the cholera death rates was used to normalize the dependent variable. Further, since the dependent variable is measured as a rate, a weighted least squares regression was selected. The weights in this analysis were defined as the reciprocal of the variance of each district's cholera mortality rate. This strategy was taken to ensure that those districts with few cholera deaths contributed little weight to the overall regression. A correction factor of 0.5 was added to the overall district mortality rate as well as to the weighting factors to adjust for a zero rate and avoid the arbitrary weighting of zero for districts with no cholera deaths.

The correlation matrix of these variables indicated that cholera mortality was positively correlated with the district measure of diarrhea mortality (0.490) and inversely related to the proportion of wells with each district (−0.508). The proportion of wells was in turn negatively correlated with both the proportion of cisterns (−0.662) and our overall density measure (−0.459). This finding

Patio Housing in
the Upper part
of the Town

Fig. 9.5. Patio housing among the poor working classes of Gibraltar (after Sutherland 1867).

suggests that those residing in more crowded districts relied more heavily on cisterns and less on wells to fulfill their water needs. Private patio wells were simply not practical in the upper districts owing to difficulty and the huge costs of drilling great distances through the thick limestone to the water table. Given Gibraltar's urban landscape and the unusual topography, these observations confirm the obvious – that poorer inhabitants occupied the less desirable areas in the upper Rock which were, in turn, devoid of wells and more densely populated (see Fig. 9.5).

The results of the regression analysis indicate that only three of the five independent variables were significant predictors for district-level cholera mortality rates. Collectively, the regression model accounted for 67.5% of the observed variability in cholera mortality. The population density measure, or square yards per person in a district, showed the strongest significant effect on cholera mortality with a standardized β coefficient of 0.65 ($t = -4.7$, $p = 0$). The proportion of wells also displayed a significant impact on cholera mortality with a standardized β coefficient of -0.596 ($t = -3.07$, $p = 0.006$). The negative coefficient indicates that districts with higher proportions of wells were correlated with lower cholera mortality rates, while a deficiency in wells implied higher levels of mortality. The patio crowding measure, or the proportion of district patios with ten or more families, proved to be another significant predictor with a β coefficient of 0.333 ($t = 2.641$, $p = 0.016$). As district patio-level crowding increased, therefore, so too did the district mortality rate. Diarrhea mortality, only borderline significant at $p = 0.05$, showed a positive correlation with cholera mortality. These results suggest that population density, patio crowding, and access to wells were all important mitigating factors in the expression of cholera mortality during the 1865 epidemic. Though well water in Gibraltar was not potable, it is possible that the hygienic use of this water may have been an important factor in reducing environmental contamination associated with the cholera epidemic.

Cholera mortality at the patio level

Narrowing the focus from district-level analyses, it is possible to evaluate how specific patio-level features influenced mortality during the cholera epidemic. An important element in the Gibraltarian ethos was long-term residence and membership in a specific patio (Sawchuk 1993). The physical structure of the patio was such that each building was viewed as a self-contained community sharing intimate communal living space and, importantly, water resources. The close proximity of life in a patio could foster friendship, group identification, and the development of strong bonds among its residents which cut across both religious and socioeconomic lines. In encouraging a sense of community spirit, patios expanded the support network beyond the immediate family and could act as adaptive organs in the hostile, impersonal, and imperfect urban environment. It was during a crisis period, such as an epidemic or shortage of rainfall, that membership in a patio became particularly important as a means of overcoming hardship.

Two different measures were developed to assess the impact of patio form on health: (1) patio size, or the number of resident families; and (2) patio water, or the type of water supply made available to households within the patio. In order to capture the relationship of patio form with health during the cholera epidemic of 1865, a relatively simple strategy was adopted. The relationship between patio size and patio water with cholera mortality was assessed through Fleiss' (1981) standardization method, which adjusts for two factors simultaneously when examining a rate or proportion, as is the case here with cholera mortality. Adjustment for the two factors based on direct standardization is useful in identifying and measuring their separate effects even when there are complex associations between the independent variables both with each other and with the dependent rate or proportion (Fleiss 1981).

Patio size was first partitioned into four categories based on the number of families occupying each patio: (1) single family, (2) two to four families, (3) five to nine families, and (4) ten or more families. The underlying expectation is that multi-family dwellings offer more potential susceptibles than buildings with fewer families and that, all else being equal, those residents should eventually experience higher cholera mortality rates. Localized neighborhood or patio outbreaks, based on chains of person-to-person infection, typically began in one-room apartments where the sick resided in close contact with other members of the household. Interestingly, comments from Gibraltar's Medical Officer of Health suggest that the number of families living together was also an important factor in the control of water resources:

> In private houses occupied by just one family, who presumably jealously guard the purity of their drinking water, the connection of the terraces is not likely to affect the health of the community at large. The case is, however, very different in dealing with tenements let out to separate families. It is a matter of common experience, that, where separate families have conjoint responsibility in maintaining the cleanliness of common stairs, courtyards, terraces, etc., there especially will be an unsatisfactory state of affairs to be found.
> (MacPherson 1892: 20)

The second element of the analysis included a four-fold classification of patio water resources: (1) well water, (2) cistern water, (3) neither a well nor a cistern, and (4) both a well and a cistern.

Patio resources, household size and cholera mortality

The pattern of cholera mortality according to patio form during the 1865 epidemic revealed that (1) the highest mortality rates (34.9 per 1000) occurred among the residents of patios without access to either a well or a cistern after adjusting for building size, (2) the next highest death rates (24.4 per 1000) were found among those patios with cistern access but lacking a well, and (3) a considerable drop in cholera mortality was seen among those resident in patios with only wells (18.7 per 1000) and those with access to both a cistern and a well (9.4 per 1000). Examination of cholera mortality rates according to building size shows a clear gradient of increasing mortality from single family dwellings (11.2 per 1000) to those holding ten or more families (26.4 per 1000) after adjusting for patio water resources.

The underlying expectation that multi-family patios can offer more potential susceptibles than buildings with fewer families, and that those more densely populated patios should eventually experience higher cholera mortality rates, appears to have some validity. The lingering question, however, is why access to cisterns, which are the primary patio sources for potable drinking water, should be less important to increasing survival than access to predominantly contaminated well water. Access to potable water, after all, should have reduced the need to search out 'other' (and possibly less reliable) sources of drinking water and, among those with cholera, should have helped to combat the diarrhea and dehydration.

Survival during the cholera epidemic appears to be far more complex than can be captured by the simple presence or absence of potable water resources alone. Part of the problem, at least in Gibraltar, could have been the manner in which water from cisterns was distributed among those in the patio. George Alton, the Secretary to the Sanitary Commissioners, provides some answers:

> ... it ought to be distinctly kept in view that one-third of the houses in the Central District of Gibraltar have neither cistern nor well water, the occupiers of which can only obtain such supply as their means enable them to purchase from water carriers. The occupiers in common of houses having cisterns, but no well, probably another third in number, are only slightly better off, for whilst they are enabled to obtain a partial supply from the storage on the premises, the actual quantity is exceedingly small, and is generally on the scale of one bucket full, or three gallons twice in the week for each dollar of rent. That is to say, a family occupying rooms which let for four dollars per month, or say 10 pounds per annum, would be allowed from the house cistern twenty-four gallons of water per week, or an average of nearly three and a half gallons per day, for the entire household, and so in proportionate to rent.
> (Alton 1870)

A number of important points are highlighted in this passage. First, it is clear that from the immediate perspective of this local health authority that residents in buildings with cisterns did not appear to fare much better than those in complete denial of any patio water resources. Second, without information on the rent paid by each household in the patio, it is impossible to gauge the actual amount of potable water potentially available to those households. Finally, the amount of water that an entire family, typically numbering about five individuals, would draw from the communal source was limited to less than 3.5 gallons per day.

Another complicating factor is the fact that the mere presence of a cistern did not guarantee absolute access to drinking water. A report of House 10, District 25, in the upper part of the town, by E. Roberts, Surveyor to the Sanitary Commissioners, is insightful in this respect:

> There are three cesspools to this House all full of sewage and soil, and in two cases running over ... The privy is in a dilapidated and filthy condition, and is used by a large number of persons, there being but three privies for 76 inhabitants who occupy the house ... This house, No. 10, is like a small

walled village, built on irregular rocky ground; the separate houses or apartments being approached by rough uneven steps and ramps all badly paved. All the houses are bad, and several totally unfit for human dwellings, the walls and floors being very damp, and the roofs in very many places admitting rain; they have no chimneys, and no ventilation except what is afforded by the doors and windows which are almost invariably on one side, and are of course closed during the night . . . **There is a large tank upon the premises, apparently in good condition, capable of holding 41,000 gallons of water, but it is never filled.**

(Roberts 1866: 8–9) (Emphasis added.)

Census returns indicate that 26 separate households yielding a total of 118 individuals occupied this large building. The majority of these households were headed by men employed as general laborers and coalheavers. Although the house is designated as one with a cistern and no well, the surveyor's description clearly indicates that these residents were lax in their communal responsibility to this important patio resource. A total of six cholera deaths was registered among the households of this patio.

Having described why the simple presence of a cistern did not hold paramount significance in the distribution of patio-level cholera mortality, it is important to now consider why the presence or absence of a patio well could impact on health in the time of cholera. While unsuitable for drinking purposes, water drawn from patio wells appears to have played a vital role in maintaining higher levels of personal and public hygiene. Sanitary well water was used for cleaning, flushing, rinsing, and, if uncontaminated, for personal ablution, cooking food, washing up cooking utensils, and so forth. Given the absolute necessity of provisioning the household with potable drinking water, poorer families without access to a well were often left with little money for the purchase of sanitary water. The transmission of infectious diseases such as cholera would be greatly facilitated in a 'dirtier' environment where communal and personal space received little or no cleansing. Furthermore, if not cautious in guarding basic hygienic principles such as hand-washing, individuals nursing the sick suffering from diarrheal disease could easily spread the infection to others in the household or patio. Recent studies in developing countries suggest that the simple ability to wash one's hands plays a more important role than many others in preventing diarrhea (Gorter *et al.* 1998; Mirza *et al.* 1997). In a review of 26 studies examining the health impact of sanitation, Esrey and Habicht (1986: 121) conclude that 'of the studies that compared the relative importance of water and sanitation, most reported that sanitation was a more important determinant of child health than was water. This was true for morbidity, growth, and mortality indicators.' It is quite possible that access to sanitary well water may have played a similarly important preventative role in the transmission of cholera in nineteenth century Gibraltar.

Conclusions

We have opportunistically exploited Gibraltar's colonial legacy of a richly detailed set of historical and demographic records to reconstruct the forces and factors that affected the health of the residents of this unusual urban community. The research paradigm employed here relies heavily on archival resources to situate the community within the prevailing conditions of the times as well as within its long-term sociocultural milieu.

Set in the 1860s, the Gibraltar case study captures a dark period when its residents were subjected to numerous hostile environmental factors (such as overcrowding, inadequate health care, poor sanitation) as well as frequent insults from a wide range of infectious diseases coupled with inadequate nutrition. Under a regime of high background mortality, a cluster of endemic infectious diseases took a consistent heavy toll of lives. Of these, weanling diarrhea exerted the largest and most consistent negative impact year after year on survivorship and the health of Gibraltar's children. Sporadic epidemics of smallpox, scarlet fever and measles further complemented the battery of insults that faced the citizens of the Rock. Of these, measles could prove to be the most serious in terms of the potential years lost to life.

The cholera epidemic of 1865 punctuated Gibraltar's dark times, severely damping survivorship and leaving among the survivors misery and hardship. The impact of this short-term episode of crisis mortality was expressed through a dramatic drop in life expectancy, a significant rise in the stillbirth rate, and an elevated Hollingsworth index value. Despite the rise in the cholera-induced mortality rate, the shock to the demographic structure of the community was not homogeneous, varying markedly according to status, residential location, and available communal resources. From a longer temporal vantage point, however, the true demographic impact of cholera was of a modest magnitude, paling in comparison to the deaths arising from the everyday pathogenic load of weanling diarrhea, tuberculosis and other respiratory infections.

References

Aaby, P. (1992). Overcrowding and intensive exposure: major determinants of variations in measles mortality in Africa. In *Mortality and Society in Sub-Saharan Africa*, ed. E. van de Walle, G. Pison and M. Sala-Diakanda, pp. 319–48. Oxford: Clarendon Press.

Aldrich, R. and Connell, J. (1998). *The Last Colonies*. Cambridge: Cambridge University Press.

Alter, G. and Carmichael, A. (1996). Studying causes of death in the past. *Historical Methods* **29**, 44–8.

Alton, G. (1870). *Remarks on the Water Supply for Gibraltar*. Gibraltar: Gibraltar Library Printing Establishment.

Antonovsky, A. (1967). Social class, life expectancy and overall mortality. *Milbank Memorial Fund Quarterly* **45**, 31–73.

Atkins, P.J. (1992). White poison? The social consequences of milk consumption, 1850–1930. *Social History of Medicine* **5**, 207–26.

Barua, D. (1992). History of cholera. In *Cholera*, ed. D. Barua and W.B. Grenough, pp. 1–36. New York: Plenum Medical Book Company.

Beaver, M.W. (1973). Population, infant mortality and milk. *Population Studies* **27**, 243–54.

Benady, S. (1994). *Civil Hospitals and Epidemics in Gibraltar*. Gibraltar: Gibraltar Books Ltd.

Black, F.L. (1997). Measles. In *Viral Infections of Humans: Epidemiology and Control*, ed. A.S. Evans and R.A. Kaslow, pp. 507–30. London: Plenum Medical Book Company.

Black, F.L., de Pinheiro, F., Hierholzer, W.J. and Lee, R.V. (1977). Epidemiology of infectious diseases: the example of measles. In *Health and Disease in Tribal Societies* (CIBA Foundation Symposium 49, London, 1976), pp. 115–35. Amsterdam: Elsevier.

Bouckaert, A. (1989). Crisis mortality: extinction and near extinction of human populations. In *Differential Mortality: Methodological Issues and Biosocial Factors*, ed. L. Ruzicka, G. Wunsch and P. Kane, pp. 217–30. Oxford: Clarendon Press.

Buffler, P.A., Cooper, S.P., Stinnett, S., Constant, C., Shirts, S., Hardy, R.J., Agu, V., Gehan, B. and Burau, K. (1988). Air pollution and lung cancer mortality in Harris County, Texas, 1979–1981. *American Journal of Epidemiology* **128**, 683–700.

Cherry, S. (1980). The hospitals and population growth: part 1 and part 2. The voluntary general hospitals, mortality and local populations in the English provinces in the eighteenth and nineteenth centuries. *Population Studies* **34**, 59–75; 251–65.

Chiang, C. (1984). *The Life Table and Its Applications*. Malabar: Krieger Publishing Co.

Collins, R. (1885). Memorandum from the Surgeon Major, Acting Health Officer to the Sanitary Commissioner's Office. Gibraltar, 20 June, Gibraltar Government Archives.

Cox, J.P. (1877). Report of an Inquiry into the Working and Administration of the Gibraltar Sanitary Commission. Gibraltar, 12 November, Gibraltar Government Archives.

Curtin, P.D. (1989). *Death by Migration: Europe's Encounter with the Tropical World in the Nineteenth Century*. Cambridge: Cambridge University Press.

De Bevoise, K. (1995). *Agents of Apocalypse: Epidemic Disease in the Colonial Philippines*. Princeton: Princeton University Press.

De Vries, R.R.P., Meera Khan, P., Bernini, L.F., van Loghem, E. and van Rood, J.J. (1979). Genetic control of survival in epidemics. *Journal of Immunogenetics* **6**, 271–87.

Dobson, M.J. (1989). Mortality gradients and disease exchanges: comparisons from old England and colonial America. *Social History of Medicine* **2**, 259–97.

Dopico, F. (1987). Regional mortality tables for Spain in the 1860s. *Historical Methods* **20**, 173–9.

212   *L.A. Sawchuk & S.D.A. Burke*

Dwork, D. (1987). The milk option: an aspect of the history of the infant welfare movement in England 1898–1908. *Medical History* **31**, 51–69.

Elkington, H.P.P. (1901). *Annual Report on the Public Health of Gibraltar for the Year 1900*. Gibraltar: Gibraltar Garrison Printing Press.

Esrey, S.A. and Habicht, J.P. (1986). Epidemiologic evidence for health benefits from improved water and sanitation in developing countries. *Epidemiologic Reviews* **8**, 117–28.

Evans, R.J. (1987). *Death in Hamburg: Society and Politics in the Cholera Years 1830–1910*. Oxford: Clarendon Press.

Exchange Committee and Chamber of Commerce (1865). Memorial to the Secretary of State, Earl Grey. Gibraltar, 12 August, Gibraltar Government Archives.

Fairchild, A.L. and Oppenheimer, G.M. (1998). Public health nihilism vs. pragmatism: history, politics and the control of tuberculosis. *American Journal of Public Health* **88**, 1105–17.

Fleiss, J.L. (1981). *Statistical Methods for Rates and Proportions*. New York: John Wiley and Sons.

Flood, F.S. (1866). Correspondence from the Police Magistrate, Police Office, Gibraltar, to S. Freeling, Colonial Secretary, 13 April. Gibraltar Government Archives.

Friedlander, D., Schellekens, J., Ben-Moshe, E. and Keysar, A. (1985). Socio-economic characteristics and life expectancies in nineteenth-century England: a district analysis. *Population Studies* **39**, 137–51.

Gentleman's Magazine (1762). London: D. Henry. March. Map Insert between pages 102 and 103.

Gibraltar Chronicle (1822). 4th January, Supplement. Gibraltar Government Archives.

Gibraltar Chronicle (1865). 9th November. Gibraltar Government Archives.

Glass, R.I. and Black, R.E. (1992). The epidemiology of cholera. In *Cholera*, ed. D. Barua and W. B. Greenough, pp. 129–54. New York: Plenum Medical Book.

Gorter, A.C., Sandiford, P., Pauw, J., Morales, P., Perez, R.M. and Alberts, J.H. (1998). Hygiene behaviour in rural Nicaragua in relation to diarrhea. *International Journal of Epidemiology* **27**, 1090–100.

Government of Gibraltar (1906). Public Health Report. Gibraltar Government Archives.

Government of Gibraltar (1939). Dietary survey of 26 households. Gibraltar Government Archives.

Grigg, E.R.N. (1958). The arcana of tuberculosis. *American Review of Respiratory Disease* **78**, 151–72; 583–603.

Gruenwald, P. (1969). Stillbirth and early neonatal death. In *Perinatal Problems: The Second Report of the 1958 British Perinatal Mortality Survey*, ed. N.R. Butler and E.D. Alberman, pp. 163–83. Edinburgh, London: E. and S. Livingstone.

Haines, M.R. (1994). *Estimated Life Tables for the United States, 1850–1900*. Cambridge, MA: National Bureau of Economic Research.

Hardy, A. (1993). *The Epidemic Streets: Infectious Disease and the Rise of Preventative Medicine, 1856–1900*. Oxford: Clarendon Press.

Hennen, J. (1830). *Sketches of the Medical Topography of the Mediterranean comprising an account of Gibraltar, the Ionian Islands, and Malta; to which is prefixed, A Sketch of a Plan for Memoirs on Medical Topography*. Fleet Street, London: Thomas and George Underwood.

Horrocks, W.H. (1904). *Annual Report on the Public Health of Gibraltar for the Year 1903*. Gibraltar: Gibraltar Garrison Printing.

Huston, J.L. (1991). Weighting, Confidence Intervals, and Ecological Regression. *Journal of Interdisciplinary History* **21**, 631–54.

Kelaart, E.F. (1846). *Flora Calpensis: Contributions to the Botany and Topography of Gibraltar and its Neighbourhood: With Plan, and Views of the Rock*. London: John Van Voorst.

Lancaster, H.O. (1990). *Expectations of Life: A Study in the Demography, Statistics, and History of World Mortality*. New York: Springer-Verlag.

Longmate, N. (1966). *King Cholera: The Biography of a Disease*. London: Hamish Hamilton.

Lunn, P. (1991). Nutrition, Immunity and Infection. In *The Decline of Mortality in Europe*, ed. R. Schofield, D. Reher and A. Bideau, pp. 131–45. Oxford: Clarendon Press.

MacPherson, R. (1890). *Annual Report on the Health of Gibraltar for 1889*. Gibraltar: Gibraltar Printing Press.

MacPherson, R. (1892). *Annual Report on the Health of Gibraltar for 1891*. Gibraltar: Gibraltar Printing Press.

MacPherson, R. (1895). *Annual Report on the Health of Gibraltar for 1894*. Gibraltar: Gibraltar Printing Press.

McGrew, R.E. (1965). *Russia and the Cholera*. Madison: University of Wisconsin Press.

McFarlane, N. (1989). Hospitals, housing, and tuberculosis in Glasgow. *Social History of Medicine* **2**, 59–85.

McKeown, T. (1976). *The Modern Rise of Population*. New York: Academic Press.

McKeown, T. (1979). *The Role of Medicine: Dream, Mirage, or Nemesis?* Princeton, NJ: Princeton University Press.

McNeil, W.H. (1977). *Plagues and Peoples*. New York: Anchor Press.

Matzarakis, A. and Mayer, H. (1997). Heat Stress in Greece. *International Journal of Biometeorology* **41**, 34–39.

Mirza, N.M., Caulfield, L.E., Black, R.E. and Macharia, W.M. (1997). Risk factors for diarrheal duration. *American Journal of Epidemiology* **146**, 776–85.

Moring, B. (1998). Motherhood, milk, and money. *Social History of Medicine* **11**, 177–96.

Morley, D.C. (1973). *Paediatric Priorities in the Developing World*. London: Butterworths.

Morris, R.J. (1976). *Cholera, 1832*. London: Croom Helm.

Neaderland, R. (1952). Teething – a review. *Journal of Dentistry for Children* **19**, 127–32.

Omran, A.R. (1971). The epidemiologic transition: a theory of the epidemiology of population change. *Milbank Memorial Fund Quarterly* **49**, 509–38.

Omran, A.R. (1983). The epidemiologic transition theory: a preliminary update. *Journal of Tropical Medicine* **29**, 305–16.

Patterson, K.D. (1994). Cholera diffusion in Russia, 1823–193. *Social Science and Medicine* **38**, 1171–91.

Pollitzer, R. (1959). *Cholera*. [With a chapter on world incidence, written in collaboration with S. Swaroop, and a chapter on problems in immunology and an annex, written in collaboration with W. Burrows.] Geneva: World Health Organization.

214   *L.A. Sawchuk & S.D.A. Burke*

Preston, S.H. (1976). *Mortality Patterns in National Populations*. New York: Academic Press.
Preston, S.H., Keyfitz, N. and Schoen, R. (1972). *Causes of Death: Life Tables for National Populations*. New York: Seminar Press.
Rabbani, G.H. and Greenough, W.B. (1992). Pathophysiology and clinical aspects of cholera. In *Cholera*, ed. D. Barua and W.B. Greenough, pp. 209–28. New York: Plenum Medical Book.
Radbill, S.X. (1965). Teething as a medical problem. *Clinical Pediatrics* 4, 556–9.
Raufman, J.P. (1998). Cholera. *American Journal of Medicine* 104, 386–94.
Reves, R. (1985). Declining fertility in England and Wales as a major cause of the twentieth century decline in mortality: the role of changing family size and age structure in infectious disease mortality in infancy. *American Journal of Epidemiology* 122, 112–26.
Richardson, S.H. (1994). Host susceptibility. In *Vibrio cholerae and Cholera: Molecular to Global Perspectives*, ed. I.K. Wachsmuth, P.A. Blake and O. Olsvik, pp. 273–89. Washington, D.C.: ASM Press.
Roberts, E. (1866). *Report of the Sanitary Commissioners of Gibraltar for the Year 1866*. Gibraltar: Gibraltar Garrison Printing Press.
Robinson, D. and Pinch, S. (1987). A geographical analysis of the relationship between early childhood death and socio-economic environment in an English city. *Social Science and Medicine* 25, 9–18.
Rodriquez Ocaña, E. (1992). Morbimortalidad del colera epidemico de 1833–35 en Andalucia. *Boletin de la Associacion de Demografia Historica* 10, 87–111.
Romero, R., Guijarro, J.A., Ramis, C. and Alonso, S. (1998). A 30-year (1964–1993) daily rainfall data base for the Spanish Mediterranean regions: first exploratory study. *International Journal of Climatology* 18, 541–60.
Rose, E.P.F. and Rosenbaum, M.S. (1991). *A Field Guide to the Geology of Gibraltar*. Gibraltar: Gibraltar Museum.
Sawchuk, L.A. (1993). Societal and ecological determinants of urban health: a case study of pre-reproductive mortality in 19th century Gibraltar. *Social Science Medicine* 36, 875–92.
Sawchuk, L.A. (1996). Rainfall, patio living, and crisis mortality in a small-scale society: the benefits of a tradition of scarcity. *Current Anthropology* 37, 863–7.
Sawchuk, L.A. (2001). *Deadly Visitations in Dark Times: A Social History of Gibraltar*. Gibraltar: Gibraltar Government Heritage Publications.
Sawchuk, L.A. and Burke, S.D.A. (1998). Gibraltar's 1804 yellow fever scourge: the search for scapegoats. *Journal of the History of Medicine and Allied Sciences* 53, 3–42.
Sawchuk, L.A., Burke, S.D.A. and Choong, H.C. (1997). Environment: a potential stressor in stillbirths. A case study in Gibraltar. *American Journal of Physical Anthropology* 24, S204.
Sawchuk, L.A., Herring, D.A. and Waks, L.R. (1985). Evidence of a Jewish advantage: a study of infant mortality in Gibraltar,1870–1959. *American Anthropologist* 87, 616–25.
Schwartz, S. (1994). The fallacy of the ecological fallacy: the potential misuse of a concept and the consequences. *American Journal of Public Health* 84, 819–24.

Scott, S. and Duncan, C. (1998). *Human Demography and Disease.* Cambridge: Cambridge University Press.

Scott, S., Duncan, S.R. and Duncan, C.J. (1995). Infant mortality and famine: a study in historical epidemiology in Northern England. *Journal Epidemiology and Community Health* **49**, 245–52.

Shryock, H.S., Siegel, J.S. *et al.* (1973). *The Methods and Materials of Demography.* Washington, D.C.: Bureau of the Census.

Smallman-Raynor, M. and Cliff, A.D. (1998). The Philippines insurrection and the 1902–4 cholera epidemic: part II – diffusion patterns in war and peace. *Journal of Historical Geography* **24**, 188–210.

Smith, E. (1889). *A Practical Treatise on Disease in Children.* London: J. and A. Churchill.

Snowden, F.M. (1995). *Naples in the Time of Cholera, 1884–1911.* Cambridge: Cambridge University Press.

Stokes, H. (1867). *Report of the Officer of Health of the Sanitary Commissioners of Gibraltar for the Year 1866.* Gibraltar Government Archives.

Susser, M. (1994). The logic in ecological: I. the logic of analysis. *American Journal of Public Health* **84**, 825–9.

Sutherland, I. (1949). *Stillbirths: Their Epidemiology and Social Significance.* London: Oxford University Press.

Sutherland, J. (1867). *Report on the Sanitary Condition of Gibraltar with Reference to the Epidemic Cholera in the Year 1865.* London: George Edward Eyre and William Spottiswode.

Szreter, S. (1988). The importance of social intervention in Britain's mortality decline c.1850–1914: a re-interpretation of the role of public health. *Social History of Medicine* **1**, 7–10.

Thomsett, R.G. (1902). *A Record Voyage in H.M.S. Malabar and Reminiscences of the Rock.* London: Digby, Long and Co.

Tranmer, M. and Steel, D.G. (1998). Using census data to investigate the causes of the ecological fallacy. *Environment and Planning* A**30**, 817–31.

Watts, S.J. (1997). *Epidemics and History: Disease, Power and Imperialism.* New Haven: Yale University Press.

Williams, N. (1996). The reporting and classification of causes of death in mid-nineteenth century England. *Historical Methods* **29**, 58–71.

Wohl, A.S. (1983). *Endangered Lives: Public Health in Victorian Britain.* Toronto: J.M. Dent and Sons.

Woods, R. (1982). The structure of mortality in mid-nineteenth century England and Wales. *Journal of Historical Geography* **8**, 373–94.

Woods, R. and Woodward, J. (1984). *Urban Disease and Mortality in Nineteenth Century England.* London and New York: Batsford Academic and Educational and St. Martin's Press.

Wrigley, E.A. (1969). *Population and History.* New York: McGraw-Hill.

Zusman, I. and Ornoy, A. (1990). Embryonic resistance to chemical and physical factors: manifestation, mechanism, role in reproduction and adaptation to ecology. *Biological Review* **65**, 1–18.

10 War and population composition in Åland, Finland

JAMES H. MIELKE

Introduction

Traditionally, anthropologists have focused their research on small populations, examining diversity and illuminating the complexity of many factors. Since an anthropological perspective is different from that of a demographer or historian, the analysis of historical archives by biological anthropologists provides new and different insights into human population dynamics. Using archival resources in studies of historical population structure and historical epidemiology will undoubtedly advance our understanding of demographic changes, the history and evolution of diseases, and the factors influencing population growth and decline. Thus, the temporal and biocultural perspective of anthropologists often uncovers intriguing relationships, contributing to our broader understanding of evolution and population change.

Historical demographers have often explained mortality patterns and demographic changes in terms of living standards and wages (Landers 1992a). A number of historical demographers are now shifting their focus, arguing that local and regional diversity in the spatial structure of populations affects the degree and intensity of exposure to infectious agents (Dobson 1992; Landers 1992b; Langford and Storey 1993). Studies now examine such things as population size and density, migration patterns, and economic and social variation within an area as major contributing factors to the spread and impact of diseases. Some researchers are now also shifting their focus from large continental populations to local and regional levels. Dobson (1992) even suggests that the term *epidemiological landscapes* may be appropriate to characterize the complex diversity seen in disease patterns and epidemic outbreaks.

Population crises were recurrent features of pre-industrial populations, shaping the demographic landscape over the centuries. The triggering event of a crisis – endogenous or exogenous – leads to responses in mortality, fertility, and migration (Palloni 1990). Contributing variables often include malnutrition, host resistance, infectious diseases, hygiene and crowding, and social factors (e.g. class, medical intervention). These features interact in complex ways,

216

shaping the impact of the crisis. This present study adds an anthropological perspective to the study of an historical mortality crisis that was triggered by a war. It will examine the effects of the War of Finland (1808–09) on the resident population of the Åland Islands, an archipelago lying between Sweden and Finland. The chapter details the impact and regional variation in mortality during the crisis, while exploring the effect that sex and age had on the mortality patterns. Spatial variation and complexity in the responses to the crisis reflect a combination of sociocultural features and the demographic and geographic structure of the islands. Studies of historical crises such as this also provide anthropologists with clues to the complexity and heterogeneity of a society's demographic responses to mortality crises in general, whether war-induced or caused by famine or epidemics.

Åland, the setting of the mortality crisis

The Åland archipelago, bounded to the north by the Gulf of Bothnia and to the south by the Baltic Sea, is located between Sweden and Finland (Fig. 10.1). Even though the province of Åland encompasses about 10 000 km^2, only 1450 km^2 is land area (Sjölund 1972). The archipelago contains about 6600 islands and skerries, which, since the mid-sixteenth century, have been divided into 15 rural parishes. By 1750 the area contained 9000 people, and by the turn of the twentieth century the population just exceeded 20 000. At the time of the War of Finland (1808–09) the population consisted of nearly 14 000 individuals.

Åland is an excellent location to examine the demography of a mortality crisis. Firstly, even though the inhabitants of Åland maintained lively connections with both Sweden and Finland, the location of the archipelago has meant that it has been relatively isolated for centuries. Thus, it is easy to define the population and trace changes in its composition and structure over both time and space. Secondly, the archival and historical data are very complete, extremely detailed, and relatively accurate (these sources are detailed below). Thirdly, the population has historically been too small and scattered to sustain most, if not all, types of infectious or communicable diseases in their endemic form. For example, the critical size (threshold) in order for measles to remain endemic in a population has been estimated to be between 250 000 and 500 000 (Bartlett 1957, 1960; Black 1989; Cliff *et al.* 1981, 1993). Fourthly, the archipelago's contemporary and historical population and genetic structures have been elucidated in a number of studies (see, for example, Workman and Jorde 1980; Jorde *et al.* 1982; Mielke *et al.* 1987), providing information helpful in interpreting population changes and regional variation in response to war.

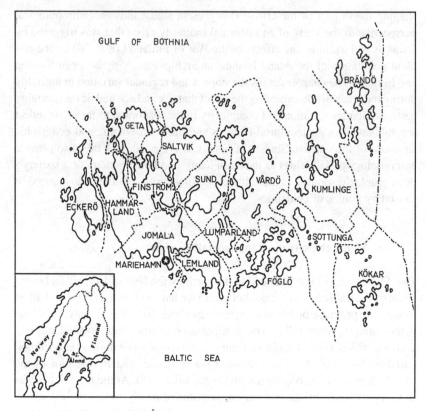

Fig. 10.1. Map of the Åland Islands, showing parish locations.

Archival sources

The data used in this study come from the Åland parish records that were kept by
the Lutheran ministers. Since the seventeenth century, Swedish Lutheran ministers have kept registries of their parishioners and various religious occurrences
(e.g. baptisms). The first uniform instructions for keeping these records were
given in the Swedish ecclesiastical law of 1686. The law required parish clergy
to maintain registers of baptisms and births, marriages, deaths and burials, in-
migration, and out-migration. It also specified that they keep records detailing
information on each parishioner's knowledge of Christianity and their participation in Communion (Communion books or *rippikirjat* or *pääkirjat*). Since
essentially everyone belonged to the Lutheran church, these records cover the
entire population. Initially, the purpose of keeping records was purely ecclesiastical. However, in the eighteenth century the Swedish government recognized

that these records could serve as basic data for population statistics covering the entire kingdom and all its inhabitants. At that time, the Swedish government, along with the Lutheran State Church, was attempting to maintain control over the populace. In 1748 the government instituted a systematic collection scheme (called *Tabellverket*) in order to collect population statistics. In 1749 the parish ministers were required to maintain two kinds of standardized statistical forms, 'Population Tables' (a census) and 'Population Change Tables.' The 'Population Tables' provide information on population size, age and sex distribution, marital status, and occupational structure by sex. These tables were compiled every three years (1751–1774) until 1775, and then changed to every five years (Pitkänen and Nieminen 1984). The 'Population Change Tables' contain yearly information on births, marriages, and deaths. Even during the war period, this registration system was not seriously compromised. So, by the early nineteenth century the parish registers and statistical tables were usually very accurate, providing a clear picture of the demography (Pitkänen and Nieminen 1984; Pitkänen 1993).

The primary data used in this study were extracted from the burial/death records (*kuolleiden*) of each of the 15 parishes. These records include the date of death and burial, the name and residence of the deceased, his or her occupation, cause of death, and age at death. The birth records (*syntyneiden*) were also examined, especially for any record of stillbirths and infant deaths. Additional and corroborative data were extracted from the 'Population' and 'Population Change' tables. For a more complete description of these Finnish archival data sets, and analysis of their accuracy, see Pitkänen (1977, 1993), Pitkänen and Nieminen (1984), Mielke and Pitkänen (1989), and Pitkänen and Mielke (1993).

Åland and the 'War of Finland'

The study of the War of Finland can provide valuable insights into the demographic impact of a war-induced mortality crisis. Firstly, the civilian population of Åland was, for the most part, not subjected to any direct military violence. They were, however, required to aid the military efforts. Hence, the increased mortality after the Swedish troops arrived in the archipelago was not related to combat. Secondly, we can study the effect that a large concentration of troops had on spreading diseases throughout the islands in a very short period of time. Thirdly, there were no major economic problems that complicated or modified the outcome of the mortality crisis. Harvests prior to the war had been relatively good, and the population could rely on fishing to buffer against any hardships. It is also probable that the Swedish army only purchased surplus foodstuffs that the population had in storage.

In July 1807, Napoleon attempted to strengthen France's supremacy in Europe by meeting with Czar Alexander I to negotiate a peace treaty near Tilsit in Eastern Prussia. As part of the conditions of the treaty, Alexander I had to convince Sweden to join the continental blockade against Britain. The Swedish government was reluctant to participate in the blockade. As a result, Alexander I issued orders to invade Finland. The 'War of Finland' started on February 12, 1808, when Russian troops invaded southeastern Finland. The Finnish troops retreated, and the Russians occupied southern Finland. Since the Åland Islands are situated at the mouth of the Gulf of Bothnia, they offered a very attractive assembly area for a direct attack on Sweden. Realizing this fact, the Russians initially occupied Åland in mid-April, 1808, with about 700 soldiers. Åland peasants, aided by a few Swedish warships and their crews, captured this occupying force in early May (Schulman 1909).

The second phase of the war for Åland occurred when Gustav IV Adolf, King of Sweden, sent Swedish troops to the islands in late May, 1808. By October there were approximately 6000–8000 troops on the islands. Since the civilian population of Åland numbered about 14 000 individuals, the population density in some parishes doubled, making conditions conducive to the spread of infectious diseases. As with many armies in the past (Major 1941; Urlanis 1971), the Swedish troops arrived on the islands with a number of communicable diseases. Samuelsson (1944) notes that typhoid fever and typhus were among the primary diseases. Other diseases included measles, smallpox, and relapsing fever. Samuelsson (1944) also suggests that leptospirosis or Weil's disease was present during the war.

Leptospirosis is a bacterial disease that is usually caused by exposure to water, food, or soil contaminated by animal urine (cattle, horses, dogs, and wild animals). The disease does not spread from person to person. The first phase of the disease involves a fever, chills, headaches, diarrhea, or vomiting. Patients recover but may become sick again, entering a second phase that may lead to liver or kidney failure (CDC 2001). Deaths are rare; however, without treatment, it takes several months to recover. Conditions on the farms and in the villages during the war period were conducive to the spread of this type of disease.

The communicable diseases brought in by the troops were effectively spread to the civilian population because the Swedish authorities required the Ålanders to work in close contact with the military. For example, one of the tasks the Ålanders performed was transporting sick soldiers to hospitals. Starting in October, contacts became even more intense when soldiers were quartered in the farmhouses with peasant families. Often 20–24 soldiers lived in a single room adjoining the family's room. Other factors contributing to the spread of disease were troop movements to new locations in the archipelago and the evacuation and relocation of the civilian population from the eastern parishes to other areas in Åland between November 1808 and January 1809 (Anderson 1945).

In mid-March, 1809, a Russian offensive forced the Swedish troops (and a number of Ålanders) to retreat across the sea ice to Sweden. Fearing the ice would not hold, most of the Russian troops returned to Finland toward the end of March, ending military presence in the archipelago for a few years.

The Treaty of Fredrikshamn, which ended the War of Finland, marked the end of Swedish domination in Åland. The archipelago was ceded to Russia and became part of the Grand Duchy of Finland and the western buffer of Russia's Baltic defense system. Under the orders of the Czar, a fortress (Bomarsund) was completed in the 1830s in the parish of Sund. Then in the summers of 1854 and 1855 Finland, Åland, and the rest of Russia's 'western buffer' became a theatre of war (part of the Crimean War) once again (Puntila 1974). A combined British and French fleet captured the Russian fortress of Bomarsund in August, 1854 (Mead and Jaatinen 1975). The majority of Åland's civilian population was not directly involved with these events. On the strength of a wartime alliance with England and France, Sweden made a bid to regain control of Åland. After negotiations, the archipelago was declared a demilitarized zone and granted international status (Dreijer 1968). Disregarding the demilitarized, international status, Russian troops occupied the islands in the early twentieth century. Åland avoided direct warfare, but a German infantry battalion made an excursion into the islands in February, 1918. In 1921 the League of Nations Council declared Åland a semiautonomous county of Finland (Barros 1968).

The War of Finland in Åland provides an example of a mortality crisis caused by an external population coming into contact with a resident population and introducing and spreading a number of infectious diseases. Similar population movements and contacts causing mortality crises have occurred during famines, epidemics, and initial contact of societies. For example, Bittles *et al.* (1986), Bittles (1988), and Smith *et al.* (1990) suggest that the regional differences in population structure in Ireland are due, in part, to the Irish Famine of 1846–51. However, Relethford and Crawford (1995), Relethford *et al.* (1997), and North *et al.* (1999), argue that recent changes such as the Irish Famine had little impact on the population structure at the regional or county level. They feel that the effect would be most dramatic at the local level (villages, towns).

Understanding the role and extent that recent historical events can have on shaping the population and genetic structure is important to anthropologists. This is especially important for interpreting the distribution of 'classic markers' or when using molecular data to infer relationships or to detail long- and

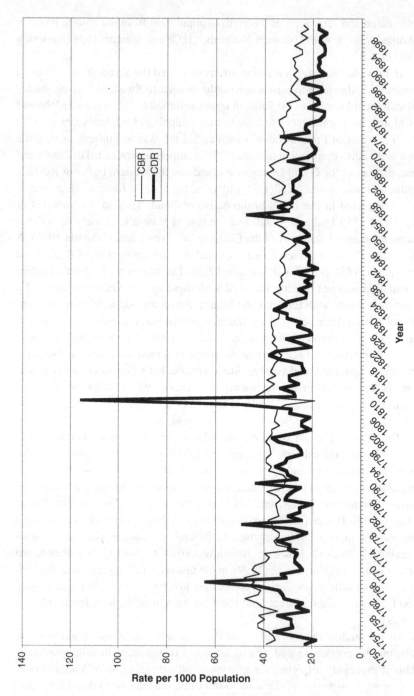

Fig. 10.2. Crude birth (CBR) and crude death (CDR) rates (per 1000 population) in Åland from 1750 to 1900. Data gleaned from Åland's death and birth records and population tables.

short-term genetic relations. By examining the impact of the War of Finland across the Åland archipelago, this present research will provide insight into the types of factors contributing to the complexity of a population crisis and potential population structure and genetic effects.

The demographic impact of the war

The general demographic impact of the War of Finland on the Ålandic population is demonstrated first by comparing the crude birth and crude death rates during the crisis to those that generally occurred in the archipelago. An examination of the age and sex differences in mortality follows this analysis. Finally, the regional diversity in mortality is detailed and discussed.

Crude birth and death rates

To gain a temporal perspective of the impact of the War of Finland on the Åland Islands, the yearly crude birth and crude death rates from 1750 to 1900 are shown in Fig. 10.2. The crude death rate increased from a prewar average of about 28 per 1000 population to 49.5 and 115.2 in 1808 and 1809, respectively. The crude birth rate dropped from about 36 to 31.1 in 1808 and then to 21 in 1809. Communicable diseases that were brought to the islands by the Swedish troops were primarily responsible for the increase in the death rate (Samuelsson 1944). At least 60% of the deaths were attributed to infectious diseases such as smallpox, typhus, typhoid fever, dysentery, and relapsing fever; while non-specific infectious diseases and miscellaneous causes were responsible for another 20%. The remaining deaths are classified as unknown, or no cause was given in the records (mainly infants). Cause of death statistics should be used with caution. Different names could be used for the same disease, or a single disease could have multiple designations (von Bonsdorff 1975). Diagnosis of the cause of death was often done by kin of the deceased and then conveyed to the Lutheran ministers who kept the records. Thus, diagnosis was far from satisfactory.

Only two deaths can be directly attributed to combat injuries between Russians and Ålanders, and these occurred in April, 1808, before the main contingent of Swedish troops arrived. There is no reason to suspect that combat-related deaths were censored or suppressed for political reasons. Civilians were simply not part of the military or fighting forces during this time period in history, nor were they targeted by armies for annihilation.

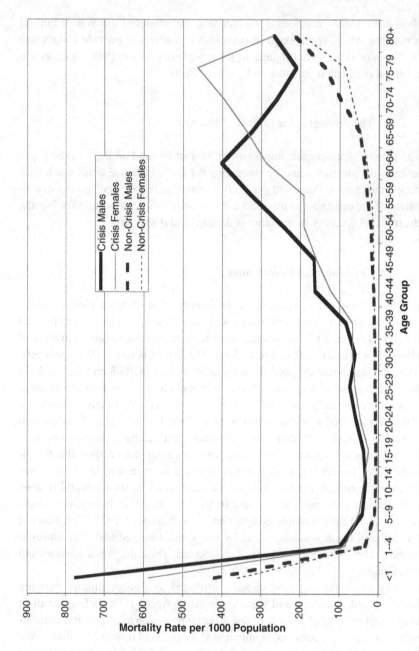

Fig. 10.3. Age-specific and sex-specific death rates in Åland during the War of Finland compared to a 'normal,' non-crisis period (1791–1805).

Age differences in mortality

The mortality pattern during the war (Fig. 10.3) generally resembles the pattern seen during a non-crisis time for individuals under the age of 40 (Åland from 1791 to 1805). Death rates were very high for young children, and then the rates declined for older children and young adults (to about 40 years of age). In *absolute* terms, the death rates for adults of both sexes were very high during the crisis after age 40, deviating from the pattern seen in non-crisis times. An alternative method to measure mortality excess is to calculate *proportional* increases by age. Proportional changes (Mielke and Pitkänen 1989) were greatest for those aged 40 to 64 (10–13 times the 'normal' rate). The main reproductive segment of the population (ages 20–39) also showed a considerable proportional increase (from 8 to 9 times 'normal'). These results suggest the potential for genetic effects, especially if the reproductive segment of the population is differentially affected by mortality. For a more detailed discussion of proportional changes in the mortality rates and the effects of the crisis on fertility, see Mielke and Pitkänen (1989).

A lack of consistent sex differences in mortality patterns

For a variety of biological and cultural reasons, we often expect to see differences in the mortality patterns between sexes (see, for example, Keyfitz and Flieger 1971; Preston 1976; Wingard 1984; Boyle and Ó Gráda 1986; Kane 1989; Langford and Storey 1993; Waldron 1993; Vallin 1995; Corsini and Viazzo 1997). As can be seen in Fig. 10.3, there are no *consistent* sex differences in the age-specific rates in Åland during the War of Finland. There is some excess male mortality in ages 55–65, and female mortality exceeds male mortality from 65 to 79 years. However, these 'cross-overs' represent variation in small samples at these older age groups and should not be considered biologically or culturally significant.

Cox's Proportional Hazards Survival Models allow one to predict survival times from one or more independent or predictor variables (covariates). A proportional hazards analysis demonstrated that there was no significant difference in the timing of mortality between the sexes, while controlling for the effect of age at death. The hazards ratio for males was almost identical to that of females (hazard ratio of 1.0145 with a 95% confidence interval of 0.9251–1.1126; females as reference at 1.0). The hazard curves for males and females are essentially identical (Fig. 10.4). Some caution should be exercised, however, since there is some overlap of the hazards distributions. This fact could indicate some sex differences in mortality at specific ages. However, all the other analyses

226   J.H. Mielke

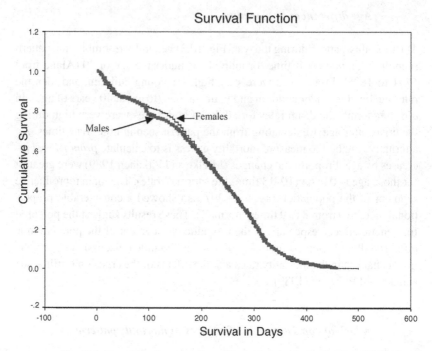

Fig. 10.4. Hazard function curve for males and females during the War of Finland.

suggest that the impact of the War of Finland on male and female mortality was essentially similar.

International data on infectious disease mortality show that male death rates are consistently higher than female rates (Preston 1976; Waldron 1983). Specifically, the death rates from infectious diseases were higher for males above 40 years of age and between birth and one year of age. Deaths were more variable between ages one and 40. Genetic factors may contribute to these differences (Waldron 1983); however, environmental factors such as nutrition and differential health care appear to be most important in causing differences among the sexes (Preston 1976; D'Souza and Chen 1980, Waldron 1983). Åland mortality data were analyzed using four age groupings (0–<1, 0–5, 5–40, 40+). Again, no significant differences between the sexes in hazards ratios were found for any of the age grouping. These results indicate that everyone, regardless of sex, was equally subjected to the adverse conditions of the war and exposed to the same array of diseases.

In a comparison of the mortality patterns during the War of Finland (1808–09) and the Great Finnish Famine (1860s)[1] Pitkänen and Mielke (1993) found that there were no significant sex differentials in mortality in mainland Finland during the 1808–09 war. However, there was considerably higher excess

Table 10.1. *Crude death rates per 1000 population during the War of Finland (1808–09) for Åland parishes*

In the parish of Finström the death records are not complete, so we have not included this parish and its companion parish Geta.

| | Crude death rate | | | | | |
| | 1808 | | 1809 | | Total | |
Parish	Males	Females	Males	Females	1808	1809
Sund/Vårdö	80	47	131	138	63	135
Saltvik	58	78	123	110	68	117
Eckerö/Ham	62	65	140	124	64	132
Jomala	53	41	144	135	46	139
Lem/Lump	47	39	147	123	43	134
Fö/So/Kö	46	48	97	84	47	91
Kum/Brändö	37	31	86	81	34	83
Total	52	46	118	108	49	115

mortality for males (especially those between 15 and 39) during the height of the famine in 1868. During both crises, the major causes of death (typhoid fever, typhus, dysentery, measles, smallpox, and relapsing fever) were essentially the same (Samuelsson 1944; von Bonsdorff 1975). The sex differences in mortality during the famine were attributed to temporary, large-scale migration of males (Pitkänen and Mielke 1993). Thus, social dislocation and the accompanying stress (e.g. living in a new location, crowded living conditions, looking for work) selectively and significantly exacerbated the crisis-related mortality (especially among the more mobile young male segment of the population) during the famine (also see Pitkänen 1993). It is difficult to disentangle the effects of economic deprivation, nutritional stress, social dislocation, and biological factors, such as differential immunity to disease, genetic makeup, or physiological factors. These findings, especially those for the War of Finland, help corroborate the finding that there were no substantial differences in death rates between the sexes in Åland during the same war-induced crisis. They also demonstrate that environmental and cultural factors can easily modify the age and sex patterns of death during a mortality crisis.

Spatial diversity in mortality rates

Even though Åland is a relatively small area (about 10 000 km^2), the increase in crude death rates was not consistent across the archipelago (Table 10.1). The Main Island parishes (e.g. Eckerö/Hammarland and Jomala) experienced higher mortality rates than the more isolated outer island parishes

Table 10.2. *Hazards regression estimates*

Adjusted for the effects of age and sex.

Parish	Hazards ratio	95% C.I.
Outer Islands	1.000	Reference
Peripheral Islands	1.2094	1.0481–1.3955
Main Island	1.2863	1.1377–1.4543

Table 10.3. *Median survival times (in days) by parish location and age group for those individuals who died during 1808–1809*

Survival times are calculated from the arrival time of the Swedish troops on the archipelago to the end of the war.

Parish location	Age 0–15	Age 15–45	Age 45+
Main island	94	219	239
Peripheral parishes	123	244	254
Outer parishes	183	260	257

(e.g. Föglö/Sottunga/Kökar and Kumlinge/Bråndö) (see Fig. 10.1 for location of parishes). Cox's proportional hazards analysis also elucidates the geographic heterogeneity of mortality. These models permit one to predict survival times from one or more independent or predictor variables. A proportional hazards analysis using parish location (Main Island, peripheral, outer islands) as a co-variate demonstrates that the impact of the crisis was disproportionately felt throughout the archipelago (Table 10.2). Survival on the Main Island was significantly shorter ($p < 0.0001$) than in the parishes surrounding the Main Island, which was, in turn, less than in the outer island parishes. The median survival times, in days, for those who were exposed to the crisis and died also clearly show the heterogeneity of death across the archipelago (Table 10.3). The median survival times are much less on the Main Island than on either the peripheral parishes surrounding the Main Island or the outer island parishes to the east. This diversity in survival times is probably related to troop concentrations, which were greater on the Main Island than in the other parishes. Travel by land was also easier on the Main Island than travel by water in the outer and peripheral parishes.

Some concluding remarks

By combining demography, history, human behavior patterns, and genetics, biological anthropologists have made important contributions to our understanding of the structure and evolution of human populations. Many analyses have effectively incorporated archival resources in these studies. The history, geography, demography, and structural composition of a population often shape the distribution and scope of hereditary diseases and the genetic structure in an area (see, for example, Norio 1984; Bittles *et al.* 1986; Smith *et al.* 1990; Relethford and Crawford 1995). Analyses of the population structure of the Åland Islands using genetic data (Jorde *et al.* 1982; Workman and Jorde 1980) and matrimonial migration matrices (Eriksson *et al.* 1973a,b; Mielke *et al.* 1976, 1982; Mielke 1980) demonstrated that there was regional genetic heterogeneity among the fifteen rural parishes. This spatial heterogeneity among the parishes was also reflected in smallpox mortality patterns during the eighteenth and nineteenth centuries (Mielke *et al.* 1984). Vaccination and geographic location were the two most important factors in the spread of smallpox epidemics in the archipelago (Jorde *et al.* 1989, 1990). Likewise, the war-induced mortality pattern reflects the geography and behavior patterns of the inhabitants of the archipelago.

The heterogeneous impact of the War of Finland on the mortality patterns in the Åland archipelago was caused by a number of factors working in concert. The primary differences were due to differential troop concentrations and movements, civilian contact with sick soldiers, crowding or increases in population density, hygiene, and the geographical dispersal of the population throughout the archipelago. The 'epidemiological landscape' (Dobson 1992) of the archipelago produced the spatial diversity in mortality patterns. It was also responsible for the similarity of mortality rates for both males and females. Thus, the geographic, demographic, and cultural structure of the archipelago played a major role in shaping the effects that epidemics and infectious diseases have historically had on the area. These results also suggest that one should be cautious when making general inferences based on the demographic responses of one or two parishes. It also argues against using one parish to make generalizations about others.

Acknowledgements

This research could not have been accomplished without the generous assistance of many people. My thanks go out to Mr Bjarne Henriksson and his staff at the Landskapsarkivet, Åland, for their time, expertise, and patience. I am indebted to Åland's Lutheran ministers who allowed our research team to examine and copy selected archival sources. I thank the staff and director of Åland's Museum for their help. This research was supported by grants from the Sigrid Jusèlius Foundation, Helsinki, Finland.

Notes

1 The Great Finnish Famine of the 1860s started in 1862 with repeatedly poor harvests, particularly in the northern and central regions of Finland. Economic conditions worsened, and after the 1865 harvest failure health conditions started to deteriorate. Mortality levels increased in the spring of 1866 and 1867. In the fall of 1867 there was a total crop failure in most parts of Finland. By 1868 the crude mortality rate was about 80 per 1000 population, and in some regions the crude death rate exceeded 200 per 1000. The population of Finland was reduced by more than 100 000 over the course of the famine. The major causes of the elevated death rates were multiple infectious diseases such as measles, smallpox, typhoid fever, typhus, dysentery, and relapsing fever (see Turpeinen 1986; Pitkänen 1993; Pitkänen and Mielke 1993 for more detail).

References

Anderson, S. (1945). Östaländska Skärgården Under Kriget 1808–1809. *Budkavlen* **24**, 1–45.

Barros, J. (1968). *The Åland Islands Question: Its Settlement by the League of Nations.* New Haven: Yale University Press.

Bartlett, M.S. (1957). Measles periodicity and community size. *Journal of the Royal Statistical Society* A **120**, 48–70.

Bartlett, M.S. (1960). The critical community size for measles in the United States. *Journal of the Royal Statistical Society* A **123**, 37–44.

Bittles, A.H. (1988). Famine and man: demographic and genetic effects of the Irish famine, 1846–1851. In: *Anthropologie et Historie ou Anthropologie Historique?*, ed. L. Buchet (Actes des Troisièmes Journées Anthropologiques de Valbonne, Notes et monographies techniques no. 24), pp. 159–75. Paris: Centre National de la Recherche Scientifique.

Bittles, A.H., McHugh, J.J. and Markov, E. (1986). The Irish famine and its sequel: Population structure changes in the Ards Peninsula, Co. Down, 1841–1911. *Annals of Human Biology* **13**, 473–87.

Black, F. L. (1989) Measles. In *Viral Infections in Humans: Epidemiology and Control*, 3rd edn, ed. A.S. Evans, pp. 451–69. New York: Plenum Medical Book Company.

Bonsdorff, B. von (1975). *The History of Medicine in Finland 1828–1918.* Helsinki: Societas Scientiarum Fennica.

Boyle, P.P. and Ó Gráda, C. (1986). Fertility trends, excess mortality, and the Great Irish Famine. *Demography* **23**, 543–62.

CDC (2001). *Leptospirosis.* Health Topics, Disease Information, Division of Bacterial and Mycotic Diseases, National Center for Infectious Diseases, Centers for Disease Control (http://www.cdc.gov/ncidod/dbmd/diseaseinfo/leptospirosis_g.htm).

Cliff, A.D., Haggett, P., Ord, J.K. and Versey, G.R. (1981). *Spatial Diffusion: An Historical Geography of Epidemics in an Island Community.* Cambridge: Cambridge University Press.

Cliff, A.D., Haggett, P. and Smallman-Raynor, M. (1993). *Measles: An Historical Geography of a Major Viral Disease from Global Eradication to Local Retreat, 1840–1990.* Oxford: Blackwell Publishers.

Corsini, C.A. and Viazzo, P.P. (eds) (1997). *The Decline of Infant and Child Mortality: The European Experience: 1750–1990.* The Hague, The Netherlands: Kluwer Law International.

Dobson, M.J. (1992). Contours of death: Disease, mortality and the environment in early modern England. In *Historical Epidemiology and the Health Transition*, ed. J. Landers, pp. 77–95. *Health Transition Review.* Supplement to Volume 2.

Dreijer, M. (1968) *Glimpses of Åland History.* Mariehamn, Åland: Ålands Museum.

D'Souza, S. and Chen, C.L.C. (1980). Sex differentials in mortality in rural Bangladesh. *Population and Development Review* 6, 257–70.

Eriksson, A.W., Fellman, J.O., Workman, P.L. and Lalouel, J.M. (1973a). Population studies on the Åland Islands I. Prediction of kinship from migration and isolation by distance. *Human Heredity* 23, 422–33.

Eriksson, A.W., Eskola, M.-R., Workman, P.L. and Morton, N.E. (1973b). Population studies on the Åland Islands. II. Historical population structure; inference from bioassay of kinship. *Human Heredity* 23, 511–34.

Hofsten, E. (1970). The availability of data about the Swedish population. *Särtryk Statistisk Tidskrift* 1, 19–26.

Jorde, L.B., Pitkänen, K.J. and Mielke, J.H. (1989). Predicting smallpox epidemics: A statistical analysis of two Finnish populations. *American Journal of Human Biology* 1, 621–9.

Jorde, L.B., Pitkänen, K.J., Mielke, J.H., Fellman, J.O and Eriksson, A.W. (1990). Historical epidemiology of smallpox in Kitee, Finland. In *Disease in Populations in Transition*, ed. A.C. Swedlund and G.J. Armelagos, pp. 183–200. New York: Bergin & Gravey.

Jorde, L.B., Workman, P.L. and Eriksson, A.W. (1982). Genetic microevolution in the Åland Islands, Finland. In *Current Developments in Anthropological Genetics*, Volume 2, ed. M.H. Crawford and J.H. Mielke, pp. 333–65. New York: Plenum Press.

Kane, P. (1989). Famine in China 1959–61: Demographic and social implications. In *Differential Mortality: Methodological Issues and Biosocial Factors*, ed. A. Bouckaert, pp. 231–53. Oxford: Clarendon Press.

Keyfitz, N. and Flieger, W. (1971). *Population: Facts and Methods of Demography.* San Francisco: W.H. Freeman & Company.

Landers, J. (ed.) (1992a). *Historical Epidemiology and the Health Transition. Health Transition Review*, Supplement to Volume 2.

Landers, J. (1992b). Historical epidemiology and the structural analysis of mortality. In *Historical Epidemiology and the Health Transition*, ed. J. Landers, pp. 47–75. *Health Transition Review*, Supplement to Volume 2.

Langford, C. and Storey, P. (1993). Sex differentials in mortality early in the twentieth century: Sri Lanka and India compared. *Population and Development Review* 19(2), 263–82.

Major, R.H. (1941). *Fatal Partners, War and Disease.* Garden City, NY: Doubleday, Doran & Company, Inc.

Mead, W.R. and Jaatinen, S.H. (1975). *The Åland Islands*. London: David and Charles.

Mielke, J.H. (1980). Demographic aspects of population structure in Åland. In *Population and Genetic Disorders*, ed. A.W. Eriksson, H. Forsius, H.R. Nevanlinna, P.L. Workman and R.K. Norio, pp. 471–86. London: Academic Press.

Mielke, J.H., Devor, E.J., Kramer, P.L., Workman, P.L. and Eriksson, A.W. (1982). Historical population structure of the Åland Islands, Finland. In *Current Developments in Anthropological Genetics: Ecology and Population Structure*, ed. M.H. Crawford and J.H. Mielke, pp. 255–332. New York: Plenum Press.

Mielke, J.H., Jorde, L.B., Trapp, P.G., Anderton, D.L., Pitkänen, K.J. and Eriksson, A.W. (1984). Historical epidemiology of smallpox in Åland, Finland, 1751–1890. *Demography* 21, 271–95.

Mielke, J.H. and Pitkänen, K.J. (1989). War demography: the impact of the 1808–09 war on the civilian population of Åland, Finland. *European Journal of Population* 5, 373–98.

Mielke, J.H., Pitkänen, K.J., Jorde, L.B., Fellman, J.O. and Eriksson, A.W. (1987). Demographic patterns in the Åland Islands, Finland, 1750–1900. *Yearbook of Population Research in Finland* 25, 57–74.

Mielke, J.H., Workman, P.L., Fellman, J. and Eriksson, A.W. (1976). Population structure of the Åland Islands, Finland. *Advances in Human Genetics* 6, 241–321.

Norio, R. (1984). The Finnish disease heritage as a product of the population structure. *Yearbook of Population Research in Finland* 22, 7–14.

North, K.E., Crawford, M.H. and Relethford, J.H. (1999). Spatial variation in anthropometric traits in Ireland. *Human Biology* 71(5), 823–45.

Palloni, A. (1990). Assessing the levels and impact of mortality in crisis situations. In *Measurement and Analysis of Mortality: New Approaches*, ed. J. Vallin, S. D'Souza and A. Palloni, pp. 194–228. Oxford: Clarendon Press.

Pitkänen, K. (1977). The reliability of the registration of births and deaths in Finland in the eighteenth and nineteenth centuries: Some examples. *Scandinavian Economic History Review* 25, 138–59.

Pitkänen, K. (1993). *Deprivation and Disease: Mortality During the Great Finnish Famine of the 1860s*. Helsinki: Publications of the Finnish Demographic Society, 14.

Pitkänen, K. and Mielke, J.H. (1993). Age and sex differentials in mortality during two nineteenth century population crises. *European Journal of Population* 9, 1–32.

Pitkänen, K. and Nieminen, M. (1984). The history of population registration and demographic data collection in Finland. In *National Population Bibliography of Finland 1945–1978*, pp. 11–22. Helsinki: Finnish Demographic Society.

Preston, S.H. (1976). *Mortality Patterns in National Populations*. New York: Academic Press.

Puntila, L.A. (1974). *The Political History of Finland, 1809–1966*. Helsinki, Finland: The Otava Publishing Company.

Relethford, J.H. and Crawford, M.H. (1995). Anthropometric variation and the population history of Ireland. *American Journal of Physical Anthropology* 96, 25–38.

Relethford, J.H., Crawford, M.H. and Blangero, J. (1997). Genetic drift and gene flow in post-famine Ireland. *Human Biology* 69(4), 443–65.

Samuelsson, G. (1944). *Lantvärnet 1808–1809.* Uppsala, Sweden.

Schulman, H. (1909). *Taistelu Suomesta 1808–1809.* Porvoo, Suomi-Finland: WSOY.

Sjölund, F. (1972). *Åland-Mariehamn Guide.* Helsinki: Kuultokuva.

Smith, M.T., Williams, W.R., McHugh, J.J. and Bittles, A.H. (1990). Isonymic analysis of post-famine relationships in the Ards Peninsula, N.E. Ireland: Effects of geographical and politico-religious boundaries. *American Journal of Human Biology* **2**, 245–54.

Turpeinen, O. (1986). *Nälkä vai tauti tappoi? Kauhunvuodet 1866–1868.* (Historiallisia Tutkimuksia, 136). Helsinki: Societas Historica Finlandiæ.

Urlanis, B.T. (1971). *Wars and Population.* Moscow: Progress Publishers.

Vallin, J. (1995). Can sex differentials in mortality be explained by socio-economic mortality differentials? In *Adult Mortality in Developed Countries: From Description to Explanation,* ed. A.D. Lopez, C. Graziella and T. Valkonen, pp. 179–200. Oxford: Clarendon Press.

Waldron, I. (1983). Sex differences in human mortality: The role of genetic factors. *Social Science and Medicine* **17**(6), 321–33.

Waldron, I. (1993). Recent trends in sex mortality ratios for adults in developed countries. *Social Science and Medicine* **36**(4), 451–62.

Wingard, D.L. (1984). The sex differential in morbidity, mortality, and life style. *Annual Review of Public Health* **5**, 433–58.

Workman, P.L. and Jorde, L.B. (1980). The genetic structure of the Åland Islands. In *Population Structure and Genetic Disorders,* ed. A.W. Eriksson, H.R. Forsius, H.R. Nevanlinna, P.L. Workman and R.K. Norio, pp. 487–508. London: Academic Press.

11 *Infectious diseases in the historical archives: a modeling approach*

LISA SATTENSPIEL

Introduction

Human biologists, especially those studying demography and disease in past populations, have long used historical archives to learn more about the populations they are studying. A variety of historical and statistical techniques have been used to analyze and interpret data derived from the archives. Mathematical and computer modeling are approaches that have only rarely been used, but they can provide interesting and valuable insights not generally possible with other methods. In the following I first discuss the nature of models in general and mathematical models in particular. This is followed by an introduction to several of the most commonly used approaches to mathematical and computer modeling in the social sciences and a selective overview of the use of these approaches to study infectious diseases in human populations. Finally, I illustrate in some detail the process of model development and analysis, drawing upon work I have done in conjunction with Ann Herring on the spread of the 1918–19 influenza epidemic in central Canada.

What is a model?

People often use the word 'model' very loosely and in different ways. To some degree, the lax usage of the word reflects both real ambiguities in the concept and the use of one word to designate several related ideas. At the most general level, any model can be thought of as an object or concept that is used to represent something else. Models simplify reality and aid in determining the role and importance of factors that influence the real world. Some models are physical reproductions of a real world object such as a train or an influenza virus. Other models are drawings of objects or ideas, such as a drawing of a tree or a flowchart of a decision-making process. A third kind of model is a verbal model – a set of statements that are logically connected to one another and that illustrate a particular process or situation. The models that are the

234

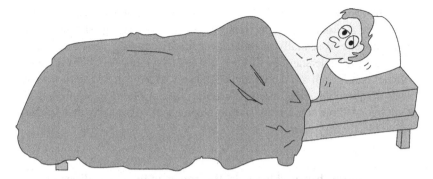

Fig. 11.1. A model of a miserable person that can be interpreted very differently by different observers. Inspiration for this model came from a drawing and discussion of 'Bessie the Cow' in Cross and Moscardini (1985).

focus of this chapter are mathematically based models, which use an explicit or implicit set of mathematical equations to characterize some aspect of the real world.

Models are not exact reproductions of a real system. No matter what type of modeling approach is used, simplifying assumptions are made that reduce real world complexities to a more manageable form. Most situations being modeled do not have a unique representation; the same model can mean different things to different people. For example, Fig. 11.1 is a model of a person lying in a bed. To one observer, this figure may bring to mind an overextended college student who has just realized that he overslept and missed his final exam, while a second observer sees in the figure a critically ill and miserable person who needs the attention of health personnel. The first observer might use this figure to build and illustrate a model that could be used to consider questions such as how to improve a college student's ability to manage time, or how to predict the consequences of negligent behavior on success in school. The second observer might use the same figure to build and illustrate a model of illness behavior, of access to health care resources, or of the natural history of a specific disease.

Mathematical models can provide valuable insights into the nature of many different types of anthropological problems. These problems are often highly complex and consist of numerous interrelated parts. Mathematical models can help to elucidate the relative importance of each of the parts, as well as the impact of connections between one or more parts. They provide a way to experiment on human populations without actually doing invasive research that would be ethically unacceptable or technically unfeasible. For example, mathematical models that take into account physiological and pharmacological processes in the human body can be used to help determine the consequences of increasing dosages of a curative, but potentially toxic drug before experimenting with living humans. Essentially, mathematical models are best used as a way to enhance basic understanding of how a complex system works. Furthermore, a model that is well structured and adequately tested can be used to answer 'what-if' questions about the behavior of a system that may aid in both the prediction of future behavior and the development of ways to alter that behavior, if desired.

Approaches used to model the impact of infectious diseases in human populations

Just as the same model can mean different things to different people, a particular question can often be modeled using a variety of approaches. Choice of a particular approach is most frequently a matter of a researcher's familiarity with and expertise in a particular method, although sometimes the specific questions asked or characteristics of the situation under study determine the most appropriate modeling technique to use. Several of the approaches most commonly used in social science research will be described here, with links, if any, to studies that have used these approaches to study the impact of infectious diseases in human populations. Neither the approaches described nor the examples mentioned are an exhaustive review of available resources. Rather, the intent is simply to provide an introduction to relevant areas of research.

Modeling approaches can be divided into three partially overlapping types: statistical approaches, mathematical models, and computer simulations. Generally speaking, statistical models begin with a set of data and try to find mathematical relationships and social, environmental, or biological factors that are likely to be the underlying cause of patterns in the data. Mathematical models work from the other direction; they begin with assumptions about and characteristics of the system that generates a set of data and try to reproduce patterns that are similar to the observed data. Computer simulations usually begin with assumptions about and characteristics of the system that generates a set of data, but mathematical equations that characterize these features are often not made explicit.

Statistical approaches

A number of statistical approaches have been used in studies of modern and historical epidemics in human populations, including regression and correlation analysis, time series analysis, generalized linear models, and spatial statistics. For example, Bartlett (1957) and Black (1975) used simple correlation or regression analysis to derive mathematical relationships between population size and the frequency and severity of epidemics. These studies clearly demonstrated with historical data on recurrent epidemics that there exists for most infectious diseases a critical population size below which an infectious disease cannot be maintained indefinitely in a population. This was a key result that has also been shown theoretically and that motivates many current efforts to determine effective control strategies for infectious diseases.

The majority of statistical modeling approaches within the anthropology of infectious disease focus on time series analyses of a set of historical data. Time series (or spectral) analysis is used to filter noise out of a detailed data set so that underlying cycles or other patterning can be observed in the data. These patterns are often related to the biology of the infectious organism or to historical events in the populations under study. Mielke *et al.* (1984) demonstrated strong periodicity in smallpox deaths in the Åland Islands, Finland, during the period 1751–1890 and tied specific mortality peaks to events such as the War of Finland in 1808–09. Scott and Duncan (1998, 2001) have used time series analysis to detect periodicity in deaths from smallpox, measles, whooping cough, scarlet fever, diphtheria, and plague in England from the seventeenth through the nineteenth centuries. Cliff and colleagues investigated a number of diseases throughout the world, including measles (Cliff *et al.* 1983, 1993; Cliff and Haggett, 1985, 1988) and influenza (Cliff *et al.* 1986). These and other diseases are discussed further by Cliff *et al.* (1981, 1998), Cliff and Haggett (1988) and Smallman-Raynor, *et al.* (1992).

A number of researchers have used generalized linear models to analyze large disease data sets and draw out essential factors explaining variation in the data so that more general patterns can be derived. These derived patterns are often then used to estimate parameters in explanatory mathematical models. Examples of this approach include Morris (1995), who used generalized linear models to analyze social network data, and Finkenstädt and Grenfell (1998), who used this method in the analysis of spatiotemporal data on measles incidence in England and Wales between 1944 and 1966. The use of generalized linear models to analyze infectious disease data is also discussed by Cliff *et al.* (1981, 1998).

In addition to these techniques, several other statistical methods have been developed by geographers to analyze the spatial distribution of infectious diseases. Many of these methods are described and illustrated in Robinson (2000) and in the books by Cliff and colleagues mentioned above. Robinson (2000) also includes a discussion of the use of GIS (global information systems) technology in infectious disease research.

Mathematical approaches

The work to be illustrated in the latter half of this chapter falls under the category of mathematical approaches to modeling infectious disease distributions. This category covers an extensive body of research, especially in mathematics and ecology, and several textbooks have been written on the topic, including those of Bailey (1975) and Anderson and May (1991). However, anthropologists have contributed relatively little to this area. In addition, the bulk of the research that focuses on human diseases is primarily theoretical in orientation. The illustrations that follow, however, will be drawn from the smaller number that deals with actual epidemics and disease data sets.

Ultimately the question being asked and the modeler's experience with different methods govern choice of the type of mathematical model to use, but there are several other issues to consider. First, a modeler must decide whether to focus at the population or individual level. In population-level models, all individuals are generally assumed to be identical with characteristics corresponding to the 'average' or 'mean' individual in the population being modeled. These models are generally easier to develop and analyze than individual-based models, and their use has resulted in substantial increases in understanding the systems being modeled. Individual-level models, by contrast, treat everyone differently and can be used to examine the impact of heterogeneity in individual behavior and biology.

A second consideration is whether a population is structured or unstructured. A structured population is one that is divided into a number of smaller groups on the basis of some structural property, such as age, sex, risk behaviors, geographic locality, etc. Individuals within each of the subgroups are assumed to be equal, but the subgroups differ in their characteristics. Individual-level models can be thought of as a subclass of structured population models with each subgroup numbering a single individual.

A third consideration important in mathematical approaches is how time is measured. Models in which time is measured in distinct steps of measurable length are called discrete-time models; those in which the time unit is arbitrarily small are called continuous-time models. The majority of continuous-time epidemic models are based on one or a system of differential equations. Discrete-time models, however, are often based on one or a system of difference equations.

Mathematical modelers also consider whether the random nature of real events is to be incorporated into a model. These random effects are most important when populations are small, which is often the case for anthropological groups. Models that ignore random effects are said to be deterministic. In this case, parameters that are likely to vary in the real world are assumed to be constant at some mean value that is (or should be) estimated from data. Like population-based models, deterministic models tend to be easier to develop and analyze and have resulted in significant insights into how real systems operate. However, since most anthropological populations are small, models that take random effects into account (called stochastic models) tend to be more realistic.

Mathematical approaches that have been used in studying the spread of infectious diseases in past and present human populations fall into two basic types: life table models and population projections, and compartmental epidemic models. Life table models are probably the most commonly used within the anthropological community and are derived from classic demography. They are discrete-time, deterministic models for populations that are generally structured according to age. Most often these models use the overall fertility and mortality schedules of a population to predict population growth and changes in age structure over time. Because infectious diseases are a common cause of death, life table models have also been used to address how specific diseases or groups of disease may affect the growth and age distribution of a population. Examples of this approach include studies looking at the effects of smallpox on Native American communities (Thornton *et al.* 1991) and the effects of a variety of diseases on Arikara populations in the North American plains (Palkovich 1981) and on colonial Mexican populations (Whitmore 1992).

Compartmental epidemic models compose the second major type of mathematical approach to disease modeling. These models explicitly follow the progression of a disease within individuals forming a population by considering the different biological stages of a disease (e.g. susceptible, infected, recovered, exposed, etc.) and rates of progression from one stage to the next. A detailed example of the formulation of a model of this type is described below. Many different types of compartmental models have been used to study human disease, including both discrete-time and continuous-time models, both deterministic and stochastic models, both individual- and population-based models, and models for both structured and unstructured populations.

The simplest compartmental epidemic models are unstructured, population-based, deterministic models in continuous time. The SIR epidemic model, which is described more fully below (see p. 247), is an example of this kind of model. Ramenofsky (1987) used insights from the analysis of the SIR model to evaluate whether infectious diseases could have been responsible for catastrophic population declines in central and northeast US aboriginal populations. Upham (1986) used the SIR model to explore the impact of smallpox on native populations in the American Southwest.

Several studies have also looked at continuous-time, deterministic, population-based models in structured populations. For example, Sattenspiel has used models of this type to study the spread of hepatitis A in day care centers (Sattenspiel 1987a; Sattenspiel and Simon 1988), the spread of measles on a Caribbean island (Sattenspiel and Powell 1993), and the spread of influenza in central Canada (Sattenspiel and Herring 1998; Sattenspiel et al. 2000; Herring and Sattenspiel 2002). Rvachev and Longini (1985) and Longini (1988) used airline data between Russian cities in a structured population model to predict patterns of spread of influenza throughout the country. Numerous models of this type have also been applied to the spread of HIV (see for example, Jacquez et al. 1988; Castillo-Chavez 1989; Kaplan and Brandeau 1994) and many other pathogens.

Discrete time, stochastic population-based models have also been used to study infectious disease spread in human groups. One of the simplest of these models is the Reed–Frost or chain-binomial model. This model defines the risk of transmission as a *probability* given contact between susceptible and infectious individuals, rather than as a constant proportion of those contacts (which is an assumption of compartmental models). It predicts the expected number of new cases as a function of the existing cases and the probability of transmission. Bartlett (1960) used this model in a classic study of measles and chickenpox in the United States. Milner (1980) used a Reed–Frost model to assess whether ethnohistoric accounts of disease among native groups in the southeastern United States reflect accurately the likely impact of infectious diseases on population change. McGrath (1988) used a similar model to study the spread of tuberculosis within prehistoric populations in the central United States. Cliff et al. (1981) used a structured chain-binomial model to study the geographic spread of measles in Iceland.

A few examples exist of composite models that follow the population-level spread of a disease across individual-based structures. Most of these models consider individuals to be nodes on a graph and to have a particular disease status (e.g. susceptible, infectious, etc.). Rules govern contact between individuals and resultant change in status. The difference between models of this type and population-based models in structured populations is that contact occurs between individuals themselves rather than between groups. In fact, as mentioned above, these models could be considered limiting cases of structured population models where all groups consist of one individual only. Much of the work in this area uses lattice-based models (sometimes called cellular automata) and the body of mathematics called percolation theory. In these models individuals are assumed to exist on the nodes of regular 1-, 2-, or 3-dimensional lattices and to have a disease status. Infection of susceptible individuals occurs as a consequence of transmission by nearest neighbors who are infected with the disease. Although most work with lattice-based epidemic models tends to be more theoretical than practical, there are a few exceptions. For example,

Rhodes and Anderson (1996) applied a lattice-based model to the spread of measles in the Faroe Islands.

Some of the most interesting work on disease spread across individual-based structures concerns the 'small-world phenomenon' (Milgram 1967; Pool and Kochen 1978). This phenomenon is the idea that any two individuals selected at random anywhere in the world can be linked by a very small number of intermediate acquaintances. Essentially these models use graph theory to define a particular type of network that includes both highly clustered and highly random associations between individuals. This combination of associations seems to be what is responsible for the unexpectedly close links among individuals. Watts (1999) and Moore and Newman (2000) applied this model to the spread of infectious diseases by introducing disease onto the small-world network of social contacts, and following the patterns of spread of the disease. Kretzschmar and Morris (1996) and Morris and Kretzschmar (1997) have also considered the connection between individual contact networks and the population level spread of infectious diseases.

Computer-based approaches

Mathematical and statistical approaches have resulted in numerous important insights into the impact of infectious diseases on human groups and the importance of various factors influencing disease transmission. However, the majority of these approaches assume large populations and/or little heterogeneity among individuals or populations. These assumptions often are not met for anthropological populations, which places boundaries on the value of mathematical and statistical approaches for anthropological questions. Computer-based approaches provide an important body of techniques that allow these limitations to be surmounted.

It is important to note the distinction between computer-based approaches that involve individual-based stochastic simulation models and the computer-based numerical analysis of a mathematical model. Historically, mathematical models were analyzed using traditional mathematical methods, including stability analysis and proofs of theorems. Attention was focused on the identification of equilibrium conditions under which a disease became either endemic or extinct and on the conditions necessary to reach these equilibrium states. The advent of inexpensive, powerful computers led to the development of several computer programs that approximate the mathematical models and that are used to explore their behavior numerically. Technically speaking, this kind of analysis is not computer simulation, since it is fully deterministic, meaning that a given set of input data always leads to the same output results. It is nonetheless an important technique used to analyze mathematical models and understand the dynamics of the systems they represent. Detailed illustrations of the numerical analysis of a compartmental epidemic model are provided below.

Four approaches to true computer simulation are in common use in social science research today: microsimulation, agent-based models, genetic algorithms, and neural networks. All of these approaches result in the development of individual-based, stochastic models. Microsimulation is the only one of these techniques that has been used substantially in epidemiology, but the importance of all others is increasing rapidly. Consequently, all four will be described in the hopes that readers will recognize the potential of each for research on historical and modern epidemics.

Individual-based computer simulation models are usually very difficult, if not impossible, to analyze mathematically and usually do not have explicit mathematical equations governing modeled behaviors. Each individual in a population possesses a set of attributes that most often vary from one individual to the next. Furthermore, most of these models are fully stochastic, in that events occur with a certain probability that varies from one run of a simulation to another. Because each simulation is different, the models are run repeatedly so that underlying patterns in the data can be observed.

Microsimulation models have been in use longer than the other computer-based approaches and nearly all epidemiological applications are of this type. Sometimes these models are called Monte Carlo simulations, in reference to the key role played by random numbers in determining the outcomes of model events. However, because the phrase 'Monte Carlo' is used in several different contexts involving random number generators, including parameter estimation procedures, it is best to avoid using it to refer to the microsimulations themselves. Microsimulation essentially involves defining the attributes of the individuals within a population and setting up rules for interactions among individuals that may result in a change in one or more of those attributes. Changes occur with predefined probabilities and when the condition for a potential change is met, a random number is generated. If the random number is equal to or lower than the probability, the change occurs; if it is higher the attribute remains as it was. For example, in a disease model the probability of transmission from an infectious individual to a susceptible individual may be set at 0.15. When two such individuals meet during the course of a simulation, a random number is generated. If the generated number is less than or equal to 0.15, then the susceptible individual becomes infectious; otherwise that individual remains susceptible. At each time unit of the simulation all individuals are evaluated and attributes are changed, if appropriate. Sometimes simulation models measure time in terms of events rather than conventional time units. In other words, changes in attributes occur whenever an event under study occurs, which results in variability in the time scale of the simulation. Models incorporating this idea are called discrete event models.

Several examples of microsimulations of the spread of infectious diseases are present in the literature. Bartlett (1961) used microsimulation to extend his groundbreaking work on critical community size (Bartlett 1957, 1960) to questions about causes of the seasonality of measles epidemics. A research group at the University of Minnesota designed detailed microsimulation models of community structure, including schools, work activities for adults, and many other features of daily life, and applied these models to the spread of several different infectious diseases, including influenza. The models and many of their results are described by Ackerman *et al.* (1984). Sattenspiel (1987b) used a microsimulation model that modeled neighborhoods and day care facilities to study the spread of hepatitis A among preschool aged children in New Mexico. Kretzschmar and Morris (1996) and Morris and Kretzschmar (1997) designed a microsimulation to help explore the behavior of their model combining social network structure and HIV transmission. Fix (1984) included infectious diseases as a primary selective factor in his simulations of the demographic and genetic structure of Semai Senoi populations in Malaysia. In addition, biologists, parasitologists, and infectious disease specialists have developed detailed microsimulation models focused on particular diseases, including schistosomiasis (de Vlas *et al.* 1996; Habbema *et al.* 1996), onchocerciasis (Plaisier *et al.* 1990), and sexually transmitted diseases (Bernstein *et al.* 1998; Van der Ploeg *et al.* 1998).

The remaining three computer-based approaches (agent-based models, genetic algorithms, and neural networks) are becoming increasingly common and valuable in social science research, but have only rarely been applied to infectious disease transmission. Agent-based models are also called individual-oriented, distributed artificial intelligence-based, or artificial societies (Gilbert and Conte 1995; Epstein and Axtell 1996; Kohler 2000). The models consist of three basic components: agents, the environment or social space, and rules that govern interactions among agents (Epstein and Axtell 1996). Agents are entities that collect information about the environment and make decisions or take action based on that information (Doran *et al.* 1994; Russell and Norvig 1995; Kohler 2000). They can be programmed either to be reactive, which means decisions are 'hard-wired' and based on a small number of relatively simple rules, or to be deliberative, which means that they have associated goals rather than hard and fast rules and can make their decisions in response to conditions in the environment (Doran 2000). The aim of agent-based models is to identify local-scale mechanisms that generate large-scale phenomena, such as social structure or group-level behaviors (Epstein and Axtell 1996).

Epstein and Axtell (1996) provide one of the few examples of the use of an agent-based model in infectious disease research. Their approach is similar to the traditional microsimulations, but instead of using decisions determined by reference to a random number generator (the Monte Carlo approach), they build a changeable immune system into each agent. If a neighboring agent is infected with a disease, the model checks to see whether the immune system is sufficient to provide immunity. If the agent does not have sufficient immunity, the disease is transmitted with 100% probability. Allowing the agents to move to neighboring unoccupied cells from which they can interact with new neighbors facilitates transmission throughout a population.

Several other agent-based models have been used to address anthropological questions. Some of these have involved development of computer code by the researchers themselves; others have involved adaptations of general programs for agent-based simulations. Several recent anthropological studies in this area are described in Kohler and Gumerman (2000). Examples include the EOS project, which uses an artificial society to explore the development of social hierarchies and centralized decision-making (Doran *et al.* 1994; Doran and Palmer 1995); the MAGICAL simulation of hunters and gatherers in the Scottish Mesolithic, which incorporates GIS technology into an agent-based model (Lake 2000); the work of Kohler *et al.* (2000) on modeling settlement patterns in Mesa Verde, which uses a freely available program called SWARM that has been developed by the Sante Fe Institute; and the Artificial Anasazi project of Dean *et al.* (2000), which uses the Sugarscape model originally developed by Epstein and Axtell (1996).

Genetic algorithms and neural networks are both examples of learning or evolutionary models. They are similar to agent-based models, but they incorporate evolutionary mechanisms that lead to alterations in parameters and/or model structure in response to the environment. As the names might suggest, genetic algorithms are based on the mechanisms of natural selection, whereas neural networks are based on structures thought to be analogous to mechanisms used by the brain to process and respond to information from the environment. Genetic algorithms have been used extensively to model rational action and associate different fitnesses with different types of actions (Gilbert and Troitzsch 1999). An extensive review of the literature failed to turn up examples of the use of either genetic algorithms or neural networks in infectious disease modeling.

Using mathematical models to supplement historical research on the spread of infectious diseases in past populations

No matter what kind of modeling is attempted, a model is only as good as the data available to use in estimating basic parameters. Historical archives are often a rich source of data that can fill this purpose. However, the useful links between historical archives and mathematical and computer models are not just limited to the archives as a source of data. Models themselves can help to direct the search through archives so that useful data are culled from the records, they can help to interpret the records, and they can stimulate the formulation of new research questions that may be answered by using the archives.

In the remainder of this chapter, the general process of using a mathematical model in conjunction with more traditional methods of archival research will be illustrated. Although the illustration is limited to the development of a compartmental epidemic model, many of the issues addressed are common to all kinds of modeling. The discussion outlines the process of working with a mathematical model. The details of the process are explained by following the path that Ann Herring (an historical demographer) and Lisa Sattenspiel (a mathematical modeler) took in setting up and working with a project to study the spread of an historical epidemic, the 1918–19 influenza epidemic in central Canadian fur-trapping populations. The discussion includes personal experiences, successes, and occasional difficulties that were encountered along the way. It is hoped that by seeing this process from the beginning readers can understand better the contributions that mathematical and computer modeling can make to archival research.

The process of building a mathematical model

The development and analysis of a mathematical model is best considered as a series of steps. These steps reflect the feedback process characteristic of the mathematical modeling enterprise, whereby a setting and questions are chosen, an appropriate model is developed, data are gathered with which to estimate model parameters, the model is run on a computer and results are analyzed. As a consequence of this process, new questions are devised, the model is adapted, and new data are collected to estimate additional parameters. The cycle can continue indefinitely by running the new model, analyzing data, readjusting the questions and models, etc.

Step 1: Identifying a project that will benefit from mathematical modeling

Mathematical modeling is essentially a technique that aids in developing a more structured way to think about a problem. As such, it can be used in almost any research project. However, it is best suited to looking at situations where there is some phenomenon that changes over time. For example, Sattenspiel and Herring's work focuses on the spread of the 1918–19 influenza (flu) epidemic in central Canada.

Work on this project was stimulated by Herring's decision to conduct research on the demographic and social consequences of epidemics of infectious disease in Aboriginal populations in Canada. One of her first thoughts was to check into the availability of suitable data from the Hudson's Bay Company (HBC) archives. The HBC was a large business that originally focused on fur-trapping and later expanded to include exploration and retail. Eventually the company reached across the entire North American continent and stretched from the Arctic Ocean south to Oregon. It also had agents in Chile, Hawaii, California, and Siberia. The company required its employees to keep meticulous records, many of which survive to the present day. These archives provide a treasure trove of information on culture, environment, economics, climate, biology, and many other features of life in the northern half of North America from the 1670s into modern times (HBCA 2000). Thus, many Canadian and northern US historians, demographers, geographers, and biologists have looked towards these archives for interesting and useful data.

After perusing the holdings of the HBC archives, Herring decided to focus her research on the community of Norway House, which is located at the north end of Lake Winnipeg in Manitoba, Canada. Initial investigations of the Norway House archives indicated a strong mortality peak in the winter of 1918–19, which further investigation clarified as being due to the Spanish flu epidemic (Herring 1994). This epidemic was a major pandemic affecting populations in every corner of the world and causing over 20 million deaths within one year (Graves 1969). Furthermore, age groups normally minimally affected by infectious disease epidemics, children aged 5–14 and adults aged 20–40, were especially hard hit by the 1918–19 flu (Walters 1978; Patterson 1986). Norway House, with a death rate of about 200 deaths per 1000 individuals, was severely affected by the epidemic, but it largely missed neighboring communities.

The degree of heterogeneity in flu mortality within a very small area was so surprising that Herring began to think about the effects of this virgin soil epidemic more generally on the demographic and social history of the community. Exposure to Ray's work on epidemics and how they moved in concert with fur trade activities (Ray 1974, 1976) and Sattenspiel's work on modeling the geographic spread of infectious diseases motivated her to consider how population mobility could have influenced the transmission of the 1918 flu across time and space in the Canadian north. Consequently, in 1992 Sattenspiel and Herring began a fruitful collaboration to study this problem.

Step 2: Identifying a specific problem or question

Once the stage is set for mathematical modeling research, it is necessary to formulate one or more interesting questions that are specific enough to be manageable, but general enough that the answer is not obvious prior to the start of the research. For example, questions that underlie our research on the geographic spread of infectious diseases in central Canada include:

- How does population mobility influence the geographic spread of infectious diseases?
- Why was there a huge spike in mortality at Norway House in 1918, especially in comparison with mortality levels at nearby communities?
- Can this variability in mortality levels among communities be explained by rates and patterns of mobility of fur-trappers working for the Hudson's Bay Company?

The first of these questions is a general question that is independent of a specific disease or locality. It has been used to form the backbone of several projects, including a study of the spread of hepatitis A in Albuquerque, New Mexico daycare centers (Sattenspiel 1987a,b; Sattenspiel and Simon 1988) and a study of the spread of measles on the Caribbean island of Dominica (Sattenspiel and Powell 1993; Sattenspiel and Dietz 1995), as well as work on the spread of the 1918–19 flu in central Canada that is discussed throughout this chapter (Sattenspiel and Herring 1998; Sattenspiel *et al.* 2000).

The second question was stimulated by Herring's archival research on the demography of populations in the Norway House district (Herring 1994), which, as mentioned above, identified significant differences in the mortality experiences of communities that were neighbors both in terms of geographic distance and in terms of social connectedness. The third question homed in on a particular explanation for the observed mortality spike. These two questions provided a clear framework to use in initial model development analysis.

Step 3: Designing the model

Most mathematical models used in epidemiology are compartmental models, in which a population is distributed into a series of disease stages with specific rates of flow that determine the numbers of individuals that move from one stage to another at any particular time. The particular stages chosen depend on the disease that is being modeled and the questions that are to be answered by using the model. A disease like influenza, with permanent immunity and a short incubation period, is usually modeled using three disease stages.

- Susceptible (S): individuals who are at risk of becoming infected
- Infectious (I): individuals who are capable of transmitting infectious particles to susceptible individuals

- Recovered or removed (R): individuals who are capable neither of becoming infected nor of transmitting infection, usually because of temporary or permanent immunity to infection.

In the simplest model with these three stages, commonly known as an SIR model because of the letters representing each included stage, a closed population with no births and deaths is assumed. Such a model is shown in Fig. 11.2, with the stages represented as boxes. Diseases with significant rates of mortality may be modeled by allowing infectious individuals to proceed to either recovery or death. Diseases with incubation periods longer than a few days, such as AIDS or measles, usually include a fourth stage, 'exposed,' in which people who are infected but are not yet capable of transmitting a disease are placed, thus generating an SEIR model. Most models of HIV transmission also include a number of distinct infectious stages to better model variability in the rates of transmission over the long course of the infectious period. Models for diseases with temporary immunity must allow removed individuals to return to a susceptible state (generating, for example, SIRS or SEIRS models).

Once the stages are defined, rates of flow between them must be determined. The simplest SIR model uses one parameter to govern the flow from the susceptible stage to the infectious stage and a second parameter to govern the flow from the infectious stage to the removed stage. Flow from the susceptible stage to the infectious stage occurs as a consequence of transmission of the disease, which is assumed to occur at a constant rate, β. Flow from the infectious stage to the removed stage occurs, for influenza and many other diseases, as a consequence of recovery followed by permanent immunity. Recovery is assumed to occur at a constant rate, γ (see Fig. 11.2).

This structure of disease stages and assumptions about flow between stages generates a single population model for the spread of an infectious disease with permanent immunity, but geographic spread is not meaningful unless multiple communities are considered. Such models use the structure illustrated in Fig. 11.2 to model the disease process within each community, but the models must also incorporate a mobility process that links the communities to each

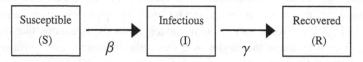

Fig. 11.2. Structure of a basic SIR epidemic model. β, the flow rate from the susceptible to infectious class, is the fraction of contacts between susceptible and infectious individuals that result in transmission of the disease; γ, the flow rate from the infectious to recovered class, is the rate of recovery from the disease; $1/\gamma$ gives the length of the infectious period.

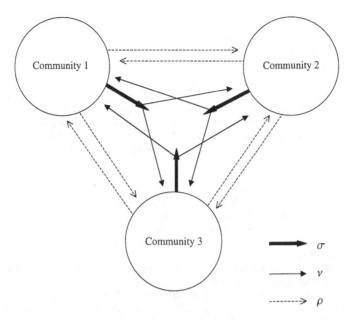

Fig. 11.3. Structure of the mobility component of the model for the geographic spread of infectious diseases. Residents of each community leave on trips at a rate, σ, represented by large arrows. The distribution of travelers among possible destinations is governed by ν and represented by small, solid arrows in the figure. Travelers return home from a visit at a rate ρ, represented by dashed arrows.

other. A simple, yet reasonable, model of such a mobility process is given in Fig. 11.3. In this model, a proportion, σ_i, of residents at home in, say, community i travel out of that community during a given time unit. These travelers then have a probability ν_{ij} of traveling from community i to community j. They return home from community j to community i at a rate ρ_{ij}.

The epidemic model and the mobility model are then combined to generate a set of mathematical equations describing the spread of an infection within and among communities (see Sattenspiel and Dietz (1995) or Sattenspiel and Herring (1998) for details of the model). The mobility process is as described above. The term β in the simple SIR formulation described above, however, represents both the number of contacts per person and the proportion of those contacts that result in transmission. The number of contacts per person is an external factor related to the social activities of a population that can potentially be observed ethnographically, whereas the proportion of contacts resulting in transmission is an internal, biological factor that is generally not directly observable. For this reason, the model used in the work described below separates the two factors into different parameters (κ and τ), even though mathematically

Table 11.1. *Parameters and variables of an SIR model for the geographic spread of infectious diseases*

A. *Structural Variables*
  S_{ii} = the number of susceptible individuals from community i who are at home
  S_{ij} = the number of susceptible individuals from community i who are visiting community j
  I_{ii} = the number of infectious individuals from community i who are at home
  I_{ij} = the number of infectious individuals from community i who are visiting community j
  R_{ii} = the number of recovered individuals from community i who are at home
  R_{ij} = the number of recovered individuals from community i who are visiting community j
  N_i = the total population size of community i ($= S_{ii} + \sum_j S_{ij} + I_{ii} + \sum_j I_{ij} + R_{ii} + \sum_j R_{ij}$)
B. *Epidemiological Parameters*
  τ = the fraction of contacts between a susceptible person and an infectious person that result in transmission of the infectious organism
  κ = the average number of contacts made per unit time
  γ = the rate of recovery from the disease; $1/\gamma$ is the average length of the infectious period
C. *Mobility Parameters*
  σ_i = the daily rate at which individuals from community i leave on trips; $1/\sigma_i$ is the average number of days between trips
  ν_{ij} = the probability that a traveler from community i goes to community j
  ρ_{ij} = the rate at which a community i resident visiting community j returns home; $1/\rho_{ij}$ is the average length (in days) of a trip

they can be and usually are represented as one parameter. Table 11.1 shows the parameters and variables found in the full model.

Step 4: Generating initial estimates of parameters

Before a model can be simulated, it is necessary to estimate initial values of all parameters and variables. Whenever possible, these estimates should be derived from analysis of actual data. For example, in the model described above, mobility and demographic parameters (population sizes, estimates of σ for each community, and all values of ν) were derived from a variety of archival sources. The primary sources of data on travel among communities were Hudson's Bay Company post journals, which are detailed journals from trading posts operated by the HBC. These journals recorded a variety of details on life around the posts, including weather, daily activities, noteworthy events, arrivals and departures of visitors, and trips taken by HBC officials into the surrounding hinterland. In addition, many of the journals include copies of accounts and correspondence of the post official and some may also include medical journals kept by the post surgeon (HBCA 2000). Additional information on modes and routes of travel was garnered from an analysis of historical maps and photographs from the HBC and National Archives of Canada.

Initial values of the rates of travel out of communities (σ) were derived by counting the number of daily arrivals and departures recorded in the Hudson's Bay Company Post Journal [hereafter, HBC or HBCA] for each location (HBCA B.154/a/87; HBCA B.156/a/44; HBCA B.283/a/8) during the epidemic period (December 1918 through February 1919). Values of σ_i were calculated by dividing the total number of visits made by residents of community i to all other communities by the product of the population size of the community and the number of days in the period covered by the visiting data (89 days). Values for the distribution of visits among communities (v) were derived from the same three HBC post journals as estimates of σ. For these parameters, the origins and destinations of travelers were noted and then the number of visits by residents of one specific community to a second specific community was divided by total number of trips anywhere made by residents of the first community.

Demographic data were collected from three primary sources: Anglican Church of Canada parish records, which provided information on births, deaths, and marriages; Norway House Cree First Nation Treaty Annuity Lists, which provided lists of people who were paid an annuity under the terms of treaties between the native populations and the Canadian government; and Government of Canada Sessional Papers, which recorded census data. Population size was estimated from data from a 1916 census recorded in the Government of Canada Sessional Papers for 1917 (Government of Canada 1917: 18–19), since this was the year closest to the epidemic in which there was a census of the study communities. Deaths dating to the time of the flu epidemic were determined from a comparison of parish records, which listed people who died, and treaty annuity lists, which listed people living in the region at a particular time.

Estimates of some basic epidemiological parameters are available from epidemiological sources. For example, the recovery rate can be estimated using readily available data. The model assumes a constant recovery rate (γ) per unit time. The recovery rate governs the flow of individuals between the infectious and recovered stages. The model assumes that this rate is constant per unit time, a condition that results in the rate being inversely proportional to the average duration of time spent in the initial (infectious) stage. In other words, the recovery rate is equal to the reciprocal of the average duration of the infectious period. Consequently, the observed value for the duration of the infectious period, about 5 days (Chin 2000), can be used to estimate γ, which was accordingly set at 0.2 for the simulations. Since there is no reason to believe individuals from different study communities varied in their reaction to the disease, this value was assumed to be the same for each community.

As is the case with most models, less adequate information is available for some parameters. This was especially the case for the two components of the transmission rate, the number of contacts per unit time (κ) and the fraction of all contacts between susceptible and infectious individuals that resulted in transmission of the disease (τ). In addition, the historical archives contained little information on the return rates from visits to other communities (ρ). To overcome limitations like this, modelers must use knowledge and insights from both archival resources and the scientific literature to choose reasonable first estimates. Then it is essential to run simulations varying these parameters to see how sensitive model outcomes are to their variation.

Initial estimates of these relatively unknown parameters were determined as follows. Although direct data on the duration of visits to a community (which are needed to estimate return rates) were not available, the archives did indicate that traders and trappers usually left a community within one day of arrival, especially if they had come from a trapline or fishing camp. Since the vast majority of winter travel involved these people, the return rates were set equal to one. The fraction of contacts that result in disease, τ, is not well estimated for most infectious diseases. Consequently, anecdotal evidence about the infectiousness of influenza was used to choose an initial value for this parameter. Since influenza is highly infectious, τ was set at 0.5 (one-half of all contacts result in infection) and was assumed to be equal for each community. It was also assumed initially that, because Norway House was the primary HBC post in the Keewatin District, visitors to that community would experience twice as many contacts per unit time as would visitors to the other communities.

Step 5: Run model simulations

Once the initial parameter estimates are determined, the stage is set to begin simulations. At the outset of the simulations all members of the population are assumed to be at home. The disease begins with one infected individual and then is followed as it spreads throughout the communities until it either dies out or reaches an equilibrium state where the numbers of individuals in each category remain constant (or predictably fluctuate). Simulations can be run to explore the consequences of varying any of the model parameters, and although it is possible to vary all of the parameters at once, it is usually best to begin by varying a single parameter at a time to assess the impact of each parameter on disease patterns. This makes it easier to evaluate the impact of interactions among two or more factors.

Ideally simulations should be run to evaluate the consequences of varying both the initial conditions of a model and the values of each parameter of the model. In reality, this process, although important, is curtailed in favor of simulations addressing specific questions of interest to the researcher. For example, the Norway House project has included explorations of the general relationship between population mobility and disease spread and the effectiveness of the mobility of fur trappers as a means of carrying the influenza virus among communities, with the ultimate goal of inferring explanations for the observed patterns of influenza mortality across communities. In order to be able to generate these broader explanations, model simulations began with very simple questions that would provide a base from which to answer the more complex and general questions.

Population mobility, for example, includes issues about where an epidemic starts as well as the frequency of travel, where the travelers go during their trips, how long the trips last, etc. To simplify this relatively complex process, the first simulations centered on whether the location of the starting point of an epidemic had an important influence on patterns of disease transmission. To address this, the initial population distribution of each community was varied. At the start of simulations all individuals in the three communities were assumed to be susceptible to the disease except for the initial case. Thus, in communities which were not home to the initial case, $S_{ii} = N_i$ and all other stages had no people in them (i.e. everyone from community i was susceptible and present at the home community), while the population of the initial case's home community had $N_j - 1$ susceptibles who were all present at the home community j (i.e., $S_{jj} = N_j - 1$) and one infected person (who could be present at any of the communities; i.e., $I_{jk} = 1$ for some $k \neq j$). In addition, all people were assumed to be at home except, in some cases, the initial case. In the three-community model for the Norway House region, there were nine different starting points for an epidemic – the initial infectious person could be a resident of community 1 who was at home, a resident of community 1 who was visiting community 2, a resident of community 1 who was visiting community 3, a resident of community 2 who was visiting community 1, etc.

To date, simulations have been run varying other basic parameters as well, including the relative numbers of contacts within communities, the rates of travel out of communities (σ), and the distribution of those travelers among communities (ν). Results of some of these simulations will be discussed further below.

Step 6: Analyze results

The most important step in mathematical modeling is probably the analysis of simulation results, which usually involves identification of patterns in the results that give insights into how different factors influence epidemic outcomes. To answer the initial question about the effects of changing the location of the initial case on the outcome of an epidemic, simulations were run using each of the nine different starting points described above. Figure 11.4 shows data from four of these simulations. The remaining five simulations give similar patterns, and when taken together the simulation results clearly indicate that the location of the starting point has a strong effect on the timing of epidemic spread among the communities. However, the number of cases within a community did not vary significantly across any of the simulations, indicating that variation in the starting point of an epidemic did not affect severity within a community.

This result led to the hypothesis that the number of contacts per unit time that an individual had within a community would have a significant effect on severity within communities. To test this hypothesis, simulations were run varying the contact rates within communities. Figure 11.5 shows that this parameter not only has a strong influence on the number of cases within communities, but

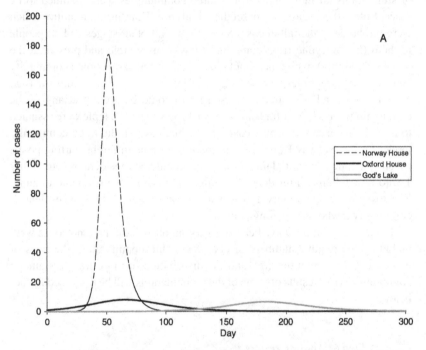

Fig. 11.4. Results of simulated epidemics with different starting points. Note that the sizes of the epidemic peaks in each community do not change significantly, but that the timing of the peaks can change substantially. (A) Starting case occurred in a person from Oxford House who was at Oxford House; (B) starting case occurred in a person from God's Lake who was visiting Norway House; (C) starting case occurred in a person from Norway House who was visiting Oxford House; (D) starting case occurred in a person from Oxford House who was visiting Norway House.

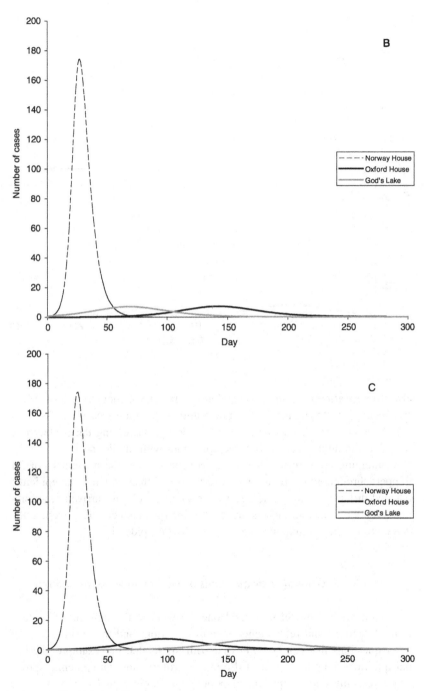

Fig. 11.4. (*cont.*)

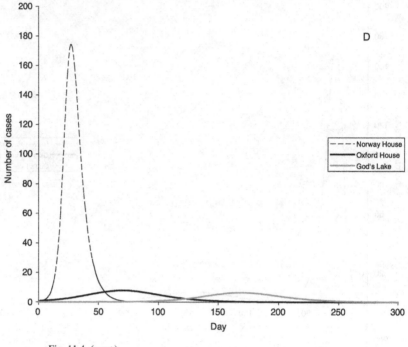

Fig. 11.4. (*cont.*)

also strongly affects the timing of epidemic spread throughout the region. More specifically, increasing the contact rate within a community causes earlier epidemics with higher epidemic peaks. Conversely, decreasing the contact rate within a community leads to delayed epidemics with smaller peaks.

Combining these results led to the conclusion that the movement of fur trappers throughout the region had a very strong effect on when the epidemic reached different communities. However, once the disease arrived at a community, internal factors such as the level of interpersonal contact governed the degree to which a community was affected by the epidemic.

Step 7: Generate new questions or modify previous questions

If a mathematical model is a good one, answers to the first questions lead to modifications and refinements of those questions and the development of other questions that can be addressed with the model. For example, the initial question about the effect of different starting points on epidemic spread addressed only one component of population mobility, and results of those

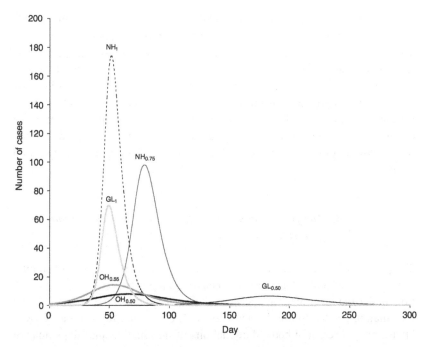

Fig. 11.5. Comparison of simulations with different rates of contact. The contact rates within each community are given in the subscript of each label. NH, Norway House; OH, Oxford House; GL, God's Lake. Note that changing the contact rate changes both size of epidemic peaks and timing of those peaks.

simulations suggested that mobility did not have a large effect on the numbers of cases in an epidemic. This led to speculations about whether other aspects of mobility influenced the case numbers or whether mobility generally affected only epidemic timing. Simulation results also stimulated thoughts about the kinds of within-community factors that might influence rates of contact and whether information on those might be available from the historical record.

Analysis of the initial simulation results generated several new questions to address with the simulations. For example,

- What is the importance of seasonal changes in mobility patterns? Would the epidemic have been different if it had arrived in the summer rather than in the winter and if so, how would it have differed?
- What was the likely impact of control strategies suggested by the authorities (e.g. informal quarantine)?
- How does the settlement structure affect patterns of disease transmission within and among the three communities?

The first of these questions was stimulated by the observation that the subarctic environment imposed on the population very different patterns of mobility and activity at different times of the year. The second question arose because there was some evidence in the historical archives of informal quarantine policies, which would have reduced rates of mobility among communities. The third question derived from the assumption that differences in contact rates were a consequence of the social organization and settlement structure of the communities. Simulations addressing this question might help to tease out more refined information about the specific factors that affect epidemic severity within communities.

Step 8: Modify the model, generate estimates of new parameters, and run new simulations

The process of addressing new questions stimulated by the analysis of simulation results is a feedback process between designing and modifying the simulation model and digging through historical archives to flesh out the ethnographic information needed to run the new model. The new questions in the Norway House project required both additional bits of information and adaptations of the existing model to better address those questions.

By emphasizing the potential importance of seasonal interactions and stimulating questions about the impact of differences between summer and winter mobility patterns, simulations helped direct the return to the historical archives to those sources that provided information on summertime mobility patterns. This provided new information on the number of trips taken in the summer and the source and destinations of those trips. However, because the travel patterns differed only in frequency and distribution of trips, no changes to the underlying model structure were required. Results relating to this work are presented by Herring and Sattenspiel (2002).

Studies of the impact of quarantine practices required both a change in model structure to allow short-term disruptions of normal travel and new information from the historical record on, for example, the types of quarantine practices implemented, their timing, and any evidence of their effectiveness. Results from numerous simulations of the new model suggest that the quarantine practices in effect, although laudable, were not likely to have altered epidemic patterns significantly (L. Sattenspiel and D.A. Herring, unpublished).

Historical data needed to address questions about the impact of settlement structure included information about where on the landscape people lived, the number of people per household, the number of households at each geographic location, the types of activities of the people that led to social contact, etc.

Simulations of the original model have been run comparing the settlement structure evident in the historical record to other anthropologically motivated theoretical patterns (Sattenspiel *et al.* 2000). However, because of small population size and increased details about daily life, effective modeling of the impact of settlement and social structure requires the use of the types of individual-based computer approaches described above. These are qualitatively different approaches that involve substantial restructuring of the underlying model.

How does mathematical modeling aid in the analysis of historical archives (and how have the archives helped the modeler)?

Mathematical modeling is a powerful tool to use in conjunction with more traditional methods of studying historical archives. Probably its most important contribution is that it provides direction and focus to the search through the archives. Models help to point out what factors are most influential in determining observed demographic patterns and what factors are less important or even insignificant. Although traditional analyses may lead to some of the same conclusions, sometimes the results are counterintuitive and wholly unexpected. For example, the initial hypothesis of the study was that patterns of mobility among the study communities would explain all or nearly all of the observed data on the Spanish flu epidemic. Yet simulations clearly showed that mobility only helped to explain the timing of the epidemic, but did not help in understanding the differential severity among communities. Model results pointed out the importance of within-community factors related to social organization and activities, which then stimulated new searches of the historical archives directed towards that aspect of life in the study populations.

A second major use of mathematical modeling is that it provides a way to 'experiment' on human populations and allows a researcher to ask and generate answers to 'what-if' questions. This is especially important in studies of health and disease, because it is unethical to introduce a disease into a real population and observe how and why it spreads. Well-designed models with parameter estimates that are grounded in reality can take the place of unethical clinical or field studies and can give answers to questions about how, for example, changes in the infectivity of a disease or changes in the nature of social contact might affect patterns of disease spread. These kinds of experiments help both archival researchers and scientists studying modern health crises to understand whether an epidemiological event represents an unusual situation specific to a particular time and place or whether the event is an example of a more general pattern. By identifying important variables affecting the situation being modeled, modeling results can also help direct researchers to other populations with which to compare the original study.

Archival research is also a useful accompaniment to mathematical modeling efforts. Issues faced regularly by archival researchers (for example, determination of whether needed data are available, the quality of existing data, or the presence of holes in the historical record) can stimulate a modeler to develop better models and new techniques of modeling. The historical details collected on the social structure of Aboriginal communities in central Manitoba, for example, have underscored the need to switch from models based on systems of differential equations, which are technically valid only for large populations, to individual-based stochastic simulations, which encompass random effects associated with small populations.

Historical archives contain untapped sources of data relevant to daily life in many different populations, and these data can provide the firm grounding to reality needed to estimate model parameters. This is extremely valuable to the mathematical modeler, because a model is only as good as the data that are used to estimate its parameters. Without these data sources, mathematical modeling is simply an academic exercise that can easily provide any picture a researcher desires. The real data available from historical archives and recent data sources are a crucial component that allows a modeler to find the picture that is most reasonable. This helps to guarantee that answers to modeling questions not only make sense but also reflect the real world, because even a faked picture can make sense. Furthermore, even authentic pictures can be interpreted in different ways by rational people (for example, remember our bedridden young man in Fig. 11.1).

Mathematical modeling thus both helps to interpret and understand the historical record and is also helped by the information available in that record. The abundance of historical information from many parts of the world guarantees that there will continue to be many opportunities for the development of new models that, in conjunction with careful historical analysis, can help the dead to speak up and tell their story.

References

Ackerman, E., Elveback, L.R. and Fox, J.P. (1984). *Simulation of Infectious Disease Epidemics.* Springfield, IL: C.C. Thomas.
Anderson, R.M. and May, R.M. (1991). *Infectious Diseases of Humans: Dynamics and Control.* Oxford: Oxford University Press.

Bailey, N.T.J. (1975). *The Mathematical Theory of Infectious Diseases and its Applications*. New York: Hafner Press.

Bartlett, M.S. (1957). Measles periodicity and community size. *Journal of the Royal Statistical Society* A**120**(1), 48–70.

Bartlett, M.S. (1960). The critical community size for measles in the United States. *Journal of the Royal Statistical Society* A**123**(1), 133–41.

Bartlett, M.S. (1961). Monte Carlo studies in ecology and epidemiology. *Proceedings of the 4th Berkeley Symposium on Mathematics, Statistics, and Probability* **4**, 39–55.

Bernstein, R.S., Sokal, D.C., Seitz, S.T., Auvert, B., Stover, J. and Naamara, W. (1998). Simulating the control of a heterosexual HIV epidemic in a severely affected East African city. *Interfaces* **28**(3 May–June), 101–26.

Black, F.L. (1975). Measles endemicity in insular populations: critical community size and its evolutionary implication. *Journal of Theoretical Biology* **11**, 207–11.

Castillo-Chavez, C. (ed.) (1989). *Mathematical and Statistical Approaches to AIDS Epidemiology*. Berlin: Springer-Verlag.

Chin, J. (ed.) (2000). *Control of Communicable Diseases Manual*, 17th edn. Washington, D.C.: The American Public Health Association.

Cliff, A.D. and Haggett, P. (1985). *The Spread of Measles in Fiji and the Pacific: Spatial Components in the Transmission of Epidemic Waves through Island Communities*. Publication HG18. Canberra: Australian National University, Research School of Pacific Studies, Department of Human Geography.

Cliff, A.D. and Haggett, P. (1988). *Atlas of the Distribution of Diseases: Analytic Approaches to Epidemiological Data*. Oxford: Blackwell.

Cliff, A.D., Haggett, P., Ord, J.K. and Versey, G.R. (1981). *Spatial Diffusion: An Historical Geography of Epidemics in an Island Community*. Cambridge: Cambridge University Press.

Cliff, A.D., Haggett, P. and Ord, J.K. (1983). Forecasting epidemic pathways for measles in Iceland: the use of simultaneous equation and logit models. *Ecology of Disease* **2**, 377–96.

Cliff, A.D., Haggett, P. and Ord, J.K. (1986). *Spatial Aspects of Influenza Epidemics*. London: Pion.

Cliff, A.D., Haggett, P. and Smallman-Raynor, M. (1993). *Measles: An Historical Geography of a Major Human Viral Disease, From Global Expansion to Local Retreat, 1849–1990*. Oxford: Blackwell.

Cliff, A.D., Haggett, P. and Smallman-Raynor, M. (1998). *Deciphering Global Epidemics: Analytical Approaches to the Disease Records of World Cities, 1888–1912*. Cambridge: Cambridge University Press.

Cross, M. and Moscardini, A.O. (1985). *Learning the Art of Mathematical Modeling*. New York: Halsted Press.

Dean, J.S., Gumerman, G.J., Epstein, J.M., Axtell, R.L., Swedlund, A.C., Parker, M.T. and McCarroll, S. (2000). Understanding Anasazi culture change through agent-based modeling. In *Dynamics in Human and Primate Societies: Agent-based Modeling of Social and Spatial Processes*, ed. T. Kohler and G. Gumerman, pp. 179–205. Oxford: Oxford University Press.

Doran, J.E. (2000). Trajectories to complexity in artificial societies: rationality, belief, and emotions. In *Dynamics in Human and Primate Societies: Agent-based*

Modeling of Social and Spatial Processes. ed. T. Kohler and G. Gumerman, pp. 89–105. Oxford: Oxford University Press.

Doran, J.E. and Palmer, M. (1995). The EOS project: integrating two models of Paleolithic social change. In *Artificial Societies: The Computer Simulation of Social Life*, ed. N. Gilbert and R. Conte, pp. 103–25. London: UCL Press.

Doran, J.E., Palmer, M., Gilbert, N. and Mellars, P. (1994). The EOS project: modeling Upper Paleolithic social change. In *Simulating Societies: The Computer Simulation of Social Phenomena*, ed. N. Gilbert and J. Doran, pp. 195–221. London: UCL Press.

Epstein, J.M. and Axtell, R. (1996). *Growing Artificial Societies: Social Science from the Bottom Up*. Washington, D.C.: Brookings Institution Press.

Finkenstädt, B. and Grenfell, B. (1998). Empirical determinants of measles metapopulation dynamics in England and Wales. *Proceedings of the Royal Society of London* B **265**, 211–20.

Fix, A.G. (1984). Kin groups and trait groups: population structure and epidemic disease selection. *American Journal of Physical Anthropology* **65**, 201–12.

Gilbert, N. and Conte, R. (eds) (1995). *Artificial Societies: The Computer Simulation of Social Life*. London: UCL Press.

Gilbert, N. and Troitzsch, K.G. (1999). *Simulation for the Social Scientist*. Buckingham, UK: Open University Press.

Government of Canada (1917). *Sessional Paper No. 27, George V* (9–10): 18–9.

Graves, C. (1969). *Invasion by Virus*. London: Icon Books.

Habbema, J.D.F., de Vlas, S.J., Plaisier, A.P. and Van Oortmarssen, G.J. (1996). The microsimulation approach to epidemiologic modeling of helminthic infections, with special reference to schistosomiasis. *American Journal of Tropical Medicine and Hygiene* **55**(5), 165–9.

HBCA (Provincial Archives of Manitoba, Hudson's Bay Company Archives, Winnipeg, Manitoba) (1918–1923). HBCA B.154/a/87, Norway House Post Journal.

HBCA (Provincial Archives of Manitoba, Hudson's Bay Company Archives, Winnipeg, Manitoba) (1918–1922). HBCA B.156/2/44, Oxford House Post Journal.

HBCA (Provincial Archives of Manitoba, Hudson's Bay Company Archives, Winnipeg, Manitoba) (1917–1922). HBCA B.283/a/8, God's Lake Post Journal.

HBCA (Provincial Archives of Manitoba, Hudson's Bay Company Archives, Winnipeg, Manitoba) (2000). http://www.gov.mb.ca/chc/archives/hbca/

Herring, D.A. (1994). 'There were young people and old people and babies dying every week': The 1918–19 influenza pandemic at Norway House. *Ethnohistory* **41**(1), 73–105.

Herring, D.A. and Sattenspiel, L. (2002). Death in winter: the Spanish flu in the Canadian Subarctic. In *The Spanish Flu Pandemic of 1918*, ed. H. Phillips and D. Killingray. London: Routledge. (In press.)

Jacquez, J.A., Simon, C.P., Koopman, J., Sattenspiel, L. and Perry, T. (1988). Modeling and the analysis of HIV transmission: the effect of contact patterns. *Mathematical Biosciences* **92**, 119–99.

Kaplan, E.H. and Brandeau, M.L. (eds) (1994). *Modeling the AIDS Epidemic: Planning, Policy, and Prediction*. New York: Raven Press.

Kohler, T.A. (2000). Putting social sciences together again: an introduction to the volume. In *Dynamics in Human and Primate Societies: Agent-based Modeling of Social and Spatial Processes*, ed. T. Kohler and G. Gumerman, pp. 1–18. Oxford: Oxford University Press.

Kohler, T.A. and Gumerman, G.J. (eds) (2000). *Dynamics in Human and Primate Societies: Agent-based Modeling of Social and Spatial Processes*. Oxford: Oxford University Press.

Kohler, T.A., Kresl, J., Van West, C., Carr, E. and Wilshusen, R.H. (2000). Be there then: a modeling approach to settlement determinants and spatial efficiency among Late Ancestral Pueblo populations of the Mesa Verde region, U.S. Southwest. In *Dynamics in Human and Primate Societies: Agent-based Modeling of Social and Spatial Processes*, ed. T. Kohler and G. Gumerman, pp. 145–78. Oxford University Press.

Kretzschmar, M. and Morris, M. (1996). Measures of concurrency in networks and the spread of infectious disease. *Mathematical Biosciences* **133**, 165–95.

Lake, M.W. (2000). MAGICAL computer simulation of Mesolithic foraging. In *Dynamics in Human and Primate Societies: Agent-based Modeling of Social and Spatial Processes*, ed. T. Kohler and G. Gumerman, pp. 107–43. Oxford University Press.

Longini, I.M., Jr. (1988). A mathematical model for predicting the geographic spread of new infectious agents. *Mathematical Biosciences* **90**, 367–83.

McGrath, J.W. (1988). Social networks of disease spread in the Lower Illinois Valley: a simulation approach. *American Journal of Physical Anthropology* **77**, 483–96.

Mielke, J.H., Jorde, L.B., Trapp, P.G., Anderton, D.L., Pitkänen, K. and Eriksson, A.W. (1984). Historical epidemiology of smallpox in Åland, Finland, 1751–1890. *Demography* **21**, 271–95.

Milgram, S. (1967). The small world problem. *Psychology Today* **2**, 60–7.

Milner, G.R. (1980). Epidemic disease in the postcontact southeast: a reappraisal. *Mid-continental Journal of Archaeology* **5**, 39–56.

Moore, C. and Newman, M.E.J. (2000). Epidemics and percolation in small-world networks. *Physical Reviews* **E61**, 5678–82.

Morris, M. (1995). Data driven network models for the spread of disease. In *Epidemic Models: Their Structure and Relation to Data*, ed. D. Mollison, pp. 302–22. Cambridge: Cambridge University Press.

Morris, M. and Kretzschmar, M. (1997). Concurrent partnerships and the spread of HIV. *AIDS* **11**, 641–8.

Palkovich, A.M. (1981). Demography and disease patterns in a protohistoric Plains group: a study of the Mobridge site (39WW1). *Plains Anthropologist* **26**(Memoir 17), 71–84.

Patterson, K.D. (1986). *Pandemic Influenza 1700–1900*. Totowa, NJ: Rowman & Littlefield.

Plaisier, A.P., Van Oortmarssen, G.J., Habbema, J.D.F., Remme, J. and Alley, E.S. (1990). ONCHOSIM: a model and computer simulation program for the transmission and control of onchocerciasis. *Computer Methods and Programs in Biomedicine* **31**, 43–56.

Pool, I. and Kochen, M. (1978). Contacts and influence. *Social Networks* **1**, 1–48.

Ramenofsky, A.F. (1987). *Vectors of Death*. Albuquerque: University of New Mexico Press.

Ray, A.J. (1974). *Indians in the Fur Trade: Their Role as Hunters, Trappers and Middlemen in the Lands Southwest of Hudson Bay, 1660–1870.* Toronto: University of Toronto Press.

Ray, A.J. (1976). Diffusion of diseases in the western interior of Canada, 1830–1850. *Geographical Review* **66**(2), 139–157.

Rhodes, C.J. and Anderson, R.M. (1996). A scaling analysis of measles epidemics in a small population. *Philosophical Transactions of the Royal Society of London* B **351**, 1679–88.

Robinson, T.P. (2000). Spatial statistics and geographical information systems in epidemiology and public health. *Advances in Parasitology* **47**, 81–128.

Russell, S.J. and Norvig, P. (1995). *Artificial Intelligence: A Modern Approach.* Englewood Cliffs, NJ: Prentice-Hall.

Rvachev, L.A. and Longini, I.M., Jr. (1985). A mathematical model for the global spread of influenza. *Mathematical Biosciences* **75**, 3–22.

Sattenspiel, L. (1987a). Population structure and the spread of disease. *Human Biology* **59**, 411–38.

Sattenspiel, L. (1987b). Spread and maintenance of a disease in a structured population. *American Journal of Physical Anthropology* **77**, 497–504.

Sattenspiel, L. and Dietz, K. (1995). A structured epidemic model incorporating geographic mobility among regions. *Mathematical Biosciences* **128**, 71–91.

Sattenspiel, L. and Herring, D.A. (1998). Structured epidemic models and the spread of the 1918–1919 influenza epidemic in the central Subarctic. *Human Biology* **70**(1), 91–115.

Sattenspiel, L., Mobarry, A. and Herring, D.A. (2000). Modeling the influence of settlement structure on the spread of influenza. *American Journal of Human Biology* **12**, 736–48.

Sattenspiel, L. and Powell, C. (1993). Geographic spread of measles on the island of Dominica, West Indies. *Human Biology* **65**, 107–129.

Sattenspiel, L. and Simon, C.P. (1988). The spread and persistence of infectious diseases in a subdivided population. *Mathematical Biosciences* **90**, 341–66.

Scott, S. and Duncan, C.J. (1998). *Human Demography and Disease.* Cambridge: Cambridge University Press.

Scott, S. and Duncan, C.J. (2001). *Biology of Plagues.* Cambridge: Cambridge University Press.

Smallman-Raynor, M., Cliff, A.D. and Haggett, P. (1992). *Atlas of AIDS.* Oxford: Blackwell.

Thornton, R., Miller, T. and Warren, J. (1991). American Indian population recovery following smallpox epidemics. *American Anthropologist* **93**, 28–45.

Upham, S. (1986). Smallpox and climate in the American Southwest. *American Anthropologist* **88**, 115–28.

Van der Ploeg, C.P.B., Van Vliet, C., De Vlas, S.J., Ndinya-Achola, J.O., Fransen, L., Van Oortmarssen, G.J. and Habbema, J.D.F. (1998). STDSIM: A microsimulation model for decision support in STD control. *Interfaces* **28**(3 May–June), 84–100.

Vlas, S.J. de, Van Oortmarssen, G.J., Gryseels, B., Polderman, A.M., Plaisier, A.P. and Habbema, J.D.F. (1996). SCHISTOSIM: A microsimulation model for the epidemiology and control of schistosomiasis. *American Journal of Tropical Medicine and Hygiene* **55**(5), 170–75.

Walters, J.H. (1978). Influenza 1918: the contemporary perspective. *Bulletin of the New York Academy of Medicine* **54**, 855–64.

Watts, D.J. (1999). Networks, dynamics, and the small-world phenomenon. *American Journal of Sociology* **105**, 493–527.

Whitmore, T.M. (1992). *Disease and Death in Early Colonial Mexico.* Boulder, CO: Westview Press.

12 *Where were the women?*

ANNE L. GRAUER

Introduction

Understanding the lives of women in medieval Britain (a period spanning roughly the twelfth through the sixteenth centuries) has a been a subject of interest and research for some time. Insights into numerous aspects of women's lives have been made through the use of documentary sources. For instance, early contributions by Abram (1916), Dale (1933), and Thrupp (1948), as well as more recent contributions by Power (1975), Charles and Duffin (1985), Hanawalt (1986), Goldberg (1991, 1992, 1995), Rosenthal (1990) and Jewell (1996) explore women's socioeconomic roles and contributions through the analysis of many types of documentary evidence, including parish church records, poll tax returns, legal documents, and literary sources, to name a few. The interplay between women and law has also been evaluated for a considerable number of years (see Abram 1916; Sheehan 1963; Cannon 1999), as have medieval women's private and family roles (see Goody 1983; Goldberg 1991; Hanawalt 1986; Kowaleski 1988), women and medicine (see Greene 1989–90, 1994), and women and religion (Power 1922; Thompson 1991; Jewell 1996).

Using historical documents to understand the lives of medieval women, however, has well-recognized limitations. In England, for instance, church records of births, marriages, baptisms, and deaths were not made compulsory until 1538 (Palliser 1979). Prior to that, records were seldom kept, or were kept by particular families or wealthy parishes. Similarly, the use of tombstones did not become a popular means of marking graves and/or providing information about the deceased until the seventeenth century (Johnson 1912). Thus, they too tell us little about the medieval period. Surviving wills have been used as tools for understanding medieval life (Bartlett 1953; Heath 1984), but will-making was usually exercised by men and 'seems to have been a habit acquired by the newly wealthy, merchants and craftsmen . . .' (Kermode 1982: 10). Tremendous success in reconstructing life in rural villages has been achieved through the analyses of manorial records (see, for example, Razi 1980; Jewell 1981) but they are of limited assistance in reconstructing the demography or history of urban centers. And tax records, while providing an extraordinary amount of economic information, commonly reflected the economic lives of more financially secure

266

men (Bartlett 1953). Jewell (1996: 16) notes that 'the picture of women in medieval England cannot be based on their own words directly, and though their
own words will be used when they can be found, the "women's point of view"
from the Middle Ages is not one that is normally available.' Subsequently, the
presence of gender bias has become a significant issue to those interested in
medieval history. Goldberg (1992: 26), however, asserts that, although written
records were selectively created and selectively maintained by men, women are
not invisible in medieval sources. Now, then, is the time to ask, 'Where were
the women?'

Archeological excavation and interpretation of the medieval period has led to
a similar question. Like inquiries made using documentary sources, the archeological record has facilitated our understanding of medieval Britain, albeit with
its own limitations. Important contributions exploring the physical development and structure of medieval towns (see, for example, Schofield and Leech
1987; Schofield and Vince 1994), and rural villages (see, for example, Beresford
1972; Aston *et al.* 1989), for instance, shed light on medieval life that historical
documents poorly reflect. Archeological excavations also commonly uncover
remnants of everyday medieval life, such as the tools and development of industry (see, for example, Crossley 1981), the use and development of fields (see,
for example, Hall 1982), housing (see, for example, Grenville 1999), the creation and use of pottery (see, for example, Haslam 1984; McCarthy and Brooks
1988), diet and crafts (see, for example, Serjeantson and Waldron 1989), and
even the unfortunate presence of pubic lice (Kenward 1999), whipworm, and
roundworm (Greig 1981).

As a result of the differing data used by historians and archeologists,
medieval history, written by historians, can appear different in approach and
conclusions than the history reconstructed by archeologists. Neither method
of exploring the past holds claim to being the 'correct' approach. Rather,
a new synthesis has emerged. Researchers such as Wylie (1985, 1991a,b),
Gilchrist (1994, 1999), and Andren (1998) seek to create a contextual analogy 'in which contemporary historical and literary evidence is used together
with archeological data, not to provide illustration or explanation, but to
link themes between media' (Gilchrist 1999: 110). Issues of gender have
become a key focus of this approach. It is argued (see Conkey and Spector
1984; Gero and Conkey 1991; Moore and Scott 1997; Nelson 1997; Gilchrist
1994, 1999; Donald and Hurcombe 2000; Sorensen 2000), that gender issues have traditionally been explored and interpreted using masculine models of sexual division of labor and have emphasized (and perpetuated) the
construction of men as the center of social conditions, relations and change.
Thus, in spite of the recovery of artifacts and data associated with the lives
of medieval women, focus and interpretation of medieval material items

have predominantly emphasized male activities and male roles (Gilchrist 1999).

The intent of this chapter is to explore another avenue of research into the lives of medieval women: the analysis of human skeletal remains. This more biologically based approach, combined with documentary and archeological evidence, has the potential to provide further insight into medieval women's lives. My goal here, like others who have adopted a synthetic approach, is to shed light on complex issues of sex and gender.

Over recent decades substantial emphasis has been placed on the careful recovery and detailed analysis of human skeletal remains from archeological sites. This attention to detail has helped to insure that all skeletal components, not just crania, are carefully recorded, recovered, and preserved. It has helped to insure that contextual information, such as material artifacts and spatial relationships, are available to researchers and used in analyses. And lastly, it has helped to show professionals and the public that human skeletal remains are valuable remnants of the past.

A wealth of information can be obtained from the human skeleton. Paleo-demographic information, such as age at death and sex of the individual can be determined through the use of multiple techniques requiring the analysis of dentition, the cranium and the postcranium (see Buikstra and Ubelaker 1994 for an overview of current techniques). When this information, gathered from the individual skeleton, is examined on the population level, an understanding of both prehistoric and historic groups can be created (Swedlund and Armelagos 1976). Saunders *et al.* (1995) for instance, in their analysis of mortality in the nineteenth century St Thomas' Anglican Church Cemetery in Belleville, Ontario, Canada, compare their skeletal sample to surviving parish records from 1821 to 1874. Along with concluding, after careful consideration of potential biases, that the skeletal sample was sufficiently representative of the once living population, they were able to synthesize skeletal and historical data and gain insight into the important role that acute infectious disease played in the lives of the parishioners. In a similar study, Grauer and McNamara (1995) compare skeletal remains from the nineteenth century Dunning Poorhouse cemetery in Chicago, Illinois, with federal and local census records of 1860 and 1870. They conclude that, although inherent biases in census collection, coupled with biases inherent in skeletal analyses, render comparison of these data sets futile, insight into childhood mortality at this institution for the destitute could be made.

Paleopathological information also makes important contributions to our understanding of the past. Numerous skeletal conditions, macroscopically recognizable on human skeletal remains, can be associated with the presence of particular pathogens and/or traumatic, systemic, metabolic, and congenital events (see Ortner and Putschar 1985; Aufderheide and Rodriguez-Martin 1998; Ortner and Aufderheide 1991; Roberts and Manchester 1995). Interpreting the presence and patterns of pathological lesions within and/or between populations has helped tease out nuances of human social interaction. Storey (1999), for instance, in her examination of skeletons from the Late Classic Maya of Copan, Honduras, uses pathological conditions to explore the effects of status on health and disease, and the potential for strong son preferences in a complex patrilineal society. Similarly, Larsen (1997, 2000; Larsen *et al.* 1995) has demonstrated that subsistence and socioeconomic changes over time within and between populations can be detected by using paleopathological techniques. And Grauer and Stuart-Macadam (1999) have highlighted ways in which issues of sex and gender can be constructed and evaluated by using human skeletal remains.

However, like documentary and archeological avenues of research, paleo-demographic and paleopathological research have potent limitations (Bocquet-Appel and Masset 1982, 1985; Buikstra and Konigsberg 1985; Jackes 1992; Wood *et al.* 1992). It has become apparent that researchers must be sensitive to and account for the numerous biases inherent in demographic and paleopathological investigation. Some examples of these biases include small sample sizes and instances where the recovered skeletal remains reflect only a portion of the once living population. The determination of age at death is a relative estimate, not an absolute one, and requires comparing the growth and age-related changes of reference populations to the population under investigation. The fact that archeologically recovered skeletons represent individuals who died, not those who lived, is another important factor, as calculating the number of individuals within a population who died at a particular age or who had a particular lesion does not tell us what proportion of the population was at risk of dying at that age, nor how many individuals in the population actually suffered from a disease. And similar to a bias in medieval archeological studies, where remnants left behind by 'everyday folks,' and not the well-kept property and belongings of the wealthy or elite, are more commonly excavated and evaluated, skeletal populations are more frequently derived from urban cemeteries of the poor, or from cemeteries associated with abolished, dissolved, or disbanded religious institutions. Not surprisingly, these cemeteries lose their distinctive features over time as payments on properties are not met, or land is sold for private use. Subsequently, distinctive grave markings in the form of headstones or funerary monuments are lost as reminders of the former land use. So, ironically, the populations overlooked in historical documentation, i.e. marginalized or disempowered groups, and the poor of all ages and both sexes, frequently stand at the center of archeological recovery and skeletal analyses.

Recognition of the biases and limitations inherent in specific techniques and approaches does not, I believe, lead to the conclusion that our potential to understand the past is futile. Rather, careful evaluation of the data chosen for evaluation leads to stronger paradigms and more robust inferences. This is especially true when multiple types of data are incorporated into the research design.

Synthesizing data

How might data synthesized from archeological excavation, historical documents, and the analysis of human skeletal remains be used to assist our understanding of the lives of medieval women? Towards answering this question, I use a population with which I am most familiar, the skeletal population associated with the parish of St Helen-on-the-Walls in York, England.

The archeology of the church and cemetery

The excavation of St Helen-on-the-Walls church and cemetery began in 1973 (Dawes and Magilton 1980; Magilton 1980). Although the Ebor Brewery had formerly owned and occupied the property (causing damage to some underlying features), approved plans to redevelop the area required an archeological excavation to be carried out in an effort to preserve threatened archaeological materials. Excavation continued until 1978, successfully uncovering the church structures and an estimated two-thirds of the original graveyard (Dawes and Magilton 1980).

According to Magilton (1980), the earliest church structure, dated tentatively to the tenth century, was a small rectangular building (7 m × 6 m) with a mortar floor. The second church structure, an expansion of the first building, indicated that a rectangular chancel was added in the twelfth century. A further extension of the chancel and the rebuilding of the nave walls, constituting the third church, was tentatively assigned a fourteenth century date. The fourth church resulted from a major phase of construction, whereby the two successive chancels were encapsulated in a larger single structure (19 m × 10 m), built on a different axis. This construction phase, possibly dating between the late fourteenth and early fifteenth centuries, was followed by an even larger phase of construction and expansion, perhaps occurring in the late fifteenth to early sixteenth centuries. Material artifacts recovered from the site consisted primarily of pottery sherds.

The excavation of the skeletal remains yielded approximately 1014 individuals (Grauer 1989, 1991a,b, 1993), densely packed in the graveyard without coffins, funerary adornments, or personal accessories. It appears that bodies were commonly interred overlapping one another, and in three separate occasions evidence suggests that graves were disturbed for the placement of another body before decomposition was complete (Dawes and Magilton 1980).

Interpreting the archeological evidence is complex. The small size of the church structures, along with the lack of religious remnants often found in churches with greater wealth, suggests that the church of St Helen-on-the-Walls was not opulently adorned. Its ultimate demise, a common fate for parish churches unable to meet the steep taxation demands of the sixteenth century monarchy, further suggests that both the parish and the parishioners were economically strapped. The crowded graveyard, with bodies overlapping one another, indicates that burials were not marked for individual identification, as was custom in the wealthiest parish churches and graveyards. The lack of coffins and adornments further attest to the modest means by which the parishioners lived.

Interpreting historical documents

A wide variety of historical documents dating from the city of York's medieval period is available for study today. One series of documents, the Freemen's Rolls, was an annual register of all individuals admitted to the Freedom of York. For York, this record begins as early as 1272 and remains uninterrupted from 1289 to 1671 (Collins 1896; Dobson 1973). The value of these registers is the light they throw upon the state of trade in York, its industries, the extent of its population and the social conditions of the inhabitants (Hawkin 1955: 11). Acquiring freedom was an important economic and social achievement. Freedom was acquired by citizenship, through the completion of a lengthy apprenticeship and/or the payment of a fee. Freedom could also be acquired by patrimony or redemption (the mayor and aldermen could award freedom). It is alleged that no person could carry on any trade in the city without acquiring the status of freeman (Collins 1896).

The Freemen's Register has been cautiously used to assess the types and changes of occupations in York as clues to the city's economic structure. For instance, records for the 50 years between 1350 and 1399 indicate that 4283 individuals received enfranchisement, while records for the 100 years between 1400 and 1499 indicate that 4357 individuals were admitted (Kermode 1982). Between the years 1500 and 1603 admissions rose to 6231 (Palliser 1979). Also interesting are the changing proportions of merchants and craftsmen. Merchants, amounting to 12% of recorded occupations admitted between 1350 and 1399, fell to 7% between 1400 and 1499 (Kermode 1982). Cloth-workers, first admitted in 1319, grew steadily in number and made up 17% of all admittances by the latter half of the fourteenth century (Miller 1965).

Whether these changing percentages are actual artifacts of the changing nature of York's economy must be questioned. Were records kept with equal consistency and accuracy over the many years of their use? Did each job, trade, or skill equally require freedom? What information about women might be gleaned from these ledgers of men's enfranchisement? It appears that while the Register of Freemen helps to highlight the variety of occupations within the city, they must be used cautiously and alongside other sources of information.

For instance, critical evaluation of the Register of Freemen has led to insight into an important aspect of York's economy: immigration. Although it is difficult to reconstruct a pattern of distribution of places of origin for the immigrants into York, the Freemen's Rolls for 1535–1566 indicate that 42.2% of the immigrants came from within a 20-mile radius of the city, 29.3% came from between 20 and 50 miles, and 28.5% were from farther than 50 miles (Palliser 1979: 130). Women, neglected in these documents, may have constituted a large proportion of these immigrants. Although only 138 names of women are recorded in the Freemen's Registers between 1272 and 1500 (Dobson 1973), and only 44 names of women are recorded between 1500 and 1603 (Palliser 1979), other documentation suggests that women were often incorporated into the work force. 'The employment of women workers has always been a marked feature of the woolen industry . . . One fourth of the cloth woven in York at the end of the fourteenth century was the work of women, and they enrolled as apprentices and were admitted to the membership of the crafts' (Lipson 1921: 34). 'Parliament recognized women brewers, bakers, weavers, spinners, but they were to receive lower wages. In York, cappers, parchment makers, listers, freshwater fishers, iron-mongers, barber surgeons, fishmongers, stringers, cooks and vintners have definite ordinances dealing with women' (Sellers 1914: lxi). Therefore, to assume that the numbers of women registered on the Freemen's Roll is an accurate account of women's role in the work force is misleading. Master craftsmen of poor means may have relied upon their wives and family for labor rather than upon apprentices and servants, and widows frequently maintained the family business without taking freedom (Palliser 1979).

Other documentary sources, such as the Lay Poll Tax Returns, or Lay Subsidy, provide further pieces of information on life in medieval York. The Lay Poll Tax was levied upon a community rather than upon the individual, and the community decided how to allot the burden of taxation (Hollingsworth 1969). Within the city of York this tax was placed on lay males over the age of fourteen. Only names of household heads are recorded within the 28 parishes listed for York. The occupation and numbers of dependents are also recorded. The Lay Subsidies of 1377 and 1381 identify almost 100 different crafts and nearly 1000 household heads (Harvey 1975). Textile workers dominate the records and other craftsmen, such as tanners, appear to cluster in a section of the city, whereas other craftsmen, such as those involved in the food industry, are dispersed throughout York. The records also highlight the disparity of wealth within the city, as one-fourth of the known parishes in York were not credited with a single taxpayer in the 1524 Lay Subsidy (Palliser 1979). Similarly, omitted from the records are those individuals of no fiscal interest to the crown: clergymen, beggars, vagrants, landless poor, servants, and most women and children (Elton 1969). To complicate the use of these documents, parishes on the outskirts of the city were grouped together differently during different years to suit the convenience of the tax collector (Bartlett 1953).

However incomplete, recent reevaluation of these historical documents helps to shed light on the lives of women. Goldberg (1986: 19) asserts that 'using poll tax returns of 1377 as a source only 90.5 adult males excluding servants are found to every 100 women in York...' In his 1992 examination of these same documents he notes that although this tax was levied on lay males over of the age of 14, and that only names of household heads are recorded, 19.3% of the households in York appear to be headed by women.

As sketchy and incomplete as the documentary evidence is for the city of York, it is even more problematic for the parish of St Helen-on-the-Walls. No direct reference to the church structures exist prior to the 1389 will of John de Blaketofte, who left a small sum to the church fabric (Borthwick Institute Prob. Req.: 1, fo. 3). In all, only 27 wills bequeathing goods or money to the parish are known to exist (Palliser 1980). One surviving document indicates that St Helen-on-the-Walls parish church was rededicated on June 26, 1424 by a suffragan bishop of York. This suggests that the fourth rebuilding phase of the church, recognized by archeological excavation, was so extensive as to call for a rededication, usually reserved for a new church foundation (Palliser 1980: 8). Finally, 'on 2 January 1550 the corporation resolved "that my lord maier (George Gayle) and his heyres shall have the churche of Seynt Ellyns on the Walls, churcheyerd and personage therof"... There is no later evidence of the church continuing in use, or of any priest serving the church after Robert Acrige, who died in 1551' (Palliser 1980: 11).

The scarcity of documentary evidence pertaining to the parish and parish-ioners of St Helen-on-the-Walls can be interpreted as an indication of poverty. The low number of wills written by parishioners stands in stark contrast to other parishes. Goldberg (1992: 26) notes that in York, prior to 1500, there are over 600 surviving wills associated with female testators from other parishes. This, along with the sporadic recording of rectors of the parish existing in the archbishop's registers dating 1282–1524, may reflect the relative unimportance of St Helen-on-the-Walls parish compared to others of the city (Palliser 1980). Royal taxation assessments further support this contention. The first of the three surviving assessments (ca. 1420) lumps St Helen-on-the-Walls within a larger category of parishes on the outskirts of the city and records their combined contribution at three pounds. Parishes located within York's richer city center paid up to fourteen pounds each (Sellers 1911, 1914). Thus, although the St Helen-on-the-Walls parish was perhaps capable of financing the rebuild-ing of the church in 1424, the assessed wealth fell substantially below the majority of parishes within the city. In the second (ca. 1490) and third as-sessment (1524), St Helen-on-the-Walls was not grouped with other parishes, but rather, was entered separately. These assessments ranked St Helen-on-the-Walls twenty-fifth out of 37 taxed parishes, and twenty-eighth out of 32 taxed parishes, respectively (York Civic Records 2, 84n, 130; Public Record Office E 179/217/92). By 1535, St Helen-on-the-Walls was rated as the lowest contributing parish out of 26 parishes tallied (Palliser 1980).

The skeletal remains

The 1014 human remains recovered from the archeological excavation of the church structures and graveyard date to approximately AD 1100–1550 (Dawes and Magilton, 1980; Grauer 1989). Determining the chronology of the burials is impossible owing to the destruction and alteration of the graveyard during the multiple periods of change to the church structures, and to the centuries of land use after the church and graveyard property became privately owned. In spite of these unfortunate limitations, information about the parishioners, which is helpful but scant from archeological excavation and analyses of historical documents, can potentially be supplemented using skeletal analyses. For instance, while medieval historical sources primarily recorded tax contributions, heads of households, membership and employment, and the legal interactions of adult men, the analysis of skeletal remains indicates that the parish of St Helen-on-the-Walls had a varied demographic composition. As seen in Fig. 12.1, 641 skeletons were assigned to age at death categories ranging from birth to over 55 years old. Children under the age of 10 made up almost 30% of the population. Out of the 320 adults in the population whose age and sex could be skeletally determined, 170 (53%) were female (Fig. 12.2). Therefore, contrary to what cursory examination of taxation records would have us believe, the urban center of York was NOT inhabited primarily by adult men. In fact, Goldberg's (1986,

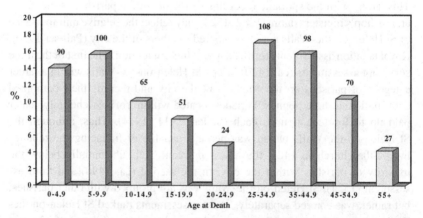

Fig. 12.1. Mortality profile of the St Helen-on-the-Walls skeletal population ($n = 641$) from York, England. Adapted from Grauer (1989, 1991a).

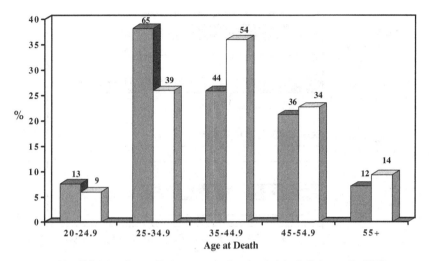

Fig. 12.2. Mortality profile by age at death and sex of the St Helen-on-the-Walls skeletal population from York, England. Females ($n = 170$) are denoted by shaded bars. Males ($n = 150$) are denoted by unshaded bars. Adapted from Grauer (1989, 1991b).

1992) assertions that there were more adult women in medieval York than there were adult men is supported by the skeletal record.

As seen in Fig. 12.3, females make up substantial proportions of other urban medieval cemeteries. The three cemeteries used here for comparison come from Lincoln and York. They were chosen for their relatively large sample sizes, the close proximity of York and Lincoln (approximately 55 miles apart), and the availability and thoroughness of the skeletal report associated with the excavation. St Mark's Church and Cemetery in Lincoln was excavated in 1976 and 1977, and yielded a total of 248 individual skeletons, of which 103 were dated between the twelfth and sixteenth centuries and could be assigned an age at death and sex (Dawes 1986). The medieval cemetery of Pennell Street, Lincoln, was excavated in 1996 and yielded 78 discrete burials, of which 46 could be assigned to an age at death and sex (Boghi and Boylston 1997). A total of 93 skeletons dating from the twelfth to the fifteenth centuries were excavated from the York Minster, yielding a total of 62 skeletons whose age and sex could be determined (J. Dawes, unpublished report). As Fig. 12.3 highlights, proportionately more females are found in the Pennell Street Cemetery and in St Helen-on-the-Walls. Males are more commonly found in St Mark's Cemetery and York Minster.

Reasons for these differences need further exploration before interpretations of the data can be made. It is possible, for instance, that the economies of different urban centers varied, causing variation in the opportunities afforded

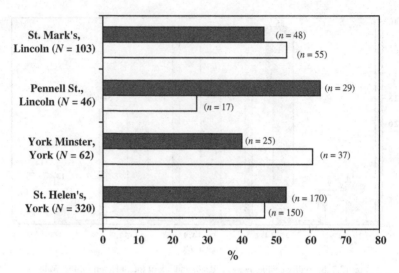

Fig. 12.3. Percentages of adult females and males in British medieval urban skeletal populations. Data for St Helen-on-the-Walls adapted from Grauer (1989, 1991b); data for York Minster adapted from J. Dawes (unpublished report); data for Pennell St, Lincoln, adapted from Boghi and Boylston (1997); data for St. Mark's, Lincoln, adapted from Dawes (1986). Females are denoted by shaded bars, males by unshaded bars.

women and men (Rigby 1995). Chronology of the cemetery might also play a role, since population density, and the proportions of females and males living in urban centers fluctuated throughout the medieval period (Bailey 1996). Lastly, careful consideration of spatial aspects of occupation, social status and gender may be warranted, as Goldberg (1992: 7) asserts that when 'evidence relating to the spatial distribution of females within the urban community is compared against the occupational topography from York poll tax and testamentary evidence . . . then single and independent females tend to be most conspicuous within the more marginal and less economically developed parishes.'

Medieval burial practices could also skew the data. Not all medieval cemeteries, or areas within the burial grounds, were equally available to women and men. Men, for instance, with their greater access to wealth and education, were more likely to have bequeathed money to the church, or have written a will (or have had a will written on their behalf) to ensure burial within the church structure or within a particularly symbolic area of the graveyard (Daniell 1997). Archeological practices and preservation issues further complicate the picture. If not all members of the congregation are buried identically, those interred within a coffin or protected by church structures might be more likely (or less likely, depending on the circumstances) to be archeologically preserved. Remains of individuals buried in areas prone to flooding, rodent activity, and/or simply buried without protection, as examples, might be more or less well preserved and more or less likely to be successfully counted centuries later when skeletal remnants are analyzed (Nawrocki 1995).

Thus, while it is apparent that the greater number of women in the St Helen-on-the-Walls skeletal population has correlates with other urban medieval populations, it is equally important that the percentages of females within the comparative populations vary. The demographic evaluation clearly shows that medieval urban centers were demographically complex and are in need of critical evaluation.

Another interesting demographic pattern needing evaluation emerges when the mortality profiles of adult females and males is examined. If we abide by history as it is more traditionally written, medieval urban centers would be seen as thriving from continued immigration of disenfranchised young men seeking wage incomes. As skeletal biologists, we could hypothesize that high immigration rates of a particular demographic segment of the population should be reflected in the mortality record. For the St Helen-on-the-Walls population, we could argue that mortality should be high for young adult males as they leave the environment within which they grew up, to live in what is commonly described as squalid conditions in urban centers. This expectation, however, is not met by the skeletal record. Females display statistically significant higher mortality than men at the age at death interval of 25–35 years old (Fig. 12.2). This suggests that substantially more women died between the ages of 25 and 35 years than men. This same pattern is found in the other urban populations. As seen in Fig. 12.4, the pattern of higher mortality within the age interval of 25–35 years old, found in the St Helen-on-the-Walls population, is also found at St Marks, Pennell St, and York Minster. Only at Pennell St does the mortality rate of females increase after this age interval.

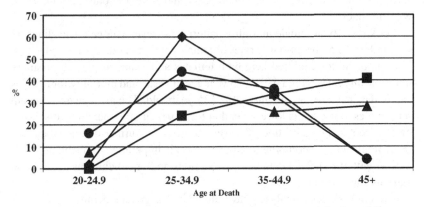

Fig. 12.4. Mortality rate of females in British medieval urban skeletal populations. Triangles, St Helen-on-the-Walls ($n = 170$); circles, York Minster ($n = 25$); squares, Pennell St, Lincoln ($n = 29$); diamonds, St Mark's, Lincoln ($n = 48$). Sources of data as for Fig. 12.3.

It is important at this juncture to highlight the limitations of this skeletal analysis, for the statistically greater number of women to men in age group 25–35 years old does not tell us everything that we should like to know. For instance, while we can see that more women of this age group were recovered from the cemetery, we do not know if there were actually more women of this age group living at the St Helen-on-the-Walls parish during the years of the cemetery's use. Alternatively, we do not know if the greater number of women in the cemetery sample aged 25–35 is due to a greater risk of women dying at that age (regardless of the number of women actually living in the parish). As pessimistic as this sounds, I argue that these are precisely the questions that might be addressed by careful synthesis of archeological, historical and skeletal data.

For example, Goldberg (1986, 1992), Rigby (1995), and Bailey (1996) have asserted that immigrants into Britain's medieval urban centers were frequently young women seeking employment in service. If this pattern holds true for cities where a substantial number of skeletal remains have been recovered, then perhaps the elevated numbers of women aged 25–35 in archeological populations might simply be due to great numbers of women of this age group living in the parishes and being buried in the parish graveyards. However, changing probabilities of death due to various risk factors might also be the cause of the noted demographic pattern. Goldberg (1986) notes that urban women sought to gain economic security through employment, and often delayed marriage until past the age of 20. Hence, the peak in female mortality between the ages of 25 and 35 years in three of the populations examined above may reflect higher probabilities of death for a number of risk factors, including childbirth and childbirth-related health complications. This is certainly possible if the average age of marriage was over the age of 20 years, and childbirth outside of marriage was infrequent.

Looking only at female mortality patterns, however, can be misleading. As seen in Fig. 12.5, the mortality rates of females and males do show similarities. In all populations explored here both females and males display surprisingly low mortality rates within the 20–25 year age interval, and mortality for females and males increases at age interval 25–35 years old. The average age at death for females, however, is lower than that for males in all populations displayed here, except at Pennell Street. Clearly, the need for further documentary and archeological data pertaining to all these sites is imperative to obtain and synthesize with the skeletal data before assertions and/or conclusions about these patterns can be offered.

Although skeletal analyses have elucidated a greater demographic complexity than that reflected in historical documents, fully satisfactory explanations for these patterns, and deviations from the patterns, have yet to be made.

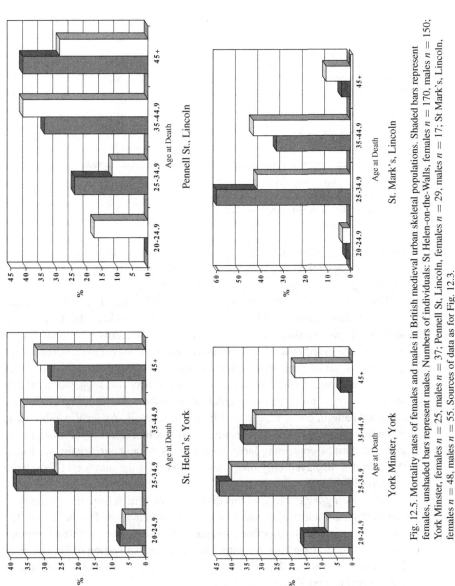

Fig. 12.5. Mortality rates of females and males in British medieval urban skeletal populations. Shaded bars represent females, unshaded bars represent males. Numbers of individuals: St Helen-on-the-Walls, females $n = 170$, males $n = 150$; York Minster, females $n = 25$, males $n = 37$; Pennell St, Lincoln, females $n = 29$, males $n = 17$; St Mark's, Lincoln, females $n = 48$, males $n = 55$. Sources of data as for Fig. 12.3.

Table 12.1. *Frequency of long bone fractures by sex in British medieval skeletal populations*

Sites	Females			Males		
	N^a	n^b	%	N^a	n^b	%
St Helen-on-the Walls	285	11	3.9	247	18	7.3
St Nicholas Shambles	71	2	2.8	90	6	6.7
Blackfriars	64	2	3.1	148	7	4.7
Fishergate	89	5	5.6	220	15	6.8
Whithorn	356	18	5.1	314	37	11.8
Chichester Hospital	70	4	5.7	128	27	21.1
Total	935	42	4.5	1147	110	9.6

Source: adapted from Grauer and Roberts (1996).
[a] Total number of individuals available for examination.
[b] Number of individuals with long bone fractures.

Optimistically I argue (Grauer *et al.* 1999) that skeletal analyses incorporating paleopathology might begin to assist in providing answers to these questions, yielding greater insight into the lives of women.

The interest and need to incorporate paleopathological research into historical studies became evident during a project completed by Grauer and Roberts (1996). In this study, long bone fracture-patterns were analyzed to assess whether medieval populations in Britain practiced the ancient, but poorly documented art of bone setting. We concluded that many complex long bone fractures had healed with minimal signs of deformity. This suggested to us that medieval communities had knowledge of immobilization techniques. An unexpected pattern, however, emerged from our analysis. As seen in Table 12.1, when long bone fractures were examined by sex in a selection of urban medieval populations containing large sample sizes throughout Britain, we found that females displayed only one fourth to one half the number of long bone fractures of men.

Determining the cause(s) of this disparity will benefit from the synthesis of skeletal/archeological data with historical documentation. Not believing that males were simply clumsier than females, or that being male grants an inherent predisposition to long bone fracture, I advocate an exploration of the historical record for insights into the activities and economic roles of females and males, and aspects of sexual divisions of labor.

In spite of the preponderance of historical evidence documenting the lives and livelihoods of men, visual images of women carrying out a variety of activities and participating in a wide range of occupations exist throughout

Europe. Women's traditional tasks, such as spinning, weaving, childcare, and cooking can be seen alongside images of women engrossed in unconventional tasks such as mining, masonry, teaching, and in commerce (Fox 1985). Written accounts also suggest that women completed a variety of tasks. In Goldberg's (1995: 167) rendition of the *Book of Husbandry*, written by Anthony Fitzherbert in 1523, an account of rural wives' tasks instructs that

> when you are up and ready, then first sweep the house, set the table, and put everything in your house in good order. Milk your cows, suckle your calves, strain your milk, get your children up and dress them, and provide for your husband's breakfast, dinner, supper, and for your children and servants, and take your place with them. . . . It is a wife's occupation to winnow all kinds of grain, to make malt, to wash and wring, to make hay, reap corn, and in time of need to help her husband fill the muck-wain or dung cart, drive the plough, load hay, corn, and such other and to go or ride to the market . . .

The variability in the scope of women's work appears tightly bound to season-ally determined agricultural tasks, and the social requirements of marriage and motherhood. Herlihy (1990: 50) asserts that 'the rural economy did not allow strict divisions between the labors of women and of men. Heavy seasonal work might require the assistance of all available hands, and crises, both personal and social, might require that women perform work normally done by men. Thus, although women did not usually drive the plow or take in the harvest, they often were enlisted in the heavy work of harvesting.'

Life in the cities and towns was different. According to Kowaleski (1988) and Rigby (1995) women's work was characterized by several features: a concentration in low-status jobs (such as spinning, washing and carding wool) or in areas such as huckstering, domestic service, and brewing, which required minimal training and capital, and relied on skills already mastered completing household duties; greatly influenced by marital and family status; often more intermittent or more impermanent than jobs occupied by men; and often invol-ved more than one economic activity, taking profits when and where circum-stances allowed. 'Herein lies a particular point of gender-difference in terms of work identity. Whereas a man might follow the same trade all his active life, a woman might have to change hers on leaving service, on marriage, and even after marriage' (Goldberg 1992: 99).

Obviously, the variability in women's occupations and roles, whether dic-tated in rural environments by agricultural season, or in urban environments by a multitude of economic and social factors, makes assessment of long bone fracture risk extremely difficult. In spite of this, clear patterns of differing risks of death due to accident can be drawn on lines of sex and gender in the his-torical documentary record. Hanawalt (1986) suggests that women in early

fourteenth-century England more often died near their home in domestic accidents, while men tended to die further from home in accidents related to work.

> The evidence of work patterns revealed by coroner's inquests into accidental
> deaths also suggests a clear division of labor. Of a sample of 2022 death of
> adults over the age of 14, women accounted for 22 per cent of cases; their
> daily routine was evidently less dangerous than that of men. That women
> spent more time in domestic tasks can be seen by the fact that 29.5 per cent
> of their deaths occurred in their own homes or closes compared with only
> 11.8 per cent of men's. Men, by contrast, were far more likely to be killed in
> fields, forests, mills, building sites and marl-pits, women's deaths were more
> likely to be connected with duties such as lighting fires, fetching water from
> wells, cooking, looking after animals, brewing or laundering.
>
> (Rigby 1995: 254–5)

Goldberg (1991) cautions against using such data to simply infer that women's
work was in the home and men's work was outside. He argues that household
work completed by women could have been more dangerous than the household
work or chores of men, and similarly, the work conducted by women outside
of the home was less dangerous than that of men. Similarly, biases in the
number of cases recorded by area, the types of accidents recorded, and reasons
for recording accidental death by the coroner might be influencing the results.

The impact of historical documentation on the analysis of medieval fracture
patterns in Britain is tremendous. Calling attention to a fracture pattern that
finds women from medieval urban centers and towns displaying one-fourth to
one-half the number of traumatic incidences than men is only the first step in the
quest to explore causes for this disparity. Since documentary evidence indicates
that cities and towns did not all share the same economic emphases (Schofield
and Vince 1994), that economic prosperity waxed and waned throughout the
middle ages (Dyer 1989), and that status and occupation played a role in where
an individual lived and died within the medieval city (Goldberg 1992), the
simple conclusion that women displayed fewer long bone fractures than men
tells us precious little and is in need of much further exploration.

Documentary evidence tantalizes us with the possibility that our results are a
feature of urban medieval life, and would not be replicated if fracture patterns in
females and males from rural environments were evaluated. Judd and Roberts
(1999), for instance, conclude in their comparison of fractures displayed in
populations from rural environments to fractures displayed in populations from
urban environments, that the higher frequency of fractures in rural (agricultural)
populations suggested that agricultural life was more dangerous than life in the
towns. The fact that the samples highlighted in this chapter are all derived
from urban environments might be providing a myopic view of medieval life.
Could our detected disparity between fracture rates in women and men be a
reflection of the division of labor predominantly found in cities and towns?

Would evaluation of rural samples find similar fracture frequencies in women and men, supporting the contention that division of labor was less recognizable or permanent in agricultural communities?

Conclusion

It has become apparent that understanding women's lives in the past requires the evaluation of women's social roles (i.e. understanding gender issues) as well as understanding women's biology (i.e. sex and reproductive capacities) (Armelagos 1999). Women's lives may have differed from men's owing to circumscribed gender roles or might in many instances have been similar. Biological differences between women and men also serve as critical variables. Differing reproductive biology does not determine the life course of a woman or man, but may play a profound role in how the body utilizes nutrients (Cook and Hunt 1999), immunologically responds to pathogens (Ortner 1999), develops and maintains bone tissue (Weaver 1999), and is susceptible to disease (Roberts *et al.* 1999; Stuart-Macadam 1999). Recognizing the complexity of human social interaction alongside human biology can guide us towards a more nuanced understanding of the past (Grauer 1995; Grauer and Stuart-Macadam 1999).

It is hoped that as paleodemographic and paleopathological analyses are synthesized with historical documents and archeological investigation, life patterns will come to light and assist our understanding of the complex roles held by medieval women. If after reading traditional histories we find ourselves asking 'Where were the women?' perhaps we might discover some answers in the archives, and others six feet underground.

References

Abram, A. (1916). Women traders in medieval London. *Economic Journal* XXVI, 276–85.

Andren, A. (1998). *Between Artifacts and Texts: Historical Archaeology in Global Perspective*. New York: Plenum Press.

Armelagos,G. (1999). Introduction: sex, gender and health status in prehistoric and contemporary populations. In *Sex and Gender in Paleopathological Perspective*, ed. A. Grauer and P. Stuart-Macadam, pp. 1–10. Cambridge: Cambridge University Press.

Aston, M., Austin, D. and Dyer, C. (eds) (1989). *The Rural Settlements of Medieval England: Studies Dedicated to Maurice Beresford and John Hurst*. Oxford: Blackwell Press.

Aufderheide, A. and Rodrigues-Martin, C. (1998). *The Cambridge Encyclopedia of Human Paleopathology*. Cambridge: Cambridge University Press.

Bailey, M. (1996). Demographic decline in late medieval England: some thought on recent research. *Economic History Review* **XLIX**, I, 1–19.

Bartlett, N. (1953). The Lay Poll Tax Returns for the City of York in 1381. *East Riding Historical Transactions*. (Offprint.)

Beresford, M.W. (1972). *Deserted Medieval Villages: Studies*. New York: St. Martin's Press.

Bocquet-Appel, J. and Masset, C. (1982). Farewell to paleodemography. *Journal of Human Evolution* 11, 321–33.

Bocquet-Appel, J. and Masset, C. (1985). Matters of moment. *Journal of Human Evolution* 14, 107–111.

Boghi, F. and Boylston, A. (1997). *The Medieval Cemetery of Pennell Street, Lincoln, Lincolnshire (SUS96). Report on the Human Skeletal Remains*. Unpublished report from the Calvin Wells Laboratory, Department of Archaeological Sciences, University of Bradford, Bradford, UK.

Buikstra, J. and Konigsberg, L. (1985). Paleodemography: critiques and controversies. *American Anthropologist* 87, 316–33.

Buikstra, J. and Ubelaker, D. (1994). *Standards for Data Collection from Human Skeletal Remains*. Proceedings of a Seminar at the Field Museum of Natural History. Arkansas Archaeological Survey Research Series No. 44. Fayetteville: Arkansas Archaeological Survey.

Cannon, C. (1999). The rights of medieval English women: crime and the issue of representation. In *Medieval Crime and Social Control*, ed. B. Hanawalt and D. Wallace, pp. 156–85. Minneapolis: University of Minnesota Press.

Charles, L. and Duffin, L. (eds) (1985). *Women and Work in Pre-Industrial England*. London: Croom Helm.

Collins, F. (1896). *Register of the Freemen of the City of York From the City Records, Vol. II, 1272–1558*. York: Publication of the Surtees Society, Volume 96.

Conkey, M.W. and Spector , J. (1984). Archaeology and the study of gender. *Archaeological Method and Theory* 7, 1–38.

Cook, D. and Hunt, K. (1999). Sex differences in trace elements: status or self-selection. In *Sex and Gender in Paleopathological Perspective*, ed. A. Grauer and P. Stuart-Macadam, pp. 64–78. Cambridge: Cambridge University Press.

Crossley, D.W. (1981). *Medieval Industry*. Council For British Archaeology Research Report 40. London: Council for British Archaeology.

Dale, M. (1933). The London silkwomen of the fifteenth century. *Economic History Review* IV, 3.

Daniell, C. (1997). *Death and Burial in Medieval England 1066–1550*. London and New York: Routledge.

Dawes, J. (1986). Human bones from St. Mark's. In *Volume XIII: Churches in Lincoln 1: St. Mark's Church and Cemetery*, ed. B.J.J. Gilmour and D.A. Stocker, pp. 33–5. Trust for Lincoln Archaeology; Council for British Archaeology.

Dawes, J. and Magilton J. (eds) (1980). *The Cemetery of St. Helen-on-the-Walls, Aldwark. The Archaeology of York: The Medieval Cemeteries*, series editor P. Addyman, Volume 12, Fascicule 1: *The Medieval Walled City Northeast of the Ouse*. Council for British Archaeology for the York Archaeological Trust.

Dobson, R.B. (1973). Admission to the freedom of the City of York in the later middle ages. *Economic History Review* 26(1), 1–21.

Donald, M. and Hurcombe, L. (eds) (2000). *Gender and Material Culture in Archaeological Perspective*. New York: St. Martin's Press.

Dyer, C. (1989). *Standards of Living in the later Middle Ages: Social Change in England c. 1200–1520.* Cambridge: Cambridge University Press.

Elton, G.R. (1969). *England 1200–1640.* London: Sources of History, Ltd and Hodder and Stoughton, Ltd.

Fox, S. (1985). *The Medieval Woman: An Illuminated Book of Days.* Boston: A New York Graphic Society Book. Little, Brown and Co.

Gero, J. and Conkey, M.W. (eds) (1991). *Engendering Archaeology: Women and Prehistory.* Oxford: Blackwell Press.

Gilchrist, R. (1994). *Gender and Material Culture: the Archaeology of Religious Women.* London: Routledge Press.

Gilchrist, R. (1999). *Gender and Archaeology: Contesting the Past.* London: Routledge Press.

Goldberg, P.J.P. (1986). Female labour, service and marriage in the late medieval urban north. *Northern History* XXII, 18–38.

Goldberg, P.J.P. (1991). The public and the private: women in the pre-plague economy. In *Thirteenth-Century England, III,* ed. P. Coss and S. Lloyd, pp. 75–81. Woodridge.

Goldberg, P.J.P. (1992). *Women, Work and Life Cycle in a Medieval Economy: Women in York and Yorkshire c. 1300–1520.* London: Clarendon Press.

Goldberg, P.J.P. (1995). *Women in England c. 1275–1525: Documentary Sources.* Manchester: Manchester University Press.

Goody, J. (1983). *The Development of Family and Marriage in Europe.* Cambridge: Cambridge University Press.

Grauer, A. (1989). *Health, Disease and Status in Medieval York.* Ph.D. Dissertation, University of Massachusetts, Amherst.

Grauer, A. (1991a). Patterns of life and death: the paleodemography of medieval York. In *Health in Past Societies: Biocultural Interpretations of Human Skeletal Remains from Archaeological Contexts,* ed. H. Bush and M. Zvelebil, pp. 67–80. British Archaeological Reports International Series, 567.

Grauer, A. (1991b). Life patterns of women from medieval York. In *The Archaeology of Gender,* ed. D. Walde and N. Willows, pp. 407–13. Chocmool Archeological Society. Calgary: University of Calgary.

Grauer, A. (1993). Patterns of anemia and infection from medieval York, England. *American Journal of Physical Anthropology* **91**(2), 203–13.

Grauer, A. (ed.) (1995). *Bodies of Evidence: Reconstructing History Through Skeletal Analyses.* New York: Wiley-Liss.

Grauer, A. and McNamara, E. (1995). A piece of Chicago's past: exploring childhood mortality in the Dunning Poorhouse Cemetery. In *Bodies of Evidence: Reconstructing History Through Skeletal Analysis,* ed. A. Grauer, pp. 91–103. New York: Wiley-Liss.

Grauer, A., McNamara, E. and Houdek, D. (1999). A history of their own: patterns of death in a nineteenth-century poorhouse. In *Sex and Gender in Paleopathological Perspective,* ed. A. Grauer and P. Stuart-Macadam, pp. 149–64. Cambridge: Cambridge University Press.

Grauer, A. and Roberts, C. (1996). Paleoepidemiology, healing and possible treatment of trauma in the medieval cemetery of St. Helen-on-the-Walls. *American Journal of Physical Anthropology* **100**(4), 531–44.

Grauer, A. and Stuart-Macadam, P. (eds) (1999). *Sex and Gender in Paleopathological Perspective*. Cambridge: Cambridge University Press.

Greene, M. (1989–90). Women's medical practice and health care in medieval Europe. *Signs: Journal of Women in Culture and Society* **14**, 434–73.

Greene, M. (1994). Documenting medieval women's medical practice. In *Practical Medicine from Salerno to the Black Death*, ed. L. Garcia-Ballester *et al.*, pp. 322–52. Cambridge: Cambridge University Press.

Greig, J. (1981). The investigation of a medieval barrel-latrine from Worcester. *Journal of Archaeological Sciences* **8**, 265–82.

Grenville, J. (1999). *Medieval Housing*. London: Leicester University Press.

Hall, D. (1982). *Medieval Fields*. Shire Archaeology Series (28). Aylesbury: Shire Publications.

Hanawalt, B. (1986). *The Ties that Bound: Peasant Families in England*. Oxford: Oxford University Press.

Harvey, John (1975). *York*. York: B.T. Batsford.

Haslam, J. (1984). *Medieval Pottery in Britain*. Shire Archaeology Series (6). Aylesbury: Shire Publications.

Hawkin, R. (1955). *A History of Freemen of the City of York*. City of York Guild of Freemen: York Surtees Society.

Heath, P. (1984). Urban piety in the later Middle Ages: The evidence of Hull wills. In *The Church, Politics and Patronage in the Fifteenth Century*, ed. B. Dobson, pp. 209–34. Stroud: Alan Sutton Press.

Herlihy, D. (1990). *Opera Muliebria: Women and Work in Medieval Europe*. McGraw-Hill Series: New Perspectives on European History, ed. R. Grew. New York: McGraw-Hill Publishing Co.

Hollingsworth, T.H. (1969). *Historical Demography*. London: The Sources of History, and Hodder and Stoughton.

Jackes, M. (1992). Paleodemography: problems and techniques. In *The Skeletal Biology of Past Peoples: Advances in Research Methods*, ed. S. Saunders and A. Katzenberg, pp. 189–224. New York: Wiley-Liss.

Jewell, H. (ed.) (1981). *Court Rolls of the Manor of Wakefield 1348 to 1350*. Wakefield Court Roll Series II. Yorkshire Archaeological Society.

Jewell, H. (1996). *Women in England*. Manchester: Manchester University Press.

Johnson, W. (1912). *Byways in British Archaeology*. Cambridge: Cambridge University Press.

Judd, M. and Roberts, C. (1999). Fracture trauma in a medieval British farming village. *American Journal of Physical Anthropology* **109**, 229–43.

Kenward, H. (1999). Pubic lice (*Pthirus pubis* L.) were present in Roman and Medieval Britain. *Antiquity* **73**(282), 911–15.

Kermode, J. (1982). The merchants of three northern English towns. In *Profession, Vocation and Culture in Later Medieval England*, ed. C. Clough, pp. 7–50. Liverpool: Liverpool University Press.

Kowaleski, M. (1988). The history of urban families in medieval England. *Journal of Medieval History* **14**(2), 47–63.

Larsen, C. (1997). *Bioarchaeology: Interpreting Behavior from the Human Skeleton*. Cambridge: Cambridge University Press.

Larsen, C. (2000). *Skeletons in Our Closet: Revealing Our Past Through Bioarchaeology*. Princeton, NJ: Princeton University Press.

Larsen, C., Craig, J., Sering. L., Schoeninger, M., Russell, K., Hutchinson, D. and Williamson, M. (1995). Cross homestead: life and death on the midwestern frontier. In *Bodies of Evidence: Reconstructing History Through Skeletal Analysis*, ed. A. Grauer, pp. 139–60. New York: Wiley-Liss.

Lipson, E. (1921). *The History of the Woollen and Worsted Industries*. London: A & C Black.

Magilton, J. (1980). *The Church of St. Helen-on-the-Walls, Aldwark. The Archaeology of York*, series editor P. Addyman, Volume 10, Fascicule 1: *The Medieval Walled City Northeast of the Ouse*. Council for British Archaeology for the York Archaeological Trust.

McCarthy, M.R. and Brooks, C. (1988) *Ancient and Medieval – Medieval Pottery in Britain AD 900–1600*. Leicester: Leicester University Press.

Miller, E. (1965). The fortunes of the English textiles industry during the thirteenth century. *Economic History Review* **18**, 64–82.

Moore, J. and Scott, E. (eds) (1997). *Invisible People and Processes: Writing Gender and Childhood into European Archaeology*. London: Leicester University Press.

Nawrocki, S. (1995). Taphonomic processes in historic cemeteries. In *Bodies of Evidence: Reconstructing History Through Skeletal Analysis*, ed. A. Grauer, pp. 49–66. New York: Wiley-Liss.

Nelson, S. (1997). *Gender in Archaeology: Analyzing Power and Prestige*. Walnut Creek, CA: AltaMira Press.

Ortner, D. (1999). Male-female immune reactivity and its implications for interpreting evidence in human skeletal paleopathology. In *Sex and Gender in Paleopathological Perspective*, ed. A. Grauer and P. Stuart-Macadam, pp. 79–92. Cambridge: Cambridge University Press.

Ortner, D. and Aufderheide, A. (1991). *Human Paleopathology: Current Syntheses and Future Options*. Washington, D.C.: Smithsonian Institution Press.

Ortner, D. and Putschar, W. (1985). *Identification of Pathological Conditions in Human Skeletal Remains*. Washington, D.C.: Smithsonian Institution Press.

Palliser, D. (1979). *Tudor York*. Oxford: Oxford University Press.

Palliser, D.M. (1980). Location and history. In *The Archaeology of York*, Volume 10, Fascicule 1: *The Medieval Walled City Northeast of the Ouse. The Church of St. Helen-on-the-Walls, Aldwark*, ed. J.R. Magilton, pp. 2–14. London: Council for British Archaeology for the York Archaeological Trust.

Power, E. (1922). *Medieval English Nunneries c. 1275 to 1535*. Cambridge: Cambridge University Press.

Power, E. (1975). *Medieval Women*. Cambridge: Cambridge University Press.

Razi, Z. (1980). *Life, Marriage and Death in a Medieval Parish: Economy, Society and Demography in Halesowen, 1270–1400*. Cambridge: Cambridge University Press.

Rigby, S. (1995). *English Society in the Later Middle Ages: Class, Status and Gender*. New York: St. Martin's Press.

Roberts, C. and Manchester, K. (1995). *The Archaeology of Disease*, 2nd edn. Stroud, Gloucester: Alan Sutton.

Roberts, C., Lewis, M. and Boocock, P. (1999). Infectious disease, sex and gender: the complexity of it all. In *Sex and Gender in Paleopathological Perspective*, ed. A. Grauer and P. Stuart-Macadam, pp. 93–113. Cambridge: Cambridge University Press.

Rosenthal, J. (ed.) (1990). *Medieval Women and the Sources of Medieval History.* University of Georgia Press.

Saunders, S., Herring, A. and Boyce, G. (1995). Can skeletal samples accurately represent the living populations they come from? The St. Thomas' Cemetery Site, Belleville, Ontario. In *Bodies of Evidence: Reconstructing History Through Skeletal Analysis*, ed. A. Grauer, pp. 69–89. New York: Wiley-Liss.

Schofield, J. and Leech, R. (eds) (1987). *Urban Archaeology in Britain.* Council for British Archaeology Research Report 61. London: Council for British Archaeology.

Schofield, J. and Vince, A. (1994). *Medieval Towns: the Archaeology of Medieval Britain.* Series Editor, Helen Clark. London: Leicester University Press.

Sellers, M. (1911). *York Memorandum Book: Part I. (1376–1419).* Surtees Society, Volume 120. York.

Sellers, M. (1914). *York Memorandum Book: Part II. (1388–1493).* Surtees Society, Volume 125. York.

Serjeantson, D. and Waldron, T. (eds) (1989). *Diet and Crafts in Towns: the Evidence of Animal Remains from the Roman to the Post-Medieval Periods.* British Archaeological Reports, British Series 199. Oxford: British Archaeological Reports.

Sheehan, M. (1963). *The Will in Medieval England.* Toronto: Pontifical Institute of Medieval Studies.

Sorensen, M.L.S. (2000). *Gender Archaeology.* Cambridge: Polity Press.

Storey, R. (1999). The mothers and daughters of a patrilineal civilization: the health of females among the Late Classic Maya of Copan, Honduras. In *Sex and Gender in Paleopathological Perspective*, ed. A. Grauer and P. Stuart-Macadam, pp. 133–48. Cambridge: Cambridge University Press.

Stuart-Macadam, P. (1999). Iron deficiency anemia: exploring the difference. In Sex and Gender in Paleopathological Perspective, ed. A. Grauer and P. Stuart-Macadam, pp. 45–63. Cambridge: Cambridge University Press.

Swedlund, A. and Armelagos, G.J. (1976). *Demographic Anthropology.* Dubuque, Iowa: Wm. C. Brown.

Thompson, S. (1991). *Women Religious: The Founding of English Nunneries After the Norman Conquest.* Oxford: Clarendon Press.

Thrupp, S. (1948). *The Merchant Class of Medieval London (1300–1500).* Chicago: Chicago University Press.

Weaver, D. (1999). Osteoporosis in the bioarchaeology of women. In *Sex and Gender in Paleopathological Perspective*, ed. A. Grauer and P. Stuart-Macadam, pp. 27–44. Cambridge: Cambridge University Press.

Wood, J.W., Milner, G.R., Harpending, H.C. and Weiss, K.M. (1992). The osteological paradox: problems of inferring prehistoric health from skeletal samples. *Current Anthropology* **33**, 343–70.

Wylie, A. (1985). The reaction against analogy. *Advances in Archaeological Method and Theory* **8**, 63–111.

Wylie, A. (1991a). Gender theory and the archaeological record: why is there no archaeology of gender? In *Engendering Archaeology*, ed. J.M. Gero and M.W. Conkey, pp. 31–54. Oxford: Blackwell Press.

Wylie, A (1991b). Feminist critiques and archaeological challenges. In *The Archaeology of Gender*, ed. D. Walde and N.D. Willows, pp. 17–23. Calgary: University of Calgary Archaeological Association.

13 Malnutrition among northern peoples of Canada in the 1940s: an ecological and economic disaster

D. ANN HERRING, SYLVIA ABONYI AND ROBERT D. HOPPA

Introduction

The Hudson's Bay region of the Canadian north captured the imagination of seventeenth-century Europeans seeking the elusive northwest passage to the Far East. The romance of the north and its promise of future riches led adventurers and agents of foreign governments to explore its land and sea. They encountered a fabulous array of fur-bearing animals – beaver, ermine, mink, fox, fisher, marten, lynx, hare and muskrat – setting the stage for the development of the European fur trade and for significant shifts in human–land relationships in the north thereafter.

A system of fur trade posts was established by the Hudson's Bay Company and the Northwest Company (which merged into a single business in 1821) that served to attract Aboriginal people to the areas around the posts and to encourage more intensive trapping of fur-bearing animals, in exchange for European goods (Fig. 13.1). Prior to sustained European contact, the Cree followed a seasonal subsistence cycle based on shifting plant and animal availability (Fig. 13.2). During the fur trade, they became enmeshed in a European industry that intensified and concentrated their subsistence activities on harvesting fur-bearing animals such as the beaver. This eventually led to abandonment of the seasonal cycle of mobility and to an increasing dependence on particular tracts of land, the economy of the fur trade, and imported resources.

The demographic and ecological consequences of the fur trade, though seemingly minor at the outset, acted incrementally through time to transform life in the Canadian north (Preston 1986). We have been investigating the implications of these changes for the health of the Cree in the Moose River region with a view to better understanding the details of the epidemiological transition in the area (Herring and Hoppa 1997, 1999; Hoppa 1998). Although the once widespread idea is fading that Aboriginal people of the Americas experienced a golden age of vigorous and universal good health followed by rapid degeneration after European contact, we still know very little about what really happened over the

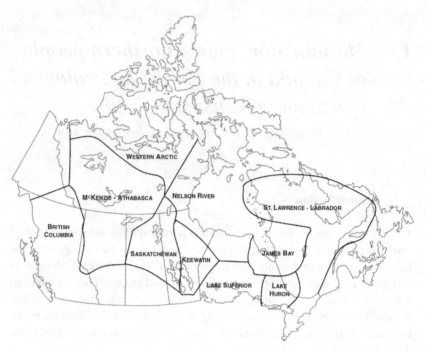

Fig. 13.1. Canada and the Hudson's Bay Company Regions (redrawn from HBCA 1922).

long term in different parts of the Canadian north, under different social and ecological circumstances. Even after the delivery of health care to Aboriginal peoples was expanded and the collection of health statistics vastly improved after the 1950s, there are few communities whose health histories are well understood (Waldram *et al.* 1995: 66).

Moose Factory, the focus of this chapter, was one of the early fur trade posts established in the Canadian subarctic, founded in 1673 by the Hudson's Bay Company (HBC) as one of three coastal posts on the southwestern shore of James Bay (Fig. 13.3). Located on an island at the head of the Moose River drainage system into Hudson's Bay, it provided a strategic location for ships to drop cargo from Europe and to load the stocks of fur harvested by Aboriginal people of the region.

One of the intriguing features of the mortality history of Moose Factory First Nation – and the impetus for this chapter – is an extraordinary increase in infant death during the early 1940s. Therein lies the beginning of the tale of how a seemingly simple question about historical infant deaths revealed a more complex history of mortality experience in the Canadian north. The purpose of this chapter is to tie together the strands of long-term ecological change,

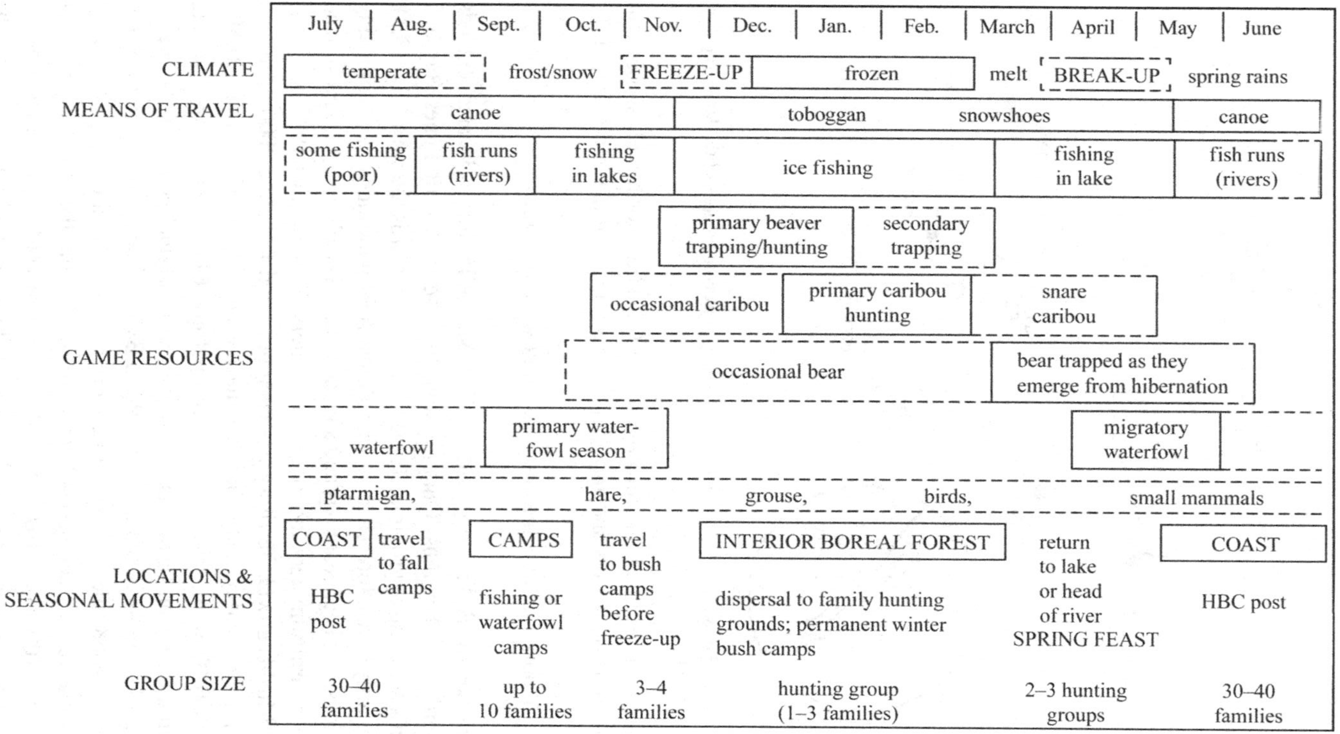

	July	Aug.	Sept.	Oct.	Nov.	Dec.	Jan.	Feb.	March	April	May	June
CLIMATE	temperate		frost/snow		FREEZE-UP	frozen			melt	BREAK-UP		spring rains
MEANS OF TRAVEL	canoe			toboggan			ice	snowshoes			canoe	

MEANS OF TRAVEL: canoe; some fishing (poor); fish runs (rivers); fishing in lakes; toboggan / ice fishing; snowshoes / fishing in lake; canoe / fish runs (rivers)

GAME RESOURCES:
- primary beaver trapping/hunting → secondary trapping
- occasional caribou → primary caribou hunting → snare caribou
- occasional bear → bear trapped as they emerge from hibernation
- primary waterfowl season → migratory waterfowl
- waterfowl, ptarmigan, hare, grouse, birds, small mammals

LOCATIONS & SEASONAL MOVEMENTS:
- COAST — HBC post; travel to fall camps
- CAMPS — fishing or waterfowl camps; travel to bush camps before freeze-up
- INTERIOR BOREAL FOREST — dispersal to family hunting grounds; permanent winter bush camps
- return to lake or head of river SPRING FEAST
- COAST — HBC post

GROUP SIZE:
- 30–40 families
- up to 10 families
- 3–4 families
- hunting group (1–3 families)
- 2–3 hunting groups
- 30–40 families

Fig. 13.2. Annual cycle of ecological and socioeconomic events among the western James Bay Cree (from Flannery 1995, reproduced with permission from McGill–Queen's University Press).

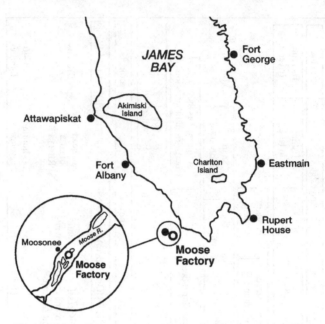

Fig. 13.3. Moose Factory and other Mushkegowuk communities of the James Bay region.

economic collapse, and malnutrition in the Canadian north with subsequent medical efforts to understand and alleviate the sources of that malnutrition.

Moose Factory and the Mushkegowuk Territory

The Mushkegowuk Territory, within which Moose Factory is located, extends northward from Timmins, Ontario, along the west coast of James Bay and Hudson's Bay, and approximately 200 km inland towards the west (Fig. 13.3). It is primarily boreal forest habitat with almost 70% of the landscape comprising a mosaic of muskeg, peat bogs and aquatic zones. The climate is characterized by short, warm summers and long, cold winters (Winterhalder 1983).

The Mushkegowuk Territory has been occupied for some 3000 years and sites with radiocarbon dates indicate the presence of the Cree from AD 800 to AD 1600 (Dawson 1983). This is well before 1611, when the first meeting between a European (Henry Hudson) and an Aboriginal person in the region was recorded (Francis and Morantz 1983). By the time Europeans returned to the area in 1668, they found the people of James Bay already participating in an extensive fur trade network that fanned out from the St Lawrence River toward the northwest. In 1673, the HBC established a post at Moose Factory that was occupied primarily by HBC personnel (Francis and Morantz 1983). The post served mainly as a point of congregation for Cree trapping families in the late spring and summer. The fragility of the resource base around this coastal post precluded the development of permanent villages because there was no continuously accessible supply of food to support a large, settled population.

By the end of the nineteenth century, a significant population of people of mixed British/Scottish and Cree descent could be found at Moose Factory. Although Christian clergy had arrived from Europe as early as 1810, their original role was to minister to the European employees of the HBC and only secondarily did they engage in missionary activity among the Cree (Francis and Morantz 1983). The missionary period, and the point at which reliable records of vital events become available in the region, dates to 1851, when Anglican missionaries began to actively seek converts in the Moose Factory area. During this period, the hazards of life were strongly affected by the changing seasons, with a marked peak in mortality occurring in the summer and a smaller peak in winter (Herring and Hoppa 1997). Epidemics of whooping cough, measles, influenza and other acute community infections swept through the region, superimposed upon a background of chronic, endemic tuberculosis (Herring and Hoppa 1999).

By 1900, Moose Factory had declined as a fur trade centre and the majority of families derived most of their income from wage labor rather than from fur trapping or trading (Ray 1996: 289). Even those still engaged in the fur trade tended to live full time in the vicinity of the post, traveling to outlying traplines in winter to collect furs, rather than managing their own traplines in the boreal forest interior. Other economic and political developments initiated further social change. With the extension of the Temiscaming and Northern Ontario Railway north in 1900, mobility was no longer as tightly constrained to the periods of freeze-up and break-up as had previously been the case. People could more easily move in and out of Moose Factory year-round and, indeed, many moved south to take advantage of new economic opportunities associated with the railroad. The period is also marked by increasing involvement of the Canadian government in the daily life of the northern Cree, notably through the enactment of a series of treaties across the nation that created the system of reserves that exist today.

The cumulative effects of these macrostructural and local contingencies eroded fur trade society at Moose Factory. The shift 'from a portable home within an ecological range, to housing in sedentary settlements' that had occurred over the span of a century (Preston 1986: 245) brought attendant changes in Cree philosophy and values, subsistence cycles, the value placed on hunting, and on virtually every aspect of social life. Preston (1986) understands the process in terms of five radical transformations, from traditional hunter's symbols to thinking in English, from the personal community to the church community, from bush homes to northern townsmen, from self-control to social control, and from an emphasis on individuals to categories.

Epidemiological patterns were transformed, too. The seasonal pattern of elevated mortality risks in summer and winter that characterized the nineteenth century was replaced by one in which the risks of death were less varied throughout the year. In addition, the frequent epidemics that punctuated the nineteenth century had largely abated by the twentieth century (Herring and Hoppa 1997). Crowded housing, the concentration of children into Residential Schools, and growing communities lacking an adequate subsistence base all contributed to a heavy burden of infectious disease and malnutrition after Treaty 9 was signed in 1905 (see Young 1988). This trend continued throughout the first half of the twentieth century (Honigmann 1948; Vivian et al. 1948). As the fur trade dwindled and the economic bases of many Aboriginal communities weakened, poor nutrition and high infectious disease loads became common by the 1930s and 1940s (Honigmann 1948; Tisdall and Robertson 1948).

The years between 1925 and 1948 were particularly harsh. In the course of conducting a survey along the west coast of James Bay in 1947–48, anthropologist John Honigmann (1948) learned of episodes of starvation in 1928, 1934 and 1939, as well as epidemics that took many lives in 1922 (unknown respiratory disease) and 1943 (measles). He observed an undernourished population that had survived a scarcity of both fur-bearing and large game animals during the winters of 1946–47 and 1947–48 (1948: 5). Access to store-bought foods, made possible by a newly created federal relief policy, alleviated the hunger of many but, ironically, made some feel less healthy, according to Honigmann (1948: 18). Vivian and colleagues' (1948) medical survey of the West Coast Cree reported caloric deficiencies, malnutrition, and an excessive burden of infectious diseases such as tuberculosis and childhood infections, as do letters and reports housed in the National Archives of Canada (NAC, RG 85, Volume 269). Tuberculosis figures prominently in local memories of the 1940s and 1950s (Abonyi 2001: 82).

All of the reports indicate a mid-twentieth-century health crisis. To evaluate the impact of the crisis, both in historical context and statistical terms, we investigated changes in survivorship at Moose Factory from 1811 to 1964. Was the first half of the twentieth century as devastating as the literature and our previous research would suggest?

Parish records as sources of information on mortality in Aboriginal communities

The survival of more than 150 years of Anglican Church of Canada parish records for Moose Factory First Nation (1811–1964) provides the opportunity to explore the features of mortality in the region surrounding this central subarctic Cree community. The records are available on microfilm at the Anglican Church of Canada Archives in Toronto, Canada (ACCA 1811–1964). With the help of several student research assistants[1] they were transcribed from printed hard copies to a database for analysis. The accuracy of the transcription was checked and errors corrected after a period of several weeks had elapsed from the time of the original data entry task.

The scarcity of available records and their often unreliable quality have meant that there have been relatively few studies of historical mortality for Aboriginal communities in Canada. Prior to the twentieth century, primary sources with quantifiable data are rare. As a result, much of the 300-year period from early contact to the present requires recourse to archeological and archival sources, such as the journals of early traders and explorers like Andrew Graham, Alexander MacKenzie, Peter Grant, David Thompson and Daniel Harmon. Parish and missionary records represent a potentially powerful source of information, allowing researchers to use both aggregate analysis (time trend in a single series of vital records) and family reconstitution methods (genealogical reconstruction from baptism, marriage, and burial records). Curiously, parish records remain relatively underexploited by students of Aboriginal population history (Herring 1992, 1994).

Information derived from parish registers, however, must be used cautiously when reconstructing demographic parameters such as mortality. A variety of factors can distort the picture created from clerical registers of vital events. While not an exhaustive list, the main factors include the loss of records, periods of underrecording, failures to register, a long gap between birth and baptism, and rapid changes in the social or economic structure of the community (Drake 1974; Lee 1977; Levine 1976; Willigan and Lynch 1982; Wrigley 1977). In particular, three factors tend to affect the overall reliability of parish records. The first relates to the degree of effort by church officials to communicate and enforce rules regarding recordkeeping, the second to the skills of the recorder, and third and perhaps most important, to the circumstances that differentially affected the recorder's ability to accurately enumerate vital events. Furthermore, the records for 'frontier' communities experiencing growth through natural increase and migration are often affected by changes in parish boundaries (Willigan and Lynch 1982). This special feature of expanding communities may lead to short-term fluctuations in the frequency of vital events. The accuracy of the records may be further constrained by transportation and communication difficulties, such as existed in many parts of the Canadian north during our study period (Hoppa 1998).

It is therefore crucial to evaluate carefully the quality of parish record data before submitting them to analysis. A detailed assessment of the quality of the Moose Factory church records can be found elsewhere (Hoppa 1998). Briefly, the burial registers consist of a total of 1459 records from 1811 to 1964, over 92% of which include age at death information. No significant differences in the recording of male versus female deaths were detected, although this cannot be taken strictly to mean that one group was not consistently underregistered compared to the other. Over the full data set, there are no differences in the precision of recorded age by sex or season of death but there is evidence of age heaping.[2]

Estimating survivorship at Moose Factory

The purpose of survival analysis is to examine the relationship between various factors and the survival time of individuals, defined by the occurrence of an event of particular interest. In this study, the event of interest is death. Using the Moose Factory burial records, survivorship was estimated via the Kaplan–Meier (1958) product-limit method. The method estimates the probability of an event occurring during a time interval, given its observation of occurrence previous to that interval. The test statistic of choice for survival analysis is the Mantel–Cox log-rank test, a modification of the Mantel–Haenzel χ^2 test (Norman and Streiner 1994). This nonparametric test is more robust than a z-test because it makes use of more data (Norman and Streiner 1994). The log-rank test is similar to a χ^2 test in that it compares the observed number of events (here, deaths) with the expected number of events, and assumes no group differences (see Hoppa 1998 for a more detailed discussion).

Figure 13.4 presents the annual number of deaths recorded in the Anglican Church of Canada parish records for Moose Factory, partitioned into three periods: 1851–1906, 1914–45 and 1946–64. The periods broadly reflect the late nineteenth century, early twentieth century, and the post-WWII eras, respectively, and capture significant shifts in socioeconomic and mortality patterns (Herring and Hoppa 1997). The first two periods roughly coincide with the 'contact-traditional era' (1821–1945) identified by ethnohistorians (Helm and Leacock 1971; Bishop and Ray 1976) and with its two components: the 'trading post dependency era' (1821–1890) and 'era of early government influence' (1890–1945). The third period represents the 'modern era' in this scheme. The annual number of deaths from 1940 to 1945 is inset in Fig. 13.4, and these are subdivided into infants (<1), children (1–14) and adult (15+) age categories.

It comes as no surprise that survivorship at Moose Factory changes significantly over time, as epidemiological transition theory would predict (Fig. 13.5). It is surprising, however, that the lowest cumulative survivorship values were observed in the first half of the twentieth century (1914–45), during the 'era of

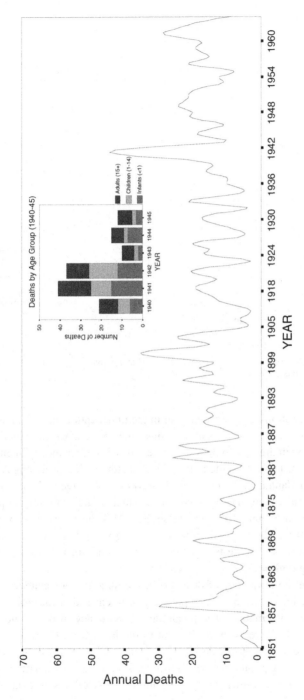

Fig. 13.4. Annual number of deaths, Moose Factory, 1851–1964. Note that deaths from 1940 to 1945 are inset and subdivided into infant (<1), child (1–14) and adult (15+) categories.

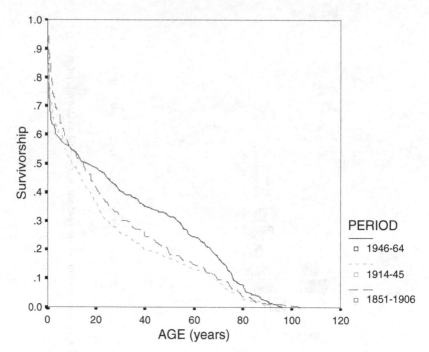

Fig. 13.5. Estimated survivorship at Moose Factory: 1851–1906, 1914–1945, and 1946–1964.

early government influence,' rather than in the nineteenth century. It has been generally accepted that the devastating consequences of European contact, expressed through malignant epidemics of infectious disease in the eighteenth and nineteenth centuries, represented the nadir of health conditions among Aboriginal people in Canada. The 'era of early government influence', by contrast, has been generally viewed as a period during which living conditions improved on the whole (cf. Young 1988: 15, Table 2.1). This clearly was not the case at Moose Factory, where the survivorship values for 1914–45 are significantly different from those for the two periods on either side of it (1851–1906 and 1946–64) (log rank = 12.44, df = 2, $p = 0.002$).

Changes in the frequency of deaths over the three periods cannot be evaluated because the duration of each period is different and because there is no reliable census information. The mean annual frequency of deaths nevertheless does seem to increase from the first to the last period (11.6 to 17.8 per annum), although not significantly so. This most likely reflects the additive effects of a growing population base at Moose Factory and the nearby community of Moosonee, complicated by population density effects associated with

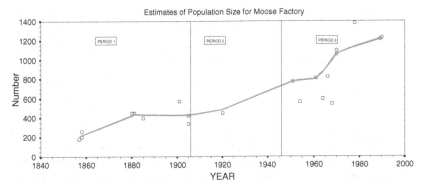

Fig. 13.6. Estimated population growth at Moose Factory: 1840–1995. Sources: Long 1995; Pugh 1972; Stephenson 1991; Piché and Romaniuk 1968; Zaslow 1960.

increasing sedentism over time. Population growth since the signing of Treaty 9 in 1905 is estimated at about 5.5% per year (Stephenson 1991), although ethnohistoric estimates of population size suggest that the most rapid growth occurred in the mid-twentieth century (Fig. 13.6).

What accounts for the drop in survivorship at Moose Factory in the early twentieth century? The burial records for Moose Factory in Fig. 13.3 show epidemic-like peaks in mortality in both 1941 and 1942. Closer inspection shows that these peaks capture a dramatic increase in infant and early childhood deaths. Childhood deaths (<15 years of age) account for 60–70% of all deaths from 1941 to 1943. Together, infant mortality in the years 1941 and 1942 boost the mean number of infant deaths for the period 1914–45 from 3.7 per annum to 13.5 per annum and significantly raise the proportional hazards rate for infants from 0.2885 (s.e. = 0.03) to 0.3110 (s.e. = 0.03).

When infant deaths from 1941 and 1942 are removed, a slight increase in survivorship is observed during the second period, but survivorship remains significantly lower than that of the post-World War II period and remains lower than that for the nineteenth century (1851–1906). This suggests that it was not just infant mortality that was depressing survivorship in the early twentieth century, but mortality at all ages (Hoppa 1998).

What was depressing survivorship at Moose Factory?

Epidemiologic transition in the Canadian Subarctic

The answer, quite simply, is poverty, expressed through the loss of food security and a reduction in the quality of the diet that, in turn, undermined the ability to resist infectious diseases such as tuberculosis.

Health and nutrition studies conducted in the 1940s by medical parties and anthropologists in northern Manitoba and along the James Bay coast (Honigmann 1948; Moore *et al.* 1946; Vivian *et al.* 1948) found communities suffering from malnutrition-related disease associated with deficient caloric intake and low levels of most nutrients. Episodic periods of dearth and starvation were familiar to northern Aboriginal communities and an expected though frightening feature of life, owing to periodic declines every few years in the natural cycles of fur-bearing animals, such as hares and squirrels, and their predators (Pugh 1972). Over and above environmental crashes that temporarily reduced the quantities of land food, however, the quality of the diet had declined since the turn of the twentieth century because of increasing reliance on store-bought food in the form of white flour, lard and sugar (Moore *et al.* 1946). Food security became an even more pressing problem toward the end of the fur trade era when the ever fluctuating food resources became further threatened by over-trapping (Honigmann 1981; Rogers 1983). In sum, ecological pressures created by the fur trade economy acted in concert with natural environmental cycles, engendering an economic crisis at Moose Factory.

By the mid-twentieth century, in fact, many northern Aboriginal communities were virtually destitute. The cumulative impact of centuries of fur harvesting, development projects such as the railway, and increased incursions of southerners into the northern economy had depleted the natural resource base in many regions and generated stiff competition for what remained. At the same time, the international demand and prices for furs crashed as the industry itself died. This conjunction of global and local developments created a two-pronged crisis of insufficient land food and the loss of the market economy for fur, threatening community sustainability. Basically, there was not enough to eat and little money with which to buy food supplies from the HBC stores. Even Aboriginal communities further south that had long since ceased to be involved in the fur trade were struggling to survive and compete with Euro-Canadian interests for access to wage labor, land, fisheries, and lumber (Moffat and Herring 1999; Tough 1984, 1990). People became undernourished and suffered from deficiency diseases, enhancing the already high rates of tuberculosis morbidity.

In other words, the depressed survivorship and increases in mortality at Moose Factory in the first half of the twentieth century are largely explained in terms of the 'un-natural history' (Santos and Coimbra 1998) of the international fur trade and the political-economic legacy of the colonial experience. A 1947 memorandum from Dr H.W. Lewis, Regional Superintendent for Eastern Arctic Indian Health Services, to Dr P.E. Moore, Director of Indian and Northern Health Services, indicates a clear understanding of these conditions:

> The chief suffering through diet deficiencies is amongst the children up to the time of second dentition. The lowered resistance leaves the child more

open to successful invasion of infectious diseases, particularly tuberculosis, impairment of skeletal and dental development and visual damage common under such conditions. It has been almost axiomatic that the general health of our isolated native races is in direct proportion to the availability of their local native foods. Supplemental scientific feeding for the young children is a constant need. Emergency feeding for all will be necessary in scattered localities when fur is scant and native foods not available in sufficient quantities. Research in the native economy should be constant and supplementary foods supplied for emergency conditions should be provided on a balanced diet basis rather than is the rule on a belly-filling basis.

(NAC, RG 29, Vol. 851-6-1, Vol. 1)

Canadian government policy and research responses 1950–1970

In view of the intertwined ecological, political and economic roots of the early twentieth-century health crisis in Aboriginal communities, we were interested in following up the recommendations of the medical parties of the 1940s to determine what national research initiatives and programs were developed and implemented over the next two decades (1950–1970). We therefore turned to the voluminous primary sources held at the National Archives of Canada and, in particular, to records from the Medical Services Branch and Indian Affairs and Northern Development (NAC, RG 29, Vol. 2987, 2989, and 2990; RG 85 Vol. 269).

The records indicate that the period from the 1950s through the 1970s can be characterized as one of active research on malnutrition and related diseases, as recommended by the survey parties. A major thrust of the work is aimed at identifying and determining the prevalence of specific nutrient deficiencies in Aboriginal communities across the country.

In a document prepared for the Panel on Indian Research in 1951, Dr L.B. Pett, then Chief of the Nutrition Division in the Department of National Health and Welfare, describes public health nutrition research as encompassing the actual state of nutrition; the foods being eaten, the analyses of these foods, their sources and the relative values placed upon them; how well the foods being used maintain a suitable state of nutrition; and methods of improving the state of nutrition that include the introduction of new (or forgotten) foods, better methods of preparation, and economic considerations (NAC, RG 85, Vol. 730, File 1003-2-3, pt. 1). All of these aspects of nutrition research were funded and undertaken, ultimately offering solutions focused on biological outcomes rather than on the ecological and environmental determinants of health, which other policy initiatives were intended to address.[3]

In 1956, for example, a nutritional survey (medical examinations, anthropometry, biochemical assays) was conducted in the northern portion of the

Province of Saskatchewan by the Nutrition Division, Department of National Health and Welfare, Ottawa (NAC, RG 29, Vol. 2990, File 851-6-x300, pt. 1). About three-quarters of the people examined showed health defects and a number of major nutritional problems were identified, including low hemoglobin (nutritional anemia), riboflavin deficiency, ascorbic acid deficiency, vitamin A deficiency and underweight. Subsequent field nurse reports on these and other communities led Dr O.J. Rath, Regional Superintendent for Indian and Northern Health Services, to note the widespread nature of nutritional deficiencies and malnutrition:

> This is a problem which continually confronts field nurses throughout the province. The same problem, I am sure, must confront field staff in other Regions as well... If we are to achieve our aim in improving the health of the Indian people, and particularly the infants and young children, the people must receive adequate nourishment. *It is realized that this problem ties in with all the social and economic problems of the Indian people, but until a solution is found for these problems in the future it is felt that there should be more immediate steps taken to improve the nutritional status of the people.* [emphasis added]
> (NAC, RG 29, Vol. 2990, File 851-6-x300, pt. 1)

The local economic and environmental circumstances faced by each community varied, even though the different pathways led to food shortages and nutritional deficiencies. In some northern locations, for example, the demise of the fur trade economy had left community members with little choice but to move to unfamiliar landscapes and to circumstances that compromised their overall well-being:

> The people at York Landing who moved from York Factory in 1957 are traditionally a coastal tribe and are not used to living in the bush environment in which they now find themselves. Many of the fur-bearing animals are new to them so they are unsure which ones are fit to eat. For instance, they are trapping a large number of lynx which are considered good eating by the other groups, but have been throwing the carcasses away because they are afraid they will become ill from eating them.
> (NAC, RG 29, File 860-8-x200, pt. 2[B] Monthly Report, The Pas Zone Headquarters, January, 1960, 860-8-C3)

Regardless of the ways in which malnutrition came to be expressed, there was general agreement at the time that the poor nutritional status of Aboriginal people constituted a national health problem. Dr L.B. Pett, Chief of the Nutrition Division, commented on the prevalence of low haemoglobin in a memorandum to Dr P.E. Moore, Director of Indian and Northern Health Services:

> ...I most certainly agree that nutritional anaemia is a commoner problem than has usually been given full credit both among Whites and Indians... I

was under the impression, perhaps a year ago, that a booklet was being prepared by Indian Health Services for Indian and Eskimo mothers. Perhaps this booklet could give a little more attention to the iron-containing foods. I am also actively discussing with Dr. Armstrong the question of an appropriate supplement for expectant mothers and for infants. It seems to me that all these different lines of approach and interests might profitably be brought together at least once to talk over most appropriate action in this field.

(NAC, RG 29, Vol. 2989, File 851-6-2, pt. 1)

As Dr Pett noted, a variety of initiatives were developed to attempt to alleviate the situation. The pressing need for food supplementation, especially for infants and children, is stressed in the early years of the correspondence (1946–61), clear recognition that many Aboriginal communities basically suffered from a lack of food. Much of the correspondence is taken up with considering the feasibility of certain regimens (e.g. the problem of juice freezing in winter in the bush); the acceptability of particular foods by Aboriginal families; the practical problems of providing baby foods in small tins; and the prohibitive cost of high-quality, nutritionally adequate food in the north. The task of developing culturally, economically and climatically suitable food supplements was extremely difficult and one that never met with much success.

As part of this initiative, serious consideration was given to creating special foods to alleviate malnutrition among children. Correspondence between November 26, 1947 and March 5, 1948, exchanged between J.O. Wharry, Vice-President of the Quaker Oats Company of Canada Limited, L.B. Pett (Chief Nutritionist, Nutrition Division), Dr P.E. Moore, Director, Indian Health Service Department, and H.A. Procter, M.D. (Assistant Director, Indian Health Services), describes efforts to develop a cereal–milk mixture that could be made into a smooth, pap-like food by adding water. While the idea was feasible in theory, the practical problems of creating a food with reduced bulk, that could be handled conveniently by people living in the bush, became major obstacles to the program (NAC, RG 29, Vol. 2989, File 851-6-2, pt. 1). This, in turn, led researchers to conduct case-control experiments with fortified foods (especially fortified flours) in Indian residential schools and in some Aboriginal communities.

Other suggestions for developing a special food for infants were modeled after international health projects:

Many countries, especially South American, have produced foods of high nutritive value and low cost using native fruits and grains with distribution subsidized by the country concerned... The development of a low cost food which with powdered milk would be adequate for the first two years of life would seem to be a very worthwhile project. A food similar to that produced

by "Meals for Millions" using soy beans as a protein source might be
feasible.
(NAC, RG 29, Vol. 2989, File 851-6-2, pt. 1, letter, W.J. Wood, Regional
Superintendent, Indian and Northern Health Service, Winnipeg, to The
Director, Indian and Northern Health Services, June 30, 1961)

Eventually, Medical Services Branch introduced a 'vitaminized' biscuit, according to a formula from the Food Division of National Health and Welfare, for teachers to distribute to schoolchildren.

Nutrition and education programs were also undertaken. In the Provinces of Manitoba and Ontario, for example, rats were shipped to Aboriginal schools (e.g. Moose Factory) for teaching demonstrations and nutrition experiments, and community-based nutrition projects were undertaken on a number of reserves, such as Berens River, Manitoba, and Moose Factory, Ontario (NAC, MSB File 851-6-4, Vol. 2). The OO-ZA-WE-KWUN Nutrition Project (1978) was especially comprehensive and attempted to use local perceptions of foods (e.g. foods characterized as strong or weak), role playing, slide presentations, and questionnaires to encourage more healthful food habits and to address identified nutritional problems, such as rickets and vitamin D deficiency (NAC, RG 29, Vol. 2990, File 851-6-x200, pt. 1).

In concert with nutritional assessments, food supplementation and education programs, a significant amount of ethnobotanical research was under way on the nutrient content of locally available plant resources and the manner in which these were incorporated into local diets (NAC, MSB Vol. 2986, File 851-6-1, Vol. 1).

Despite the intense and sincere efforts by medical and nutritional experts to alleviate malnutrition in Aboriginal communities in the two decades following World War II, the project never reached the potential envisaged by its architects. The food supplementation program for children, in the form of the vitaminized biscuit, was judged to be not only a failure, but a contributor to poor nutrition:

As a result of the use of this biscuit over the past ten years, Nutrition Canada has indicated that all age groups of school children and the school leavers in their late teens and early twenties are deficient in all the items listed in the content of this food supplement with the exception of the two Vitamins B. We simply cannot afford such a haphazard approach to food and vitamin supplementation. When the distribution is not under the surveillance of the Medical Services personnel, then the vitaminized biscuit becomes a liability, as everyone assumes that the child is getting all the essential nutrients when, in fact, this is not the case.
(Memorandum, Marcia C. Smith, Senior Consultant, Maternal and Child Health Division to M14, May 8, 1974)

Table 13.1. *Comparison of costs (in dollars) for basic staples: City of Hamilton, Moose Factory and Attawapiskat, 1996*

Item	Quantity	Hamilton	Moose Factory	Attawapiskat
white bread	1 loaf	0.99	1.09	2.40
brown bread	1 loaf	0.99	1.99	2.55
margarine	1 pound	0.99	2.79	2.35
eggs	1 dozen large	1.79	2.09	2.93
2% milk	4 litres	3.49	5.99	12.71
cooking onions	1 bag	1.25	1.99	3.14
carrots	1 bag	0.99	1.20	3.27
iceberg lettuce	1 head	0.79	1.59	2.97
tomatoes	1 package	1.74	2.09	4.26
potatoes	10 pounds	1.99	3.69	10.09
bananas	per kg	0.64	1.95	3.52
apples	per kg	2.18	3.75	5.11
oranges	per kg	2.18	5.39	5.14
lean hamburger	per kg	4.17	6.69	8.49
frozen fish	1 pound	3.99	4.99	8.05
pork chops	per kg	7.69	9.15	11.39
TOTAL COST		$35.86	$56.43	$88.37

The persistence of nutritional inadequacies prompted Otto Schaeffer to advocate regular monitoring of the nutritional status of a number of northern Aboriginal communities:

> My first choice of population groups in particular need of repeated monitoring of nutritional status would be Inuvik and Delta Indians and Eskimos, Ft. Providence and Simpson Indians, Rankin Inlet Eskimos, Frobisher Bay and Cape Dorset Eskimos, Whitehorse Indians. Others may have to be added such as Arctic Quebec Eskimos and *James Bay Indians* [emphasis added], and more traditional groups may be interesting for contrast but not urgent in regard to indicators for preventive action in particular for those living in the Northwest Territories.
>
> (RG 29, Vol. 2987, File 851-6-1, pt. 5A, Nutrition General, letter, Dr O. Schaeffer, Director, Northern Medical Research Unit to Dr Marcia C. Smith, Senior Consultant, Maternal and Child Health Division, January 10, 1974)

In fact, iron deficiency anemia continues to be highly prevalent among James Bay Cree infants today (Willows *et al.* 2000), as do other nutritional problems (Abonyi 2001) despite some sixty years of research and programs aimed at alleviating the problem. Barriers to accessing high-quality protein and fresh fruits and vegetables, including prohibitive costs (Table 13.1), continue to plague northern communities (Abonyi 2001: 121–6). It is clear that the focus, both in research and policy, on taking 'immediate steps ... to improve the nutritional status of the people' envisioned in 1959 by O.J. Rath (NAC, RG 29, Vol. 2990) has had limited impact. At the same time, there has been little progress towards advancing solutions to the ecological and economic root causes that have been recognized both locally and nationally since the 1940s.

Conclusions

This project was stimulated by our attempt to understand the increase in infant and childhood deaths among the people of Moose Factory in the 1940s. It also constitutes one facet of our study of the details of the epidemiologic transition among Aboriginal people of the Canadian Subarctic, as exemplified by the Moose Factory experience. This study shows that the first half of the twentieth century was marked not only by a reduction in the frequency of epidemics and the loss of the seasonal pattern of mortality, but also by a significant and unexpected reduction in survivorship. The depression in survivorship was not simply the product of elevated rates of infant and child mortality, but reflects an overall increase in mortality. Evidently, the conditions of life at Moose Factory in the first half of the twentieth century were in decline, perhaps more so than the literature and our previous research suggested. This reduction in the quality of life coincided with increased involvement of the Canadian Government in Aboriginal life, new development projects and increased penetration of non-natives into the northern economy. Whether this downward trajectory is unique to the Mushkegowuk Cree, or part of a larger pattern of health decline from the nineteenth to the mid-twentieth century, remains to be quantified for other groups. We suspect that such analyses will reveal that Aboriginal life expectancy in general deteriorated in the first half of the twentieth century before rebounding in the post-World War II period.

The search for explanations led us to explore post-World War II nutrition research among Aboriginal communities, as represented through National Archives of Canada document collections. The dire situation at Moose Factory was not unusual, but part of a national health and nutrition crisis among First Nations which was the cumulative effect of complex social and ecological change since the nineteenth century. In particular, the general quality of the diet had declined because of increasing reliance on store-bought food in the form of white flour, lard and sugar (Moore *et al.* 1946), aggravated by growing food insecurity toward the end of the fur trade era when the ever-fluctuating food resources became further threatened by over-trapping (Honigmann 1981; Rogers 1983). Recognition that this was a national – and not a restricted local – problem led the Government of Canada to develop major programs in the areas of health care, social assistance, economic development and education.

As part of these initiatives, energetic programs of nutrition research were funded and undertaken, but made few inroads. A number of factors contributed to this situation. The mandate of the studies carried out by Medical Services Branch essentially involved assessing the nutritional well-being of communities, the sources and quality of foods being eaten and their relative value, as well as improving the overall state of nutrition. This focus necessarily removed the study of malnutrition from its historical origins in colonialism and the European fur trade, as well as from the ways in which these forces had transformed the socioeconomic and political fabric of life for northern people. Malnutrition inevitably became medicalized, diagnosed through the specific biological conditions of individuals and treated at the community level through the use of vitamin supplements, food fortification, education programs and medication. In essence, malnutrition increasingly and necessarily was studied as a biological (or internal) problem, not as an ecological and economic (or external) problem, primarily because the treatment available to medical researchers did not include the resources needed to address the fundamental ecological and economic determinants of health.

During a year of fieldwork at Moose Factory in the late 1990s by one of us (Abonyi), it became evident that community members understand the malnutrition of the 1940s as an ecological problem arising from the disruption of human–land and human–animal relationships. They recall, sometimes with humor and often with frustration, participating in various research projects, surveys and assays, as outsiders came and went in a blur – seldom getting to know the people, their circumstances, or perspectives. Our foray into the archives not only helped to explain how the differences between local and national medical research perspectives on malnutrition arose, but also offered a new vista on human research in the north and the factors that contributed to the biological reductionism that can be said to have characterized nutritional research protocols in Aboriginal communities in the post-World War II period.

Acknowledgements

We are grateful to the Anglican Church of Canada for permission to study the Moose Factory missionary and parish records. Thanks go to John Long and Dick Preston for additional information, advice and guidance throughout the research process and to the people of Moose Factory who shared their memories and experiences. Funding for this project is gratefully acknowledged from the Social Science and Humanities Research Council of Canada (#756-95-0181),

McMaster University Arts Research Board (#5-58547), the Northern Scientific
Training Program, and a fieldwork research grant from the School for Graduate
Studies, McMaster University.

Notes

1 We are grateful to Mary-Anne Ciampini, Tracy Farmer, and Todd Garlie for their
help.
2 Age heaping is the tendency to round ages to multiples of particular numbers, such
as 2, 5 or 10.
3 National programs of family allowance and relief were implemented and extended
to Aboriginal people, with special relief rations targeted for infants (NAC, RG 10,
Vol. 7094, File 1/10-3-0). Conservation efforts were also implemented, such as the
Territorial and Provincial Migratory Bird Sanctuaries located in the James Bay
Region from the late 1930s (NAC, RG 85, Vol. 850, File 7824).

References

Abonyi, S. (2001). *Sickness and symptom: Perspectives on diabetes among the
Mushkegowuk Cree*. Ph.D. Thesis, McMaster University, Hamilton, Canada.
Anglican Church of Canada Archives (ACCA). (1811–1964). *Moose Factory burials,
Diocese of Moosonee*. Toronto, Canada: General Synod Office.
Bishop, C.A. and Ray, A.J. (1976). Ethnohistoric research in the central subarctic: some
conceptual and methodological problems. *Western Canadian Journal of Anthro-
pology* 6(1), 116–44.
Dawson, K.C.A. (1983). Prehistory of the interior forest of northern Ontario. In *Boreal
Forest Adaptations: The Northern Algonkians*, ed. A.T. Steegman, Jr., pp. 55–84.
New York: Plenum Press.
Drake, M. (1974). *Historical Demography: Problems and Projects*. Milton Keynes: The
Open University Press.
Flannery, R. (1995). *Ellen Smallboy: Glimpses of a Cree Woman's Life*. Montreal and
Kingston: McGill–Queen's University Press.
Francis, D. and Morantz, T. (1983). *Partners in Furs: A History of the Fur Trade
in Eastern James Bay, 1600–1870*. Kingston and Montreal: McGill–Queen's
University Press.
Helm, J. and Leacock, E.B. (1971). The hunting tribes of subarctic Canada. In *North
American Indians in Historical Perspective*, ed. E.B. Leacock and J. Helm, pp. 343–
74. New York: Random House.
Herring, D.A. (1992). Toward a reconsideration of disease and contact in the Americas.
Prairie Forum 17, 1–13.
Herring, D.A. (1994). 'There were young people and old people and babies dying every
week': The 1918–1919 influenza pandemic at Norway House. *Ethnohistory* 41,
73–105.

Herring, D.A. and Hoppa, R.D. (1997). Changing patterns of mortality seasonality among the western James Bay Cree. *International Journal of Circumpolar Health* **56**, 121–33.

Herring, D.A. and Hoppa, R.D. (1999). Endemic tuberculosis among nineteenth century Cree in the central Canadian Subarctic. *Perspectives in Human Biology* **4**, 189–99.

Honigmann, J.J. (1948). *Foodways in a Muskeg Community*. Ottawa: Department of Northern Affairs and Natural Resources, Northern Coordination and Research Centre.

Honigmann, J.J. (1981). West Main Cree. In *Handbook of North American Indians*, Volume 6, *Subarctic*, ed. J. Helm, pp. 217–30. Washington, D.C.: Smithsonian Institution.

Hoppa, R.D. (1998). Mortality in a northern Ontario fur-trade community: Moose Factory, 1851–1964. *Canadian Studies in Population* **25**, 175–98.

Hudson's Bay Company Archives (HBCA). (1922). *Map of the Dominion of Canada shewing* [sic] *the establishments of the Hudson's Bay Company*. HBCA G.3/556 (N6490).

Kaplan, E.L. and Meier, P. (1958). Nonparametric estimation from incomplete observations. *Journal of the American Statistical Association* **53**, 457–85.

Lee, R.D. (1977). Methods and models for analyzing historical series of births, deaths, and marriages. In *Population Patterns in the Past*, ed. R.D. Lee, pp. 337–70. New York: Academic Press.

Levine, D. (1976). The reliability of parochial registration and the representativeness of family reconstitution. *Population Studies* **30**, 107–22.

Long, J.S. (1995). Historical Context. In *Ellen Smallboy. Glimpses of a Cree Woman's Life*. R. Flannery, pp. 65–75. Montreal: McGill–Queen's University Press.

Moffat, T. and Herring, D.A. (1999). The roots of high rates of infant death in Canadian Aboriginal communities in the early twentieth century: The case of Fisher River, Manitoba. *Social Science and Medicine* **48**, 1821–32.

Moore, P.E., Kruse, H.D., Tisdall, F.F. and Corrigan, R.S.C. (1946). Medical survey of nutrition among the northern Manitoba Indian. *Canadian Medical Association Journal* **54**, 223–33.

National Archives of Canada (NAC). Ottawa, Ontario, Canada.

Norman, G.R. and Streiner, D.L. (1994). *Biostatistics: The Bare Essentials*. Toronto: Mosby.

Piché, V. and Romaniuk, A. (1968). Une enquête socio-démographique auprès des Indiens da la Baie James: 1968. *Anthropologica* **14**, 219–30.

Preston, R.J. (1986). Twentieth-century transformations of the West Coast Cree. In *Proceedings of the Seventeenth Annual Algonquian Conference/Actes du Dix-Septieme Congres des Algonquinistes*, ed. W. Cowan, pp. 238–51. Ottawa, Canada: Carleton University.

Pugh, D.E. (1972). *Cultural optimality: A Study of the rise and decline of the Cree culture of north eastern Ontario*. Ph.D. Thesis, Carleton University, Ottawa.

Ray, A.J. (1996). *I Have Lived Here Since the World Began*. Toronto: Lester Publishing Limited.

Rogers, E.S. (1983). Cultural adaptations: The northern Ojibwa of the boreal forest, 1670–1980. In *Boreal Forest Adaptations: The Northern Algonkians*, ed. A.T. Steegmann, Jr., pp. 85–142. New York: Plenum Press.

Santos, R.V. and Coimbra, C.E.A., Jr. (1998). On the (un)natural history of the Tupi-Mondé Indians: Bioanthropology and change in the Brazilian Amazon. In *Building a New Biocultural Synthesis: Political-Economic Perspectives on Human Biology*, ed. A.H. Goodman and T.L. Leatherman, pp. 269–94. Ann Arbor: The University of Michigan Press.

Stephenson, K. (1991). *The Community of Moose Factory: A Profile*. TASO Report, Second Series, No. 2. Hamilton, Canada.

Tisdall, F.F. and Robertson, E.C. (1948). Voyage of the medicine men. *The Beaver* **28**, 42–6.

Tough, F. (1984). The establishment of a commercial fishing industry and the demise of Native fisheries in northern Manitoba. *Canadian Journal of Native Studies* **4**, 303–19.

Tough, F. (1990). Indian economic behaviour, exchange and profits in northern Manitoba during the decline of monopoly, 1870–1930. *Journal of Historical Geography* **16**, 385–401.

Vivian, R.P., McMillan, C., Moore, P.E., Robertson, E.C., Sebrell, W.H., Tisdall, F.F. and McIntosh, W.G. (1948). The nutrition and health of the James Bay Indian. *Canadian Medical Association Journal* **59**, 505–18.

Waldram, J.B., Herring, D.A., and Young, T.K. (1995). *Aboriginal Health in Canada: Historical, Cultural and Epidemiological Perspectives*. Toronto: University of Toronto Press.

Willigan, J.D. and Lynch, K. (1982). *Sources and Methods of Historical Demography*. New York: Academic Press.

Willows, N.D., Morel, J. and Gray-Donald, K. (2000). Prevalence of anemia among James Bay Cree infants of northern Quebec. *Canadian Medical Association Journal* **162**, 323–6.

Winterhalder, B. (1983). History and ecology of the boreal forest zone in Ontario. In *Boreal Forest Adaptations: The Northern Algonkians*, ed. A.T. Steegmann, pp. 9–54. New York: Plenum Press.

Wrigley, E.A. (1977). Births and baptisms: The use of Anglican baptism registers as a source of information about the numbers of births in England before the beginning of civil registration. *Population Studies* **31**, 281–312.

Young, T.K. (1988). *Health Care and Cultural Change: The Indian Experience in the Central Subarctic*. Toronto: University of Toronto Press.

Zaslow, M. (1960). Rendezvous at Moose Factory, 1882. *Ontario History* **53**, 82–94.

14 *Archival research in physical anthropology*

MALCOLM T. SMITH

Introduction

My aim is to present an overview of the main trends in the development and practice of archival anthropology, reflecting on the paths by which the project has developed, and noting not only the mainstream, but also the occasional intriguing backwater. Before moving on in later sections to a description of data sources, analytical approaches, and exemplary studies, I shall briefly outline the broad range of anthropological studies that utilize archival data, and offer some thoughts on the particular qualities of the sources which lend themselves to anthropological analysis. Finally, I give some hostages to fortune in a brief statement of future directions for this endeavor. Now and then, I shall indicate a little of the parallel work done especially by historians and historical geographers, to remind us of the commonality of interest across academic disciplines, and to show what keeps the biological anthropology distinct.

Archival anthropology is in many ways no different in spirit and intent from the discipline as followed by many workers studying living populations. Often the investigators are the same people and the only difference is the provenance of the data. That said, the fact of relying on archival data imposes its own special constraints, grants its own freedoms, and inculcates a community of interest among practitioners across disciplines – the anthropologist and the historian in the record office can sometimes feel equally remote from their colleagues in the disciplines' mainstream.

The scope of archival research in biological anthropology embraces aspects of human evolution, including elucidation of the processes of microevolution and their consequences for the genetic structure of populations, and aspects of human biology, including survival and reproduction, and components thereof such as mortality, fertility, nuptiality, and the effects of nutrition and disease on these measures. The topics that are to do with evolutionary mechanisms and population structure are clearly within the purview of biological anthropology, but the work on fertility, mortality and disease is much more ambiguously situated, to some extent common and occasionally contested ground among social and medical historians, historical geographers and biological anthropologists.

Why work on data archives? There are a number of reasons, and I give the most prosaic first. The data *may* be easy to obtain, so that a lot of survey effort can be avoided. This may sound an unworthy reason, but it is a serious one, and we can see just how seriously the topic of preservation and re-analysis of data is taken, by the remarkable efforts that are being made for the curation of data, especially digitized data (see below). I must acknowledge, however, that the easy access archive is a rarity, and most archival work requires the researcher spending effort on the original data source, often scarcely indexed, perhaps never even read before. Anyone who has transcribed a manuscript marriage register, a foundling home entry book, or a dusty bundle of wills and inventories will emphatically deny that these are data easily won.

The time-span of archives offers cogent inducement to those with an interest in microevolution or the historical dimension of demography. Data taken from European and European-derived populations not infrequently span three or four hundred years, and may cover still longer periods. This offers the opportunity not only to describe and interpret secular trends, to see the dynamic as well as the static, but also to elect to study the consequences of specific phenomena, be they diseases, famines, political changes or industrialization.

A further reason for work with historical data is the increased emotional distance between the researcher and populations suffering the ill-effects of famine, disease, warfare or child abandonment. However compelling their stories, these people cannot elicit from us the same response of concern as those in a desperate plight today. Freed from the immediate concern to act, we are sometimes able thereby to take a more detached, analytical and even speculative view. These properties of historical data have been appreciated before, notably by Segalen (1991). The ethical dimension of engagement is removed, both in terms of intrusion in the lives of individual subjects, and in the absence of any possibility of changing their fate.

Data sources

The records commonly used by historical demographers, social and economic historians and biological anthropologists fall into three main classes: vital records (both ecclesiastical and civil), censuses, and the rest. The vast preponderance of data comes from the first two categories, mostly spanning the sixteenth to the twentieth centuries. Vital records reflect the standing population through sampling its vital events as they occur. Thus, one never gets actually to know how many people there are in the population, only how many get married, are born, or die in any given year. Censuses are different in nature, depicting the entire population at an instant, and their occurrence, widely spaced in time, means that the view they present is like a series of snapshots. The miscellany of other kinds of record is less comprehensive in coverage, but may provide listings of population such as tax lists (see Schurer and Arkell 1992), migration evidence (Clark 1979) or insight into social and kin relationships revealed through wills. Archival studies are of course limited to literate societies, and the records routinely exploited are from the European Early Modern and Modern periods. There are, however, records from the edges and beyond this constrained range, and I shall highlight some of these exceptional alternative sources in the following paragraphs. For example, records survive from the Roman period (Lo Cascio 1999) including censuses of Roman Egypt (see below) and sources from elsewhere in the Roman Empire (Parkin 1999), records of the Ottoman empire (Kiel 1999) and censuses of colonial rule, including British censuses of India (Barrier 1981) and Portuguese censuses from Africa (Curto 1994).

In the context of a paper on brother–sister marriage grounded in theories of the incest taboo, Hopkins (1980) used as his principal data source papyrus census returns surviving from AD 19–20 to AD 257–8, taken in Egypt under the Roman administration. Bagnall and Frier (1994) presented transcriptions in English of this primary census material together with a demographic analysis of the data. Out of 121 marriages 20 were between full sibs, in a demographic context in which only about 40% of families were likely to have had children of both sexes surviving to marriageable age.

Census returns from around 300 households survive, referring to some 1100 individuals, with about two-thirds from a single area, the town of Arsinoe in Fayum. These data were sufficiently numerous to give confidence to quantitative interpretation on matters such as mortality, fertility, age at marriage, nuptiality, and the age difference between spouses. There were also Roman censuses of Gaul, Sicily, and, as is well-known from the Nativity story in the Gospel of Luke, Judaea. It is however, the Egyptian ones that survive the best, though whether they can be used to generalize demographic behavior more widely across the Roman Empire is disputed. Shaw (1992) and Parkin (1999) are much more cautious than Hopkins or Bagnall and Frier about the wider applicability of these data. Recently Scheidel (1997) has analyzed this same material in the context of interest in the frequency and degree of consanguineous marriage and arguments about inbreeding genetic load.

Although medieval records do not contain the systematic census-taking and recording of vital events that occur in later centuries, there are a number of sources which can provide data on migration and, occasionally, fertility. Principal among these in England are the Manor Court Rolls, which record events transacted in the local community, and which may incidentally reveal the actions and relationships of individuals and their provenance. A fine exemplary study of the English provincial town of Halesowen in the thirteenth and fourteenth century indicates the wealth of information to be gained from Manor Court Rolls (Razi 1980), though the author notes a gender bias in the records, with relatively little information about women (see chapter 12, this volume). Other studies, such as those of Warboys (Raftis 1974) and Godmanchester (Raftis 1982) have yielded, among other things, estimates of migration rates and distances. Alternative medieval sources used to estimate migration include Freeman Rolls, and Lay Subsidy returns. McClure (1979) used the distribution of people with surnames derived from placenames, in the period before surnames became hereditary, to estimate lifetime migration into the cities of London, York and Norwich.

A class of records much studied by historians but relatively ignored by anthropologists is the enormous legacy of documentation from foundling homes, orphanages and other institutions (see, for example, chapters 5 and 6 this volume). Foundling hospitals developed in Europe, and in European colonial settlements, from the middle ages onward. They are said to have originated in an attempt to avert the practice of infanticide through drowning or exposure, widespread in classical antiquity (Boswell 1988). The longest series of data comes from Italy, where foundling homes were early established, and where the tradition of abandonment has perhaps been the most enduring. Viazzo *et al.* (2000) analyzed the series of records from Spedali del Innocenti in Florence, which runs from 1447 to the present day. Major studies of foundling records have been published by Kertzer (1993), Sá (1995), and Sherwood (1988). Although a correlation between rates of abandonment and grain prices has been shown (Peyronnet 1976), the marked secular trend in child abandonment seems as much cultural as economically or demographically based. An 'evolutionary' approach to the phenomenon of child abandonment has been taken by Hrdy (1992), who coined the concept of 'delegated mothering' to encompass various modes of allocare, including wet-nursing as the preferred choice of many affluent parents (Fildes 1988) and abandonment as an expedient borne of poverty and shame.

A distinctive category of records, exploited independently by a number of researchers, is historical anthropometric measurements. Relethford and collaborators have used data collected by C.W. Dupertius and H. Dawson during 1934–6 as part of the Harvard Anthropological Survey of Ireland (North *et al.* 1999; Relethford *et al.* 1980) in a series of papers exploring historical migration, genetic drift and isolation by distance as determinants of population structure (Relethford and Crawford 1995; Relethford *et al.* 1997; chapter 3, this volume).

Old anthropometric and anthroposcopic data sets are also the product of routine measures of stature and other variables, collected from recruits to the armed forces, other employment, penal institutions and schools (Beddoe 1885; Rosenbaum and Crowdy 1990; chapter 7, this volume). Such data have been analyzed within the framework of historical human biology (Tanner 1981), and of economic history by Floud *et al.* (1990), who used historical data to explore the effects of change in prices and wages on achieved stature. The result is a fascinating amalgamation of approaches from distinct disciplines converging on a single issue. Anthropometric data from East India Company recruits, prison records and obstetric records have been used by Ó Gráda (1991, 1994) to gauge the nutritional status of the population of rural Ireland before and after the Great Famine of 1845–50. Archived data from Floud *et al.* (1990) were utilized by Jordan (1998) in his survey of the status of children in Ireland. Further notable contributions to this field have been made by Komlos (1994, 1995).

Analytical approaches

The rise of historical demography had its origins both in the example of French historical demographers (Gautier and Henry 1958), and in the wider movement which gave rise to social history, the view that the history of the ordinary people was both worthy of study and tractable as a subject of enquiry (Laslett 1965; Bonfield *et al.* 1986). In England, the development of methods based on aggregative analysis and family reconstitution (Wrigley 1966, 1972) and the eventual application of these to large-scale and long-term analysis are among the most important achievements of the Cambridge Group for the History of Population and Social Structure (Wrigley and Schofield 1981; Wrigley *et al.* 1997). Historical demographic research in anthropology developed from this model, and many early publications show its influence (Dobson and Roberts 1971; Jeffries *et al.* 1976, Küchemann *et al.* 1967). The secular trend or seasonality of births, marriages and deaths, and the spatial or social distribution of marriage partners are the demographic data underpinning this kind of research (see, chapters 4 and 13, this volume).

Recent work by Scott and Duncan (1998) has combined the perspective of population biology with both aggregative and family reconstitution techniques to investigate the population of the parish of Penrith, Cumbria, from the sixteenth to the nineteenth centuries. Their interest includes the effects of prices and weather, and of social class differences in nutrition and fertility, on the demographic basis of population cycles (Scott and Duncan 1999; Scott *et al.* 1998). The same authors have also investigated the periodicity of infectious disease (Scott *et al.* 1996; Duncan *et al.* 1997, 2000), and indeed descriptive and explanatory studies of temporal and spatial distributions of non-genetic disease have frequently been a focus for anthropological studies (Jorde 1989b;

Mielke *et al.* 1984; Sattenspiel *et al.* 2000; chapters 8 and 11, this volume) an interest shared with historians, demographers, historical geographers (Dobson 1997; Landers 1993; Woods and Shelton 1997). The range of anthropological work also includes fine-scale case-studies which embrace both biological and sociocultural factors in describing the complexity of causation and association in historical communities (Sawchuk 1980; chapter 9, this volume).

Archival sources have been used to test hypotheses derived from behavioral ecology or sociobiology. For example, the Trivers and Willard (1973) hypothesis about control of offspring sex ratio was invoked in Dickemann's interpretation of childhood sex ratios from Colonial censuses of India, alongside other quantitative reports of sex ratio bias from China and Europe (Dickemann 1979). Further research in this vein is the study of female claustration by Boone (1986, 1988) based on genealogies from Medieval and Early Modern Portugal. A very distinctive style of analysis in support of similar hypotheses appears in the work of Betzig. Her analyses of the origins of European marriage systems (Betzig 1992, 1995) exemplify wide textual as well as archival analysis. In style this work can be likened to that of historians like Stone (1977) or Boswell (1988), with inference based on the evidence of writings, correspondence and diaries of literate elites rather than the vital records of the whole population. Much more explicitly historical demographic in approach is the work of Voland (1984, 1988, 1989) who has used family reconstitution studies of German parishes to test hypotheses about fertility strategies, for example, among different socioeconomic groups. Evolutionary hypotheses about parental investment in offspring have also been tested in relation to bequests in wills in Sacramento, California, by Judge and Hrdy (1992).

Studies of population genetic structure seek to measure, predict and explain the distribution of genes and genotypes in a population. The measurement may be based on genetic markers or metrics, or estimated from surname distributions. Predictions of kinship are based on observed or estimated migration, or on inbreeding calculated from genealogies or dispensations (Cavalli-Sforza and Bodmer 1971). Protocols for such analysis have been elaborated and discussed in a number of reviews (for example Jorde 1980; Swedlund 1980, 1984; Morton 1982; Morton *et al.* 1971; Mielke and Swedlund 1993).

At their most comprehensive such studies integrate a number of data sources to test hypotheses of causation: for example, that the distribution of genetic variation reflects the observed pattern of historical migration, or the geographical pattern of settlement. However, it is often the case that one or just a few parameters may be ascertained. For example, marital mobility alone, as evidenced in endogamy and marriage distance, might be used to show the breakdown of genetic isolation over time. Likewise, inbreeding alone has become a popular measure of genetic structure, especially where Roman Catholic marriage dispensations are available. This development has occurred alongside the interest in consanguinity in living populations, almost in spite of the fact that the populations for which there is historical evidence have, by global standards, low levels of inbreeding (Bittles 1998). Notwithstanding the criticisms of detractors (Rogers 1991) and the caveats of practitioners (Crow 1989, and many others) the acceptance of isonymy models, which enable the use of surnames as surrogate alleles, together with the widespread availability of surname data sets, has led to a proliferation of historical isonymy studies. The most commonly used techniques permit the estimation of random and non-random components of inbreeding within populations (Crow and Mange 1965), relationship (Lasker 1977), random isonymy (Morton *et al.* 1971) or genetic distance (Relethford 1988) between populations, and descriptors of population subdivision (Lasker and Kaplan 1985). Measures of isonymy may then be compared with kinship predicted from migration or genealogy, or relationships between isonymy and geographical distance, gene frequencies, or some other parameter may be sought. Isonymic methods overestimate kinship because many names have a number of independent origins (are polyphyletic). On the other hand, hereditary surnames encompass a greater genealogical depth than pedigree or dispensation evidence, and may present a counterbalance to generalizations about migration behavior based on short runs of data. My own view is that local studies of isonymy, in which bearers of even common surnames may plausibly be descended from a common ancestor, are less likely to be subverted by the polyphyletic origin of names than large-scale ones, for which some selective sampling of rarer surnames might be preferable. For example, the fourth most common surname in a recent study using the Italian CD ROM telephone directory was Esposito – a name given as a family name to foundling children (Barrai *et al.* 1999).

Potential mates analysis was devised by Dyke (1971), and enhances the understanding of mate choice and inbreeding by setting individual action within the context of the whole population. It depends upon the biological kin relationships of a population being known, and so requires a population whose members have ascertained genealogies. Comparison is then made between the kinship of those who actually marry with the kinship of potential mates, i.e. the array of individuals of appropriate kinship, age and marital status. A thorough review is given by Leslie (1985), and exemplary applications are by Brennan and collaborators (Brennan 1981; Brennan *et al.* 1982; Relethford and Brennan 1982).

The index of opportunity for selection was devised by Crow (1958) to indicate how much natural selection might occur if the differential mortality and fertility between individuals were due solely to the adaptation of genotypes to their environment. There has been sporadic rather than intensive interest in application of the method (see, for example, Jorde and Durbize 1986; Hed 1984, 1987), and it has recently been used more with information surveyed from living populations (Reddy and Chopra 1990; Hemam and Reddy 1998) than with historical data, for which it requires record-linked demographic data. This may be because it obviously measures only opportunity for, rather than actual occurrence of, natural selection, and so reveals more about a population's status in respect of demographic transition than about selection.

Regional coverage

In surveying studies of population structure by region, I have arranged the subject matter into three categories: (1) European mainland, including the British Isles, (2) colonizing settlement from Europe, and (3) islands.

Among biological anthropologists in Europe there is a strong tradition of studying aspects of migration and population structure, often under the blanket title of 'biodemography' (see, for example, Vienna *et al.* 1998). (This nomenclature had for many years been unproblematic, though the term 'biodemography' has recently been adopted in the United States to indicate an approach to research in ageing and longevity based on biological potentials.) Biodemography developed partly out of the profoundly influential French tradition of historical demography, but perhaps also derived inspiration from the familiarity in Catholic Europe with the issue of marriage within the prohibited degrees of consanguinity, and with the archive of records that it has generated (Cavalli-Sforza and Bodmer 1971).

The seminal paper by Barrai *et al.* (1962) exemplifies the data sources and concerns that underlay this development. This paper, like some other early works (see, for example, Sutter 1968), goes far beyond the descriptive, testing hypotheses about the demographic context of consanguineous marriage, including the effects of patrilocality and of the age difference between spouses in determining the different categories of consanguineous marriage. The preoccupation with the natural history of inbreeding is an important strand in the intellectual tradition of European biodemography, with studies in France (Sutter and Tabah 1948; Crognier 1985; Vernay 2000), Spain (Fuster 1986; Fuster *et al.* 1996; Calderon 1989; Calderon *et al.* 1998; Peña *et al.* 1997) and Portugal (Abade *et al.* 1986; Areia 1986) and Italy (Barrai *et al.* 1992; Pettener 1985, 1990; Danubio *et al.* 1999; Paoli *et al.* 1999; Rodriguez-Larralde *et al.* 1993) and elsewhere. For an analytical review of many cases, see McCullough and O'Rourke (1986).

The chronological pattern that emerges from European studies of inbreeding is gratifyingly non-intuitive. Whilst the commonly-observed decline in inbreeding throughout most of the twentieth century is just what we should predict assuming the secular breakdown of isolation, the studies with the longest runs of data show inbreeding levels increasing through the time, sometimes quite sharply, until the late nineteenth and early twentieth century (Moroni 1967; Bourgoin-Vu Thien Khang 1978). This is an observation that demands an explanation, usually offered in terms of inheritance laws, migration patterns or demographic transition (Sutter 1968). Where such hypotheses have been tested, as in the important paper by Pettener (1985), the results have been rather ambiguous.

In Britain the development of historical demography by the Cambridge Group (see above) stimulated biological anthropologists at Oxford to use the same data sources to illuminate the genetic structure of historical populations. Their focus of research was an area of rural Oxfordshire named Otmoor (Küchemann *et al.* 1967), and subsequently on a number of historical parishes in Oxford City (Küchemann *et al.* 1974). This work introduced a number of influential ideas for the analysis of the effects of geographical distance (Jeffries *et al.* 1976) and social class (Harrison *et al.* 1970, 1971) on marital migration, including an early formulation of a migration matrix model (Hiorns *et al.* 1969) which was used to show the decline of genetic isolation through time, and the Neighbourhood Knowledge model (Boyce *et al.* 1967), one of the few behavioral models used to explore genetic structure (Fix 1999). Subsequent work analyzed surname distributions (Küchemann *et al.* 1979; Lasker 1978), the distribution of classic genetic markers (Hiorns *et al.* 1977) and aspects of human biology (Harrison *et al.* 1974). The Otmoor studies were brought together recently in a single volume by Harrison (1995). Elsewhere in mainland Britain, Roberts and colleagues characterized the population structure of agricultural (Dobson and Roberts 1971; Roberts 1982,) and mining populations (Roberts 1980; Lasker and Roberts 1982), and major surveys of the spatial distribution of surnames were carried out by Mascie-Taylor and Lasker (1984, 1985, 1990), Sokal *et al.* (1992) and in Scotland by Holloway and Sofaer (1989). Smaller-scale studies have tested the hypothesis that sociocultural characteristics contribute to the genetic structuring of populations, in research into occupational communities including fishing villages on the English coast (Pollitzer *et al.* 1988; Smith *et al.* 1984), and into settlement history and religious affiliation in the Ards Peninsula, Northern Ireland (Bittles and Smith 1994; Smith *et al.* 1990). Religion as a correlate of population structure in Ireland has also been investigated by Relethford and Crawford (1998).

Away from Europe, there has been substantial research on the demography of colonizing settlement in the New World and elsewhere. Such settlement characteristically has included kin-structured migration in the first instance (McCullough and Barton 1991). In North America, classic studies of religious isolates such as the Amish and the Hutterites, among others, have demonstrated the profound influence of founder effect, genetic drift and endogamy in promoting inbreeding and unusually high frequencies of rare genetic disorders (McKusick *et al.* 1964; Crow and Mange 1965; Allen 1988; Hurd 1983a,b).

Research on the genetic demography of the Lower Connecticut Valley constitutes one of the classic studies of colonizing settlement in North America, identifying the remarkable regularity of spatial settlement as new towns were founded, and combining analytical methods based on migration matrices, isolation by distance and isonymy (Relethford 1986; Relethford and Jaquish 1988; Swedlund *et al.* 1984a,b, 1985), including Lasker's repeating pairs method (Relethford 1992). Despite the evidence of kin-structured settlement, and social stratification influencing mate choice, migration between settlements resulted in relatively low predictions of heterogeneity and a marked pattern of isolation by distance, which became weaker over time (Swedlund 1984).

Studies of Utah Mormons constitute perhaps the most complete body of work from the point of view of integration of parameters of genetic structure, and none exemplifies so comprehensively the use of kinship as the common language for the elucidation and interpretation of population structure (Jorde 1982, 1987, 1989a). The studies were made possible in part by the genealogical interest of the Church of Jesus Christ of the Latter Day Saints. The research characterizes the population structure of an open, colonizing community with a diverse founding population. An added strength is the continual testing of the historically derived kinship measures (from migration, genealogies and surnames) against different types of genetic polymorphism – classic markers and DNA polymorphisms – in the present-day population (O'Brien *et al.* 1994, 1996).

Another project based on a very extensive genealogical database is the analysis of genetic structure of the French Canadians of Quebec (Gradie *et al.* 1988). These studies have focused attention on genetic disorders, especially autosomal recessives, with a view to identifying the probable ancestors of present-day carriers of the genes (DeBraekeleer 1994). The very heterogeneous distribution of recessive disorders in Quebec seems to derive in part from a concentration of founders from Perche, France (DeBraekeleer and Dao 1994), and the Quebec population can be demonstrated to comprise two or perhaps three (Gagnon and Heyer 2001) genealogically and genetically defined components. Molecular studies combined with kinship estimates from genealogies and isonymy have suggested that founder effect and genetic drift, rather than close consanguinity, account for the high frequency of disorders in Sanguenay-Lac-St-Jean (DeBraekeleer 1995, 1996; DeBraekeleer and Gauthier 1996a,b). The genealogical database has also made possible an assessment of inbreeding effects on reproductive success (Edmond and DeBraekeleer 1993a,b).

Besides these integrated studies in North America, there has been a substantial amount of smaller-scale work examining aspects of colonial settlement in North America (see, for example, Christensen 1999; Lebel 1983; Mathias *et al.* 2000; Reid 1988), and a wealth of research on central and south America (see, for example, Dipierri *et al.* 1991; Pinto-Cisternas *et al.* 1985; Rodriguez-Larralde *et al.* 2000; Madrigal and Ware 1997, 1999; chapter 2, this volume), and some on European colonial settlement in Australia (Harding 1985; Lafranchi *et al.* 1988; Kosten and Mitchell 1984). There have also been studies of later European immigrant populations to the United States, including one on the Italian community of Boston, examining the extent to which the marriage patterns of the homeland were maintained in the new cosmopolitan context (Danubio and Pettener 1997).

Islands have long been a magnet for study, for two principal reasons. On practical grounds one sees the possibility of comprehensive coverage, even by a single or small number of researchers. Secondly, the physical isolation and small population size of islands may increase the expectation of observing inbreeding and drift. This enticement, however, may also encourage a kind of antiquarian interest.

The South Atlantic island of Tristan da Cunha presents the classic example of a small, isolated population, whose founding population has been investigated in great detail, with an ascertained genealogy leading to the modern population (Roberts 1971). Classic, too, was the demonstration of bottlenecks in the population history subsequent to foundation, and the consequent reduction of the gene pool (Roberts 1968). Computational methods have been used to calculate the probabilities of founders' gene survival (Thomas and Thompson 1984) and bioassay of genetic distributions has been linked to the genealogical structure (Jenkins *et al.* 1985; Soodyall *et al.* 1997). Studies in the Azores have used a historical database of genealogies to trace ancestors of patients with Machado–Joseph Disease (MJD). In this archipelago the world's highest frequencies of MJD are found, with two distinct MJD haplotypes associated with two unconnected genealogies, probably of independent continental origin (Lima *et al.* 1998).

Islands round Britain have attracted an enormous amount of research. In the Orkney Islands, Scotland, Brennan and collaborators achieved a detailed understanding of the dynamics of vital events, mate choice and inbreeding through record linkage and potential mates analysis of the Sanday population (Brennan 1981, 1983; Brennan and Relethford 1983; Brennan *et al.* 1982). The demography of marriage, fertility and inbreeding in the Outer Hebrides has been investigated (Clegg 1986, 1999; Clegg and Cross 1995) and inbreeding from genealogies or isonymy has been studied in Eriskay (Robinson 1983) and Orkney (Roberts and Roberts 1983; Roberts *et al.* 1979). Off the coast of England, the Scilly Isles (Raspe and Lasker 1980), the Isle of Wight (Smith 1993), Holy Island (Cartwright 1973), and Jersey (Gottlieb *et al.* 1990) have all attracted attention.

The Åland archipelago lies in the Baltic, and the historical and present-day genetic structure have been intensively studied (Eriksson *et al.* 1973a,b; Eriksson 1980; Jorde *et al.* 1982; Mielke 1980; Mielke *et al.* 1994; Relethford and Mielke 1994; Workman and Jorde 1980). These studies, along with those of historical Massachusetts and the Utah Mormons, deserve to be regarded among the classic examples integrated analysis. Against a detailed demographic background, the pattern of mortality through time and space has been linked to migration, infectious disease, medical intervention and warfare (Mielke *et al.* 1984; chapter 10, this volume). Through an analysis of migration, mate choice and genetic structure of Sottunga, the importance of founder effect and drift in the local origins and distribution of von Willibrand disease was established (O'Brien *et al.* 1988, 1989).

Progress and prospects

Studies in Europe and the New World have demonstrated the broad dependence of population structure on geographical distance as the single most general determinant of population structure, and have elucidated chronological and geographical patterns of consanguineous marriage. Additionally, landscape features (Küchemann *et al.* 1967; Swedlund 1984), altitude (Pettener 1985), population size (Pettener 1985; Relethford 1985), ethnicity (Biondi *et al.* 1990, 1993, 1996), political boundaries (Boldsen and Lasker 1996), socioeconomic status (Harrison *et al.* 1970, 1971) and religious affiliation (Bittles and Smith 1994; Crawford *et al.* 1995; Koertvelyessy *et al.* 1992), and warfare (chapter 10, this volume) have been shown as presumably causal correlates of population structure.

We know, too, how isolation, migration and small population size affect the distribution of genes and genotypes. And studies of isolates can account for the present-day distribution of particular genes, including those causing genetic diseases, in terms of founder effect, genetic drift and inbreeding, as evidenced in studies of the Amish (McKusick *et al.* 1964), Sangueney du Lac (DeBraekeleer 1994, 1995), Sottunga (O'Brien *et al.* 1988), and the Azores (Lima *et al.* 1998).

How will archival anthropology develop in the future? The scope of investigation will undoubtedly be broadened by the availability of digitized data sets. We have begun to see the publication of research based on CD-ROM data sets, including the national telephone directories of Switzerland (Barrai *et al.* 1996),

Germany (Rodriguez-Larralde *et al.* 1998), Austria (Barrai *et al.* 2000), Italy (Barrai *et al.* 1999) and the United States of America (Barrai *et al.* 2001), and the registers of electors in Venezuela (Rodriguez-Larralde *et al.* 2000). Machine-readable source material will accumulate rapidly, both under the auspices of 'scholarly' initiatives such as the UK Data Archive, and also through the agency of 'amateur' family history organizations, and through commerce. For example, the important post-Famine listing of householders in Ireland, the Griffiths Valuation, is available on CD-ROM, as is the 1881 census of England and Wales, which is also accessible through the UK Data Archive. Though details of access are not yet clear, the 1901 Census of England and Wales will be made available by the Public Record Office via the Internet.

Although the provision of data sets to an extent determines what is possible, the intellectual development of archival studies depends upon the choice of appropriate research objectives and the scale of investigation on which they bear. The current body of knowledge has resulted largely from the steady accumulation of local studies, examining a few general themes within the context of a particular locality and its circumstances. There has perhaps been an overemphasis on isolation by distance and secular trend, and we seem sometimes to be content with 'time' and 'distance' as almost self-evident and self-justifying correlates of population structure (Fix 1999). Important though they are, these concepts subsume rather than elucidate the behaviors that provide the human link between the parameter and the genetics, and so lose some of the detailed context which archive studies can reveal. Of course, from another point of view, such analysis is successful exactly because time or distance summarizes complex activities of the human heart, brain, legs and loins.

I believe that it is at an important intermediate level of agency that much is still to be gained, and that it is both feasible and desirable to attempt systematically to evaluate the effects of broader historical contingency on microevolutionary processes and genetic structure. Factors such as land tenure, inheritance system, disease incidence, famine, industrialization, urbanization and demographic transition all have predictable and testable consequences, which could be addressed through the integration of fine-grain local studies. This interdisciplinary approach would enable archival anthropology to exploit its uniquely human source material, the written records of former generations, to yield further insights into the effects of historical contingency on human biology and microevolution.

References

Abade, A., Antunes, M.A., Fernandes, M.T. and Mota P.G. (1986). Inbreeding as measured by dispensations and isonymy in Rio Do Onor, Portugal. *International Journal of Anthropology* 1, 225–8.

Allen, G. (1988). Random genetic drift inferred from surnames in Old Colony Mennonites. *Human Biology* 60, 639–53.

Areia, M.L.R. (1986). Aspects socio-culturels de l'etude de la consanguinité dans les communautés rurales du Portugal. *Revista de Antropologia* 29, 135–45.

Bagnall, R.S. and Frier, B.W. (1994). *The Demography of Roman Egypt.* Cambridge: Cambridge University Press.

Barrai, I., Cavalli-Sforza, L.L. and Moroni, A. (1962). Frequencies of pedigrees of consanguineous marriages and mating structure of the population. *Annals of Human Genetics* 25, 347–77.

Barrai, I., Formica, G., Scapoli, C., Beretta, M., Mamolini, E., Volinia, S., Barale, R., Ambrosino, P. and Fontana, F. (1992). Microevolution in Ferrara – isonymy 1890–1990. *Annals of Human Biology* 19, 371–85.

Barrai, I., Rodriguez-Larralde, A., Mamolini, E., Manni, F. and Scapoli, C. (2000). Elements of the surname structure of Austria. *Annals of Human Biology* 27, 607–22.

Barrai, I., Rodriguez-Larralde, A., Mamolini, E., Manni, F. and Scapoli, C. (2001). Isonymy structure of USA population. *American Journal of Physical Anthropology* 114, 109–23.

Barrai, I., Rodriguez-Larralde, A., Mamolini, E. and Scapoli, C. (1999). Isonymy and isolation by distance in Italy. *Human Biology* 71, 947–61.

Barrai, I., Scapoli, C., Beretta, M., Nesti, C., Mamolini, E. and Rodriguez-Larralde, A. (1996). Isonymy and the genetic structure of Switzerland. 1. The distributions of surnames. *Annals of Human Biology* 23, 431–55.

Barrier, N.G. (ed.) (1981). *The Census in British India.* New Delhi: Manohar Publications.

Beddoe, J. (1885). *The Races of Britain.* Bristol: Arrowsmith.

Betzig, L. (1992). Roman polygyny. In *Darwinian History*, ed. L. Betzig, special edition of *Ethology and Sociobiology* 13, 309–49.

Betzig, L. (1995). Medieval monogamy. *Journal of Family History* 20, 181–216.

Biondi, G., Lasker, G.W., Raspe, P. and Mascie-Taylor, C.G.N. (1993). Inbreeding coefficients from the surnames of grandparents of the schoolchildren in Albanian-speaking Italian villages. *Journal of Biosocial Science* 25, 63–71.

Biondi, G., Perrotti, E., Mascie-Taylor, C.G.N. and Lasker, G.W. (1990). Inbreeding coefficients from isonymy in the Italian Greek villages. *Annals of Human Biology* 17, 543–6.

Biondi, G., Raspe, P., Mascie-Taylor, C.G.N. and Lasker, G.W. (1996). Repetition of the same pair of surnames in marriages in Albanian Italians, Greek Italians, and the Italian population of Campobasso province. *Human Biology* 68, 573–83.

Bittles, A.H. (1998). *Empirical Estimates of the Global Prevalence of Consanguineous Marriage in Contemporary Societies.* Stanford, CA: Morrison Institute.

Bittles, A.H. and Smith, M.T. (1994). Religious differentials in post-Famine marriage patterns, Northern Ireland, 1840–1915. I. Demographic and isonymy analysis. *Human Biology* 66, 59–76.

Boldsen, J.L. and Lasker G.W. (1996). Relationship of people across an international border based on an isonymy analysis across the German-Danish frontier *Journal of Biosocial Science* 28, 177–83.

Bonfield, L., Smith, R.M. and Wrightson, K. (eds) (1986). *The World We Have Gained: Histories of Population and Social Structure.* Oxford: Basil Blackwell.

Boone, J.L. (1986). Parental investment and elite family structure in preindustrial states: a case study from late Medieval–Early Modern Portuguese genealogies. *American Anthropologist* **88**, 859–78.

Boone, J.L. (1988). Parental investment, social subordination and population processes among the 15th and 16th century Portuguese nobility. In *Human Reproductive Behaviour, a Darwinian Perspective*, ed. L. Betzig, M. Borgerhoff Mulder and P. Turke, pp. 201–19. Cambridge: Cambridge University Press.

Boswell, J. (1988). *The Kindness of Strangers.* New York: Pantheon Books.

Bourgoin-Vu Tien Khang, J. (1978). Quelques aspects de l'histoire génétique de quatre villages pyrénéens depuis 1740. *Population* **33**, 633–58.

Boyce, A.J., Küchemann, C.F. and Harrison, G.A. (1967). Neighbourhood knowledge and the distribution of marriage distances. *Annals of Human Genetics* **30**, 335–8.

Brennan, E.R. (1981). Kinship, demographic, social and geographic characteristics of mate choice in Sanday, Orkney Islands, Scotland. *American Journal of Physical Anthropology* **65**, 121–8.

Brennan, E.R. (1983). Secular changes in migration between birth and marriage. *Journal of Biosocial Science* **15**, 391–406.

Brennan, E.R., Leslie, P.W. and Dyke, B. (1982). Mate choice and genetic structure of Sanday, Orkney Islands, Scotland. *Human Biology* **54**, 477–89.

Brennan, E.R. and Relethford, J.H. (1983). Temporal variation in the mating structure of Sanday, Orkney Islands. *Annals of Human Biology* **10**, 265–80.

Calderón, R. (1989). Consanguinity in the archbishopric of Toledo, Spain, 1900–79. I. Types of consanguineous mating in relation to premarital migration and its effects on inbreeding levels. *Journal of Biosocial Science* **21**, 253–66.

Calderón, R., Peña, J.A., Delgado, J. and Morales, B. (1998). Multiple kinship in two Spanish regions: new model relating multiple and simple consanguinity. *Human Biology* **70**, 535–61.

Cartwright, R.A. (1973). The structure of populations living on Holy Island. Northumberland. In *Genetic Variation in Britain*, ed. D.F. Roberts and E. Sunderland, pp. 95–107. London: Taylor and Francis.

Cavalli-Sforza, L.L. and Bodmer, W.F. (1971). *The Genetics of Human Populations.* San Francisco: Freeman.

Christensen, A.F. (1999). Population relationships by isonymy in frontier Pennsylvania. *Human Biology* **71**, 859–73.

Clark, P. (1979). Migration in England during the late 17th and early 18th centuries. *Past and Present* **83**, 57–90.

Clegg, E.J. (1986). The use of parental isonymy in inbreeding in two Outer Hebridean populations. *Annals of Human Biology* **13**, 211–24.

Clegg, E.J. (1999). Probabilities of marriage in two Outer Hebridean islands, 1861–1990. *Journal of Biosocial Science* **31**, 167–93.

Clegg, E.J. and Cross, J.F. (1995). Religion and fertility in the Outer Hebrides. *Journal of Biosocial Science* **27**, 79–94.

Crawford, M.H., Koertvelyessy, T., Huntsman, R.G., Collins, M., Duggirala, R., Martin, L. and Keeping, D. (1995). Effects of religion, economics and geography on genetic-structure of Fogo Island, Newfoundland. *American Journal of Human Biology* **7**, 437–51.

Crognier, E. (1985). Consanguinity and social change: an isonymic study of a French peasant population, 1870–1979. *Journal of Biosocial Science* **17**, 267–79.

Crow, J.F. (1958). Some possibilities for measuring selection intensities in man. *Human Biology* **30**, 1–13.

Crow, J.F. (1989). The estimation of inbreeding from isonymy (reprinted and updated). *Human Biology* **61**, 935–54.

Crow, J.F. and Mange, A.P. (1965). Measurement of inbreeding from the frequency of marriages of persons of the same surname. *Eugenics Quarterly* **12**, 199–203.

Curto, J.C. (1994). Sources for the pre-1900 population history of Sub-Saharan Africa: the case of Angola, 1773–1845. *Annales de Démographie Historique*, 1994, 319–38.

Danubio, M.E. and Pettener, D. (1997). Marital structure of the Italian community of Boston, Massachusetts, 1880–1920. *Journal of Biosocial Science* **29**, 257–69.

Danubio, M.E., Piro, A. and Tagarelli, A. (1999). Endogamy and inbreeding since the 17th century in past malarial communities in the Province of Cosenza (Calabria, Southern Italy). *Annals of Human Biology* **26**, 473–88.

DeBraekeleer, M. (1994). Hereditary disorders in the French-Canadian population of Quebec. 1. In search of founders. *Human Biology* **66**, 205–23.

DeBraekeleer, M. (1995). Geographic distribution of 18 autosomal recessive disorders in the French-Canadian population of Saguenay-Lac-Saint-Jean, Quebec. *Annals of Human Biology* **22**, 111–22.

DeBraekeleer, M. (1996). Autosomal recessive disorders in Saguenay-Lac-Saint-Jean (Quebec, Canada): Estimation of inbreeding from isonymy. *Annals of Human Biology* **23**, 95–9.

DeBraekeleer, M. and Dao, T.N. (1994). Hereditary disorders in the French-Canadian population of Quebec. 2. Contribution of Perche. *Human Biology* **66**, 225–49.

DeBraekeleer, M. and Gauthier, S. (1996a). Autosomal recessive disorders in Saguenay-Lac-Saint-Jean (Quebec, Canada): a study of inbreeding. *Annals of Human Genetics* **60**, 51–6.

DeBraekeleer, M. and Gauthier, S. (1996b). Autosomal recessive disorders in Saguenay-Lac-Saint-Jean (Quebec, Canada): study of kinship. *Human Biology* **68**, 371–81.

Dickemann, M. (1979). Female infanticide, reproductive strategies and social stratification: a preliminary model. In *Evolutionary Biology and Human Social Behaviour: An Anthropological Perspective*, ed. N. Chagnon and W. Irons, pp. 312-67. North Scituate, MA: Duxbury Press.

Dipierri, J.E., Ocampo, S.B. and Russo, A. (1991). An estimation of inbreeding from isonymy in the historical (1734–1810) population of the Quebrada-de-Humahuaca (Jujuy, Argentina). *Journal of Biosocial Science* **23**, 23–31.

Dobson, M. (1997). *Contours of Death and Disease in Early Modern England.* Cambridge: Cambridge University Press.

Dobson, T. and Roberts, D.F. (1971). Historical population movement and gene flow in Northumberland parishes. *Journal of Biosocial Science* **3**, 193–208.

Duncan, C.J., Duncan, S.R. and Scott, S. (1997). The dynamics of measles epidemics. *Theoretical Population Biology* **52**, 155–63.

Duncan, S.R., Scott, S. and Duncan, C.J. (2000). Modelling the dynamics of scarlet fever epidemics in the 19th century. *European Journal of Epidemiology* **16**, 619–26.

Dyke, B. (1971). Potential mates in a small human population. *Social Biology* **18**, 28–39.

Edmond, M. and DeBraekeleer, M. (1993a). Inbreeding effects of fertility and sterility – a case-control study in Saguenay-Lac-Saint-Jean (Quebec, Canada) based on a population registry 1838–1971. *Annals of Human Biology* **20**, 545–55.

Edmond, M. and DeBraekeleer, M. (1993b). Inbreeding effects on prereproductive mortality – a case-control study in Saguenay-Lac-Saint-Jean (Quebec, Canada) based on a population registry 1838–1971. *Annals of Human Biology* **20**, 535–43.

Eriksson, A.W. (1980). Genetic studies on Åland: geographical, historical and archival data and some other potentialities. In *Population Structure and Genetic Disorders*, ed. A.W. Eriksson, H.R. Forsius, H.R. Nevalinna, P.L. Workman and R. Norio, pp. 459–70. London: Academic Press.

Eriksson, A.W., Eskola, M.-R., Workman, P.L. and Morton, N.E. (1973a). Population studies on the Åland Islands. II. Historical population structure: inference from bioassay of kinship and migration. *Human Heredity* **23**, 511–34.

Eriksson, A.W., Fellman, J.O., Workman, P.L. and Lalouel, J.M. (1973b). Population studies on the Åland Islands. I. Prediction of kinship from migration and isolation by distance. *Human Heredity* **23**, 422–33.

Fildes V. (1988). *Wet Nursing: a History from Antiquity to the Present*. Oxford: Basil Blackwell.

Fix, A. (1999). *Migration and Colonization in Human Microevolution*. Cambridge: Cambridge University Press.

Floud, R., Wachter, K. and Gregory, A. (1990). *Health, Height and History: The Nutritional Status of the British, 1750–1880*. Cambridge: Cambridge University Press.

Fuster, V. (1986). Relationship by isonymy and migration pattern in northwest Spain. *Human Biology* **58**, 391–406.

Fuster, V., Morales, B., Mesa, M.S. and Martin, J. (1996). Inbreeding patterns in the Gredos mountain range (Spain). *Human Biology* **68**, 75–93.

Gagnon, A. and Heyer, E. (2001). Fragmentation of the Québec population genetic pool (Canada): evidence from the genetic contribution of founders per region in the 17th and 18th centuries. *American Journal of Physical Anthropology* **114**, 30–41.

Gautier, E. and Henry, L. (1958). *La Population de Crulai*. Paris: Presses universitaires de France.

Gottlieb, K., Raspe, P. and Lasker, G.W. (1990). Patterned selection of mates in St-Ouen, Jersey, and the Scilly Isles examined by isonymy. *Human Biology* **62**, 37–647.

Gradie, M.I., Jorde, L.B. and Bouchard, G. (1988). Genetic-structure of the Saguenay, 1852–1911 – evidence from migration and isonymy matrices. *American Journal of Physical Anthropology* **77**, 321–33.

Harding, R.M. (1985). Historical population-structure of 3 coastal districts of Tasmania, Australia – 1838–1950. *Human Biology* **57**, 727–44.

Harrison, G.A. (1995). *The Human Biology of the English Village*. Oxford: Oxford University Press.

Harrison, G.A., Gibson, J.B., Hiorns, R.W., Wigley, M., Hancock, C., Freeman, C.A., Küchemann, C.F., Macbeth, H.M., Saatcioglu, A. and Carrivick, P.J. (1974).

328   *M.T. Smith*

Psychometric, personality and anthropometric variation in a group of Oxfordshire villages. *Annals of Human Biology* **1**, 365–81.

Harrison, G.A., Hiorns, R.W. and Küchemann, C.F. (1970). Social class relatedness in some Oxfordshire parishes. *Journal of Biosocial Science* **2**, 71–80.

Harrison, G.A., Hiorns, R.W. and Küchemann, C.F. (1971). Social class and marriage patterns in some Oxfordshire populations. *Journal of Biosocial Science* **3**, 1–12.

Hed, H.M.E. (1984). Opportunity for selection during the 17th–19th centuries in the diocese of Linkoping as estimated with Crow index in a population of clergymens wives. *Human Heredity* **34**, 378–87.

Hed, H.M.E. (1987). Trends in opportunity for natural-selection in the Swedish population during the period 1650–1980. *Human Biology* **59**, 785–97.

Hemam, N.S. and Reddy, B.M. (1998). Demographic implications of socioeconomic transition among the tribal populations of Manipur, India. *Human Biology* **70**, 597–619.

Hiorns, R.W., Harrison, G.A., Boyce, A.J. and Küchemann, C.F. (1969). A mathematical analysis of the effects of movement on the relatedness between populations. *Annals of Human Genetics* **32**, 237–50.

Hiorns, R.W., Harrison, G.A. and Gibson, J.B. (1977). Genetic variation in some Oxfordshire villages. *Annals of Human Biology* **4**, 197–210.

Holloway, S.M. and Sofaer, J.A. (1989). Coefficients of relationship by isonymy within and between the regions of Scotland. *Human Biology* **61**, 87–97.

Hopkins, K. (1980). Brother-sister marriage in Roman Egypt. *Comparative Studies in Society and History* **22**, 303–54.

Hrdy, S.B. (1992). Fitness tradeoffs in the history and evolution of delegated mothering with special reference to wet-nursing, abandonment and infanticide. *Ethology and Sociobiology* **13**, 409–42.

Hurd, J.P. (1983a). Comparison of isonymy and pedigree analysis measures in estimating relationships between 3 Nebraska Amish churches in central Pennsylvania. *Human Biology* **55**: 349–55.

Hurd, J.P. (1983b). Kin relatedness and church fissioning among the Nebraska Amish of Pennsylvania. *Social Biology* **30**: 59–66.

Jeffries, D.J., Harrison, G.A., Hiorns, R.W. and Gibson, J.B. (1976). A note on marital distances and movement, and age at marriage, in a group of Oxfordshire villages. *Journal of Biosocial Science* **8**, 155–60.

Jenkins, T., Beighton, P. and Steinberg, A.G. (1985). Serogenetic studies on the inhabitants of Tristan-da-Cunha. *Annals of Human Biology* **12**, 363–71.

Jordan, T.E. (1998). *Ireland's Children: Quality of Life, Stress, and Child Development in the Famine Era*. Westport, Connecticut: Greenwood Press.

Jorde, L.B. (1980). The genetic structure of subdivided human populations: a review. In *Current Developments in Anthropological Genetics*, volume 1: *Theory and Methods*, ed. J.H. Mielke and M.H. Crawford, pp. 135–208. New York: Plenum Press.

Jorde, L.B. (1982). The genetic structure of the Utah Mormons: migration analysis. *Human Biology* **54**, 583–97.

Jorde, L.B. (1987). Genetic structure of the Utah Mormons: isonymy analysis. *American Journal of Physical Anthropology* **72**, 403–12.

Jorde, L.B. (1989a). Inbreeding in the Utah Mormons: an evaluation of estimates based on pedigrees, isonymy and migration matrices. *Annals of Human Genetics* **53**, 339–55.

Jorde, L.B. (1989b). Predicting smallpox epidemics: a statistical analysis of two Finnish populations. *American Journal of Human Biology* **1**, 621–9.

Jorde, L.B. and Durbize, P. (1986). Opportunity for natural-selection in the Utah Mormons. *Human Biology* **58**, 97–114.

Jorde, L.B., Workman, P.L. and Eriksson, A.W. (1982). Genetic microevolution on the Åland Islands. In *Current Developments in Anthropological Genetics*, Volume 2: *Ecology and Population Structure*, ed. M.H. Crawford and J.H. Mielke, pp. 333–65. New York: Plenum Press.

Judge, D. and Hrdy, S.B. (1992). Allocation of accumulated resources among close kin: inheritance in Sacramento, California, 1890–1984. *Ethology and Sociobiology* **13**, 495–522.

Kertzer, D. (1993). *Sacrificed for Honor*. Boston: Beacon Press.

Kiel, M. (1999). The Ottoman Imperial Registers. Central Greece and Northern Bulgaria in the 15th–19th Century, the demographic development of two areas compared. In *Reconstructing Past Population Trends in Mediterranean Europe (3000BC to AD 1800)*, ed. J. Bintliff and K. Sbonias, pp. 195–218. Oxford: Oxbow Books.

Koertvelyessy, T., Crawford, M.H., Pap, M. and Szilagyi, K. (1992). The influence of religious affiliation on surname repetition in marriages in Tiszaszalka, Hungary. *Journal of Biosocial Science* **24**, 113–21.

Komlos J. (1994). *Stature, Living Standards, and Economic Development: Essays on Anthropometric History*. Chicago: University of Chicago Press.

Komlos J. (1995). *The Biological Standard of Living on Three Continents: Further Explorations in Anthropometric History*. Boulder, Colorado: Westview Press.

Kosten, M. and Mitchell, R.J. (1984). Family-size and social-class in 19th-century Tasmania, Australia. *Journal of Biosocial Science* **16**, 55–63.

Küchemann, C.F., Boyce, A.J. and Harrison, G.A. (1967). A demographic and genetic study of a group of Oxfordshire villages. *Human Biology* **39**, 251–76.

Küchemann, C.F., Harrison, G.A., Hiorns, R.W. and Carrivick, P.J. (1974). Social class and marital distance in Oxford City. *Annals of Human Biology* **1**, 13–27.

Küchemann, C.F., Lasker, G.W. and Smith, D.I. (1979). Historical changes in the coefficient of relationship by isonymy among the populations of the Otmoor villages. *Human Biology* **51**, 63–77.

Lafranchi, M., Mitchell, R.J. and Kosten, M. (1988). Mating structure, isonymy and social-class in late 19th-century Tasmania. *Annals of Human Biology* **15**, 325–36.

Landers, J. (1993). *Death and the Metropolis: Studies in the Demographic History of London 1670–1830*. Cambridge: Cambridge University Press.

Lasker, G.W. (1977). A coefficient of relationship by isonymy: a method for estimating the genetic relationship between populations. *Human Biology* **49**, 489–93.

Lasker, G.W. (1978). Relationships among Otmoor villages and surrounding communities as inferred from surnames contained in the current register of electors. *Annals of Human Biology* **5**, 105–11.

330   *M.T. Smith*

Lasker, G.W. and Kaplan, B.A. (1985). Surnames and genetic structure: repetition of the same pairs of names of married couples, a measure of subdivision of the population. *Human Biology* **57**, 431–40.

Lasker, G.W. and Roberts, D.F. (1982). Secular trends in relationship as estimated by surnames: a study of a Tyneside parish. *Annals of Human Biology* **9**, 299–308.

Laslett, P. (1965). *The World We Have Lost.* London: Methuen.

Lebel, R.R. (1983). Consanguinity studies in Wisconsin I: secular trend in consanguineous marriage, 1843–1981. *American Journal of Human Genetics* **15**, 543–60.

Leslie, P.W. (1985). Potential mates analysis and the study of human population structure. *Yearbook of Physical Anthropology* **28**, 53–78.

Lima, M., Mayer, F.M., Coutinho, P. and Abade, A. (1998). Origins of a mutation: genetics of Machado-Joseph Disease in the Azores (Portugal). *Human Biology* **70**, 1011–23.

Lo Cascio, E. (1999). The population of Roman Italy in town and country. In *Reconstructing Past Population Trends in Mediterranean Europe (3000BC to AD 1800)*, ed. J. Bintliff and K. Sbonias, pp. 161–71. Oxford: Oxbow Books.

Madrigal, L. and Ware, B. (1997). Inbreeding in Escazu, Costa Rica (1800–1840, 1850–1899): Isonymy and ecclesiastical dispensations. *Human Biology* **69**, 703–14.

Madrigal, L. and Ware, B. (1999). Mating pattern and population structure in Escazu, Costa Rica: A study using marriage records. *Human Biology* **71**, 963–75.

Mascie-Taylor, C.G.N. and Lasker, G.W. (1984). Geographic-distribution of surnames in Britain – the Smiths and Joneses. *Journal of Biosocial Science* **16**, 301–8.

Mascie-Taylor, C.G.N. and Lasker, G.W. (1985). Geographical-distribution of common surnames in England and Wales. *Annals of Human Biology* **12**, 397–401.

Mascie-Taylor, C.G.N. and Lasker, G.W. (1990). The distribution of surnames in England and Wales – a model for genetic distribution. *Man* **25**, 521–30.

Mathias, R.A., Bickel, C.A., Beaty, T.H., Petersen, G.M., Hetmanski, J.B., Liang, K.Y. and Barnes, K.C. (2000). A study of contemporary levels and temporal trends in inbreeding in the Tangier Island, Virginia, population using pedigree data and isonymy. *American Journal of Physical Anthropology* **112**, 29–38.

McClure, P. (1979). Patterns of migration in the late Middle Ages: the evidence of English place-name surnames. *Economic History Review*, 2nd series, **32**, 167–82.

McCullough, J.M. and Barton, E.Y. (1991). Relatedness and kin-structured migration in a founding population – Plymouth Colony, 1620–1633. *Human Biology* **63**, 355–66.

McCullough, J.M. and O'Rourke, D.H. (1986). Geographic distribution of consanguinity in Europe. *Annals of Human Biology* **13**, 359–68.

McKusick, V.A., Hostetler, J.A., Egeland, J.A. and Eldridge, R. (1964). The distribution of certain genes in the Old Order Amish. *Cold Spring Harbor Symposium in Quantitative Biology* **29**, 99–114.

Mielke, J.H. (1980). Demographic aspects of population structure in Åland. In *Population Structure and Genetic Disorders*, ed. A.W. Eriksson, H.R. Forsius, H.R. Nevalinna, P.L. Workman and R. Norio, pp. 471–86. London: Academic Press.

Mielke, J.H., Jorde, L.B., Trapp, P.G., Anderton, D.L., Pitkänen, K. and Eriksson, A.W. (1984). Historical epidemiology of smallpox in Åland, Finland: 1751–1890. *Demography* **21**, 271–95.

Mielke, J.H., Relethford, J.H. and Eriksson, A.W. (1994). Temporal trends in migration in the Åland Islands – effects of population-size and geographic distance. *Human Biology* **66**, 399–410.

Mielke, J.H. and Swedlund, A.C. (1993). Historical surveys of population structure. In *Research Strategies in Human Biology*, ed. G.W. Lasker and N.M.T. Mascie-Taylor, pp. 140–85. Cambridge: Cambridge University Press.

Moroni, A. (1967). Andamento della consanuineita nell'Italia settentrionale negli ultimi quattro secoli. *Atti Associazione Genetica Italiana* **12**, 202–22.

Morton, N.E. (1982). *Outline of Genetic Epidemiology.* New York: Karger.

Morton, N.E., Yee, S., Harris, D.E. and Lew, R. (1971). Bioassay of kinship. *Theoretical Population Biology* **2**, 507–24.

North, K.E., Crawford, M.H. and Relethford, J.H. (1999). Spatial variation of anthropometric traits in Ireland. *Human Biology* **71**, 823–45.

Ó Gráda, C. (1991). The heights of Clonmel prisoners 1845–49: some dietary implications. *Irish Economic and Social History* **18**, 24–33.

Ó Gráda, C. (1994). *Ireland: a New Economic History:1780–1939.* Oxford: Clarendon Press.

O'Brien, E., Jorde, L.B., Rönnlöf, B., Fellman, J.O. and Eriksson, A.W. (1988). Founder effect and genetic disease in Sottunga, Finland. *American Journal of Physical Anthropology* **77**, 335–46.

O'Brien, E., Jorde, L.B., Rönnlöf, B., Fellman, J.O. and Eriksson, A.W. (1989). Consanguinity avoidance and mate choice in Sottunga, Finland. *American Journal of Physical Anthropology* **79**, 235–46.

O'Brien, E., Rogers, A.R., Beesley, J. and Jorde, L.B. (1994). Genetic-structure of the Utah Mormons – comparison of results based on RFLPs, blood-groups, migration matrices, isonymy, and pedigrees. *Human Biology* **66**, 743–59.

O'Brien, E., Zenger, R. and Jorde, L.B. (1996). Genetic structure of the Utah Mormons: A comparison of kinship estimates from DNA, blood groups, genealogies, and ancestral arrays. *American Journal of Physical Anthropology* **8**, 609–14.

Paoli, G., Franceschi, M.G. and Lasker, G.W. (1999). Changes over 100 years in degree of isolation of 21 parishes of the Lima Valley, Italy, assessed by surname isonymy. *Annals of Human Biology* **71**, 123–33.

Parkin, T. (1999). Clearing away the cobwebs: a critical perspective on historical sources for Roman population history. In *Reconstructing Past Population Trends in Mediterranean Europe (3000BC to AD 1800)*, ed. J. Bintliff and K. Sbonias, pp. 153–60. Oxford: Oxbow Books.

Peña, J.A., Morales, B. and Calderon, R. (1997). New method for comparing levels of microdifferentiation: Application to migration matrices of two populations from the Basque Country (Spain). *Human Biology* **69**, 329–44.

Pettener, D. (1985). Consanguineous marriages in the upper Bologna Appenine (1565–1980): microgeographic variation, pedigree structure and correlation of inbreeding secular trend with population size. *Human Biology* **57**, 267–88.

Pettener, D. (1990). Temporal trends in marital structure and isonymy in S-Paolo Albanese, Italy. *Human Biology* **62**, 837–51.

Peyronnet, J.C. (1976). Les enfants abandonnés et leurs nourrices à Limoges au XVIII-siècle. *Revue d'Histoire Moderne et Contemporaine* **23**, 418–41.

Pinto-Cisternas, J., Pineda, L. and Barrai, I. (1985). Estimation of inbreeding by isonymy in Iberoamerican populations – an extension of the method of Crow and Mange. *American Journal of Human Genetics* **37**, 373–85.

Pollitzer, W.S., Smith, M.T. and Williams, R.W. (1988). A study of isonymic relationships in Fylingdales parish from marriage records from 1654 through 1916. *Human Biology* **60**, 363–82.

Raftis, J.A. (1974). *Warboys: Two Hundred Years in the Life of an English Mediaeval Village*. Toronto: Pontifical Institute of Mediaeval Studies.

Raftis, J.A. (1982). *A Small Town in Late Medieval England: Godmanchester, 1278–1400*. Toronto: Pontifical Institute of Mediaeval Studies.

Raspe, P.A. and Lasker, G.W. (1980). The structure of the human population of the Isles of Scilly: inferences from surnames and birthplaces listed in census and marriage records. *Annals of Human Biology* **7**, 401–10.

Razi, Z. (1980). *Life, Marriage and Death in a Medieval Parish: Economy, Society and Demography in Halesowen, 1300–1400*. Cambridge: Cambridge University Press.

Reddy, B.M. and Chopra, V.P. (1990). Opportunity for natural-selection among the Indian populations. *American Journal of Physical Anthropology* **83**, 281–96.

Reid, R.M. (1988). Church membership, consanguineous marriage, and migration in a Scotch-Irish frontier population. *Journal of Family History* **13**, 397–414.

Relethford, J.H. (1985). Examination of the relationship between inbreeding and population size. *Journal of Biosocial Science* **17**, 97–106.

Relethford, J.H. (1986). Microdifferentiation in historical Massachusetts – a comparison of migration matrix and isonymy analyses. *American Journal of Physical Anthropology* **71**, 365–75.

Relethford, J.H. (1988). Estimation of kinship and genetic distance from surnames. *Human Biology* **60**, 475–92.

Relethford, J.H. (1992). Analysis of marital structure in Massachusetts using repeating pairs of surnames. *Human Biology* **64**, 25–33.

Relethford, J.H. and Brennan, E.R. (1982). Temporal trends in isolation by distance on Sanday, Orkney Islands. *Human Biology* **54**, 315–27.

Relethford, J.H. and Crawford, M.H. (1995), Anthropometric variation and the population history of Ireland. *American Journal of Physical Anthropology* **96**, 25–38.

Relethford, J.H. and Crawford, M.H. (1998). Influence of religion and birthplace on the genetic structure of Northern Ireland. *Annals of Human Biology* **25**, 117–25.

Relethford, J.H., Crawford, M.H. and Blangero, J. (1997). Genetic drift and gene flow in post-famine Ireland. *Human Biology* **69**, 443–65.

Relethford, J.H. and Jaquish, C.E. (1988). Isonymy, inbreeding, and demographic variation in historical Massachusetts. *American Journal of Physical Anthropology* **77**, 243–52.

Relethford, J.H., Lees, F.C. and Crawford, M.H. (1980). Population structure and anthropometric variation in rural western Ireland: Migration and biological variation. *Annals of Human Biology* **7**, 411–28.

Relethford, J.H. and Mielke, J.H. (1994). Marital exogamy in the Åland Islands, Finland, 1750–1949. *Annals of Human Biology* **21**, 13–21.

Roberts, D.F. (1968). Genetic effects of population size reduction. *Nature* **220**, 1084–88.

Roberts, D.F. (1971). The demography of Tristan da Cunha. *Population Studies* **25**, 465–79.

Roberts, D.F. (1980). Inbreeding and ecological change: an isonymic analysis of secular trends in a Tyneside parish over three centuries. *Social Biology* **27**, 230–40.

Roberts, D.F. (1982). Population structure of farming communities of northern England. In *Current Developments in Anthropological Genetics*, Volume 2: *Ecology and Population Structure*, ed. M.H. Crawford and J.H. Mielke, pp. 367–84. New York: Plenum Press.

Roberts, D.F. and Roberts, M.J. (1983). Surnames and relationships – an Orkney study. *Human Biology* **55**, 341–7.

Roberts, D.F., Roberts, M.J. and Cowie, J.A. (1979). Inbreeding levels in Orkney Islanders. *Journal of Biosocial Science* **11**, 391–5.

Robinson, A.P. (1983). Inbreeding as measured by dispensations and isonymy on a small Hebridean island, Eriskay. *Human Biology* **55**, 289–95.

Rodriguez-Larralde, A., Barrai, I., Nesti, C., Mamolini, E. and Scapoli, C. (1998). Isonymy and isolation by distance in Germany. *Human Biology* **70**, 1041–56.

Rodriguez-Larralde, A., Formica, G., Scapoli, C., Beretta, M., Mamolini, E. and Barrai, I. (1993). Microevolution in Perugia – isonymy 1890–1990. *Annals of Human Biology* **20**, 261–74.

Rodriguez-Larralde, A., Morales, J. and Barrai, I. (2000). Surname frequency and the isonymy structure of Venezuela. *American Journal of Human Biology* **12**, 352–62.

Rogers, A.R. (1991). Doubts about isonymy. *Human Biology* **63**, 663–8.

Rosenbaum, S. and Crowdy, J.P. (1990). British army recruits: 100 years of heights and weights. *Journal of the Royal Army Medical Corps* **138**, 81–6.

Sá, I. dos Guimerães (1995). *A Circulação de Crianças na Europa do Sul: O Caso dos Expostos do Porto no Sécolo XVIII*. Lisbon: Fundaçao Calouste Gulbenkian.

Sattenspiel, L., Mobarry, A. and Herring, D.A. (2000). Modeling the influence of settlement structure on the spread of influenza among communities. *American Journal of Human Biology* **12**, 736–48.

Sawchuk, L.A. (1980). Reproductive success among the Sephardic Jews of Gibraltar – evolutionary implications. *Human Biology* **52**, 731–52.

Scheidel, W. (1997). Brother-sister marriage in Roman Egypt. *Journal of Biosocial Science* **29**, 361–71.

Schurer, K. and Arkell, T. (1992). *Surveying the People*. Oxford: Leopard's Head Press.

Scott, S. and Duncan, C.J. (1998). *Human Demography and Disease*. Cambridge: Cambridge University Press.

Scott, S. and Duncan, C.J. (1999). Nutrition, fertility and steady-state population dynamics in a pre-industrial community in Penrith, Northern England. *Journal of Biosocial Science* **31**, 505–23.

Scott, S., Duncan, C.J. and Duncan, S.R. (1996). The plague in Penrith, Cumbria, 1597/8: Its causes, biology and consequences. *Annals of Human Biology* **23**, 1–21.

Scott, S., Duncan, S.R. and Duncan, C.J. (1998). The interacting effects of prices and weather on population cycles in a preindustrial community. *Journal of Biosocial Science* **30**, 15–32.

Segalen, M. (1991). *Fifteen Generations of Bretons*. Cambridge: Cambridge University Press.

Shaw, B.D. (1992). Explaining incest: brother-sister marriage in Graeco-Roman Egypt. *Man* **27**, 267–99.

Sherwood, J. (1988). *Poverty in Eighteenth Century Spain: the Women and Children of the Inclusa.* Toronto: University of Toronto Press.

Smith, M.T. (1993). Migration and population structure of the Isle of Wight, UK: the antecedents of tourism. *Anthropologie et Préhistoire* **104**, 93–101.

Smith, M.T., Smith, B.L. and Williams, W.R. (1984). Changing isonymic relationships in Fylingdales parish, North Yorkshire, 1841–1881. *Annals of Human Biology* **11**, 449–57.

Smith, M.T., Williams, W.R., McHugh, J.J. and Bittles, A.H. (1990). Isonymic analysis of post-Famine relationships in the Ards Peninsula, N.E. Ireland: genetic effects of geographical and politico-religious boundaries. *American Journal of Human Biology* **2**, 245–54.

Sokal, R.R., Harding, R.M., Lasker, G.W. and Mascie-Taylor, C.G.N. (1992). A spatial-analysis of 100 surnames in England and Wales. *Annals of Human Biology* **19**, 445–76.

Soodyall, H., Jenkins, T., Mukherjee, A., du Toit, E., Roberts, D.F. and Stoneking, M. (1997). The founding mitochondrial DNA lineages of Tristan da Cunha islanders. *American Journal of Physical Anthropology* **104**, 157–66.

Stone, L. (1977). *The Family, Sex and Marriage in England 1500–1800.* London: Weidenfeld and Nicolson.

Sutter, J. (1968). Fréquence de l'endogamie et ses facteurs au XIX siècle. *Population* **23**, 303–24.

Sutter, J. and Tabah, L. (1948). Fréquence et répartition des mariages consanguins en France. *Population* **3**, 607–30.

Swedlund, A.C. (1980). Historical demography. In *Current Developments in Anthropological Genetics*, Volume 1, *Theory and Methods*, ed. J.H. Mielke and M.H. Crawford, pp. 17–42. New York: Plenum Press.

Swedlund, A.C. (1984). Historical studies of mobility. In *Migration and Mobility*, ed. A.J. Boyce, pp. 1–18. London: Taylor and Francis.

Swedlund, A.C., Anderson, A.B. and Boyce, A.J. (1984a). A comparison of marital migration and isonymy matrices with geographic distance in the Connecticut River Valley. *American Journal of Physical Anthropology* **63**, 225–35.

Swedlund, A.C., Anderson, A.B. and Boyce, A.J. (1985). Population-structure in the Connecticut Valley. 2. A comparison of multidimensional-scaling solutions of migration matrices and isonymy. *American Journal of Physical Anthropology* **68**, 539–47.

Swedlund, A.C., Jorde, L.B. and Mielke, J.H. (1984b). Population structure in the Connecticut Valley. I. Marital Migration. *American Journal of Physical Anthropology* **65**, 61–70.

Tanner, J. (1981). *A History of the Study of Human Growth.* Cambridge: Cambridge University Press.

Thomas, A. and Thompson, E.A. (1984). Gene survival in isolated populations: the number of distinct genes on Tristan da Cunha. *Annals of Human Biology* **11**, 101–11.

Trivers, R.L. and Willard, D.E. (1973). Natural selection of parental ability to vary the sex ratio of offspring. *Science* **179**, 90–2.

Vernay, M. (2000). Trends in inbreeding, isonymy, and repeated pairs of surnames in the Valserine Valley (French Jura), 1763–1972. *Human Biology* **72**, 675–92.

Viazzo, P.P., Bortolotto, M. and Zanotto, A. (2000). Five centuries of foundling history in Florence: changing patterns of abandonment, care and mortality. In *Abandoned Children*, ed. C. Panter-Brick and M.T. Smith, pp. 70–91. Cambridge: Cambridge University Press.

Vienna, A., De Stefano, G.F., Bastianini, A. and Biondi, G. (1998). Biodemography in Siena, Italy. *Journal of Biosocial Science* **30**, 521–8.

Voland, E. (1984). Human sex ratio manipulation: historical data from a German parish. *Human Evolution* **13**, 99–107.

Voland, E. (1988). Differential infant and child mortality in evolutionary perspective: data from late 17th to 19th century Ostfriesland (Germany). In *Human Reproductive Behaviour, a Darwinian Perspective*, ed. L. Betzig, M. Borgerhoff Mulder and P. Turke, pp. 253–61. Cambridge: Cambridge University Press.

Voland, E. (1989). Differential parental investment: some ideas on the contact area of European social history and evolutionary biology. In *Comparative Socioecology: The Behavioural Ecology of Humans and Other Mammals*, ed. V. Standen and R.A. Foley, pp. 391–403. Oxford: Blackwell Scientific Publications.

Woods, R. and Shelton, N. (1997). *An Atlas of Victorian Mortality*. Liverpool: Liverpool University Press.

Workman, P.L. and Jorde, L.B. (1980). Genetic structure of the Åland Islands. In *Population Structure and Genetic Disorders*, ed. A.W. Eriksson, H.R. Forsius, H.R. Nevalinna, P.L. Workman and R. Norio, pp. 471–86. London: Academic Press.

Wrigley, E.A. (ed.) (1966). *An Introduction to English Historical Demography*. London: Weidenfeld and Nicolson.

Wrigley, E.A. (ed.) (1972). *Nineteenth Century Society – Essays in the Use of Quantitative Methods for the Study of Social Data*. Cambridge: Cambridge University Press.

Wrigley, E.A. and Schofield, R.S. (1981). *The Population History of England 1541–1871*. Cambridge: Cambridge University Press.

Wrigley, E.A., Davies, R.S., Oeppen, J.E. and Schofield, R.S. (1997). *English Population History from Family Reconstitution 1580–1837*. Cambridge: Cambridge University Press.

Index

Page numbers in italics refer to figures and tables.